BK 621.381 P926E 1982
ELECTRONIC INSTRUMENTATION
/PRENSKY, S
3 C1982 22.95 FV
3000 578810 30012
St. Louis Community College

W9-DIU-146

WITHDRAWN

St. Louis Community
College

ELECTRONIC INSTRUMENTATION

3rd Edition

ELECTRONIC INSTRUMENTATION

Sol D. Prensky
Richard L. Castellucis
Southern Technical Institute
Marietta, Georgia

PRENTICE-HALL, INC., Englewood Cliffs, NJ 07632

Library of Congress Cataloging in Publication Data

Prensky, Sol D.
Electronic instrumentation.

Bibliography: p.
Includes index.
1. Electronic instruments. I. Castellucis, Richard L.
II. Title.
TK7878.4.P65 1982 621.3815′48 81–5212
ISBN 0–13–251611–X AACR2

Editorial/production supervision and interior design: BARBARA BERNSTEIN
Manufacturing buyer: GORDON OSBOURNE
Cover design: WANDA LUBELSKA DESIGN

Printed in the United States of America

10 9 8 7 6 5 4 3 2 1

ISBN: 0-13-251611-X

PRENTICE-HALL INTERNATIONAL, INC., *London*
PRENTICE-HALL of AUSTRALIA PTY. LIMITED, *Sydney*
PRENTICE-HALL of CANADA, LTD., *Toronto*
PRENTICE-HALL of INDIA PRIVATE LIMITED, *New Delhi*
PRENTICE-HALL of JAPAN, INC., *Tokyo*
PRENTICE-HALL of SOUTHEAST ASIA PTE. LTD., *Singapore*
WHITEHALL BOOKS LIMITED, *Wellington, New Zealand*

Contents

Preface to the Third Edition

Over the past few years we have witnessed a series of advancements, improvements, and applications of electronic circuits and systems that almost boggle the mind. The electronic instrumentation field has, by necessity, been obliged to keep pace with this incredible technological explosion. Whereas the basic functions of the instrumentation remain fundamentally the same, changes in technique, accuracy, and operation offering improved capabilities of measurement are constantly emerging.

This revision has attempted to accomplish several goals. First, the original style and approach of the text was maintained. This was done because, even though some techniques did change, the basic concepts of measurement have remained fairly constant. Rather than fill the text with what might have resembled a paste-up of manufacturers' catalogs, a few instruments were chosen not because they are the newest or the most modern-looking, but because they (in the opinion of this writer) describe the measurement concept of that particular set of instruments. To use only the latest instruments for examples in a text fails to acknowledge that there are still labs using some very good "old solid stand-bys." Second, to keep the text at a manageable size, a "Readers Digest" type condensation was performed. Third, new material was added where deemed appropriate.

This revision is just that—a revision of a text that covers the field of electronic instrumentation. The basic concepts still hold true. The new material is an attempt to add those fields omitted in the previous additions.

I wish to express my thanks to the staff of Prentice-Hall for asking me to do this revision. I thank all of those manufacturers who gave permission to have their instruments included in the text. I especially thank Professor Lou Covert of Southern Technical Institute for his invaluable assistance, guidance, and reading of the material with comments for its improvement. My thanks to David Boelio of Prentice-Hall for putting up with my failing to meet deadlines. But most of all my thanks to the family of Professor Sol D. Prensky for giving me the opportunity to share in his work. I dedicate this edition to the memory of Professor Prensky, a man who devoted his life to the learning experience.

RICHARD L. CASTELLUCIS

1

General Instrumentation

1-1. GENERAL DESCRIPTION OF INSTRUMENTATION

Instruments of many kinds have the common purpose of supplying information concerning some variable quantity (sometimes called a *parameter*[1]) that is to be measured. This information is generally obtained as a deflection of a pointer on a meter or in the form of a digital readout, and in this general way the instrument performs an *indicating function*. In many cases the instrument also provides a chart record of the instantaneous indications, thus performing a *recording function*. A third important instrument function, particularly in industrial-process situations, is accomplished when the information is used by the instrument to *control* the original measured quantity. We can thus classify instruments by function into three main groups: the large, general group that has only the one indicating function; another large group of indicator/recorders; and finally the specialized group that performs all three functions of indicating, recording, and controlling.

The last-named group, which includes the control function, forms the basis of automatic systems, wherein a process is automatically controlled by the information fed back by the monitoring instrument. This leads to more complicated

[1] A quantity that is experimentally varied in a series of steps.

but very useful *automated systems* (*or automation*), a specialized field that has rapidly extended to science and industry its benefits of minimizing repetitive work.

This text emphasizes the more general instrument functions of *indicating and recording*, especially in those instruments that supply information for *measurement and test purposes;* the control function will be examined in those cases where controlling enters as an integral part of the indicating and recording functions of electronic instrumentation.

General Measurement and Test Methods

Many methods are available for obtaining a given measurement; some measurements are made by *mechanical* means, as where the force exerted by a gas under pressure is measured by means of a Bourdon-tube pressure gage; other methods are primarily *electrical*, as in a measurement of solution conductivity by a current meter; and still other methods are *electronic*, as, for example, in the electronic voltmeter, where electron devices contribute amplification to provide more sensitive detection of the measured quantity.

The use of electronic amplification together with electrical measurement methods offers means of electronic measurement having significant advantages. The use of amplification to produce *highly sensitive indications* is valuable in detecting small physical changes, while the ability to obtain this *indication at a remote location* helps in monitoring inaccessible or dangerous locations. Heightening the effectiveness of the electronic method of measurement is the fact that it can be extended to a wide variety of nonelectrical fields by the action of *transducers.* These are devices that convert other forms of energy to an electrical form. The many available types of transducers enable us to measure many forms of energy in terms of electrical signals; these transducer types are described in a separate chapter.

Energy Conversion by Transducers

The basic action of transducers may be examined here in terms of energy conversion. For example, we have the familiar microphone used to convert *sound* energy to an electrical signal; the *thermocouple* for converting heat energy to an equivalent electrical voltage; the *photocell* for converting light energy to electricity; the *Geiger–Mueller tube* for the conversion of nuclear radiation into electrical pulses; and a host of transducers for obtaining electrical outputs from *mechanical forms of energy*, such as force, pressure, displacement, flow, acceleration, and the like.

Electronic versus Electrical Instruments

The newer developments in electronic instruments have been built up on the base of the older, electrical instruments, so the instrument forms at present are thoroughly mixed. We will see more clearly the role played by the electronic—as opposed to electrical—instruments if we identify the unique features the former contribute. Basically, the *electronic instrument includes in its makeup some electron device*, such as a vacuum tube, semiconductor diode, transistor, gas tube, and integrated circuit, while the purely electrical instrument does not. Thus, electronic voltmeters, whether of the vacuum-tube or solid-state variety, are obviously electronic instruments, while the common dc voltmeter, based on the moving-coil meter movement, is clearly an electrical instrument. There are instances, however, where this distinction may not be so clear-cut, as in some solid-state applications; but it is clear enough, generally speaking, to be a very useful guide.

For example, the *greater speed* inherent in the electronic switching method (using tubes or transistors) is easily observed in comparison with the electro-mechanical method using relays; similarly, electronic instruments that incorporate some degree of *amplification of the input signal* in the circuit arrangement stand out clearly as more sensitive than their purely electrical counterparts. In addition to speed and amplification, specific features contributed by electron devices in instruments are discussed more fully in Chapter 6.

Summarizing, it may be said in general that the *use of electronic principles provides instruments having higher sensitivity, faster response, and greater flexibility* than their mechanical or purely electrical counterparts in *indicating*, *recording*, and, where required, *controlling* the measured quantity.

1-2. DEVELOPMENT OF ELECTRONIC INSTRUMENTATION

A brief glance at some steps in the development of our knowledge and control of the electron will help us to understand the potentialities of the present electron devices. At the turn of the century (in 1906) a Nobel Prize was awarded to Joseph J. Thomson for his discovery of the electron. At that time, too, the very first three-element electron tube, the "audion," was introduced by its inventor, Lee DeForest. Thereafter, electronic amplifiers and circuits developed at a rapid rate; communication and radar circuit developments were spurred by wartime necessities in 1914 and 1941. Peacetime uses for entertainment purposes were first given a great impetus by the broadcast-radio receiver introduced as a "music-box" by David Sarnoff (forming the basis for the huge Radio Corporation of America); then again by Allen B. DuMont's development of the cathode-ray tube, culminating in Major Armstrong's later developments in FM reception. These were followed by developments from many manufacturers of television receivers.

The accent on communication developments in AM and FM radio, and television, suitable for mass distribution, was a powerful factor in making electronic components available at reasonable cost. As a result of these civilian and military developments, industrial applications developed concurrently.

The more recent advances in space exploration and the advent of the micro-processor have brought us to a new frontier. The significant point is the success that has been achieved in such a relatively short time in harnessing the potentialities of electron devices. Developers have been quick to sense the versatile properties of the electron and to incorporate them in devices able to respond quickly and faithfully to the slightest impulse in the form of an electrical signal.

A few examples from the abundance of possible instances may be cited to illustrate electron-device capabilities:

1. In the field of medical electronics, the faithful recording by the electroencephalograph[2] of minute electrical impulses generated in the body in the form of brain waves indicates the extremely *sensitive response* made possible by electronic amplification.
2. In the TV broadcast station, the lightweight, almost inertia-free property of the electron is employed in the transmitter oscillator and amplifier tubes to provide extremely rapid alternations, calling for electron oscillations at rates greater than 200 million times/sec. This VHF band (216 MHz for TV Channel 13) is not even close to the upper limit of *speed of electron response* employed in the microwave regions, where frequencies in the range of thousands of MHz [gigahertz (GHz)] are in common use.
3. In the field of telemetry, we have an illustration of the great *flexibility* of electronic instrumentation; for example, in a space capsule over a dozen electrical signals monitor a like number of variable quantities, and these signals are then sequentially multiplexed on a single carrier, to be received in intelligible form by a ground station and processed by computers.

Even a brief view of the field should include the important area of computers, both *analog* and *digital*, in which the electrical signals mentioned in the above examples can be further processed. Such examples serve to emphasize the great versatility that can be attained by applications of electronic instrumentation in the performance of highly sensitive and intricate tasks.

1-3. FORMS OF ELECTRONIC INSTRUMENTS

The general appearance of three electronic indicating instruments is illustrated in Fig. 1-1. Figure 1-1(a) shows the familiar pointer-type *indicating* instrument; Fig. 1-1(b) shows a digital voltmeter (DVM); and Fig. 1-1(c) shows a solid-state voltmeter. Examples of three electronic graphic recorders are shown in Fig. 1-2.

The study of electronic instruments, by its nature, must inlcude a study of essential electrical principles. However, since this text emphasizes the electronic aspects of instrumentation, the reader is referred to other texts for more detailed discussion of electrical measurement and control instrumentation. Additional references, giving further details on specific electronic instrument topics, are listed in Appendix B.

[2]Amplifier and recorder of brain waves.

(a)

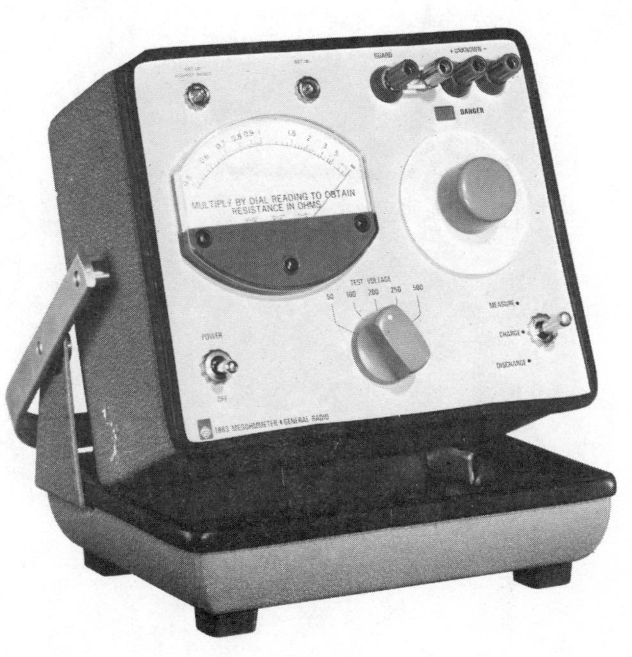

(b)

Figure 1-1. Typical electronic instruments: (a) VOM (courtesy VIZ Mfg. Co.); (b) megohmmeter (courtesy GenRad, Inc.); (c) milliohmmeter (courtesy Hewlett-Packard).

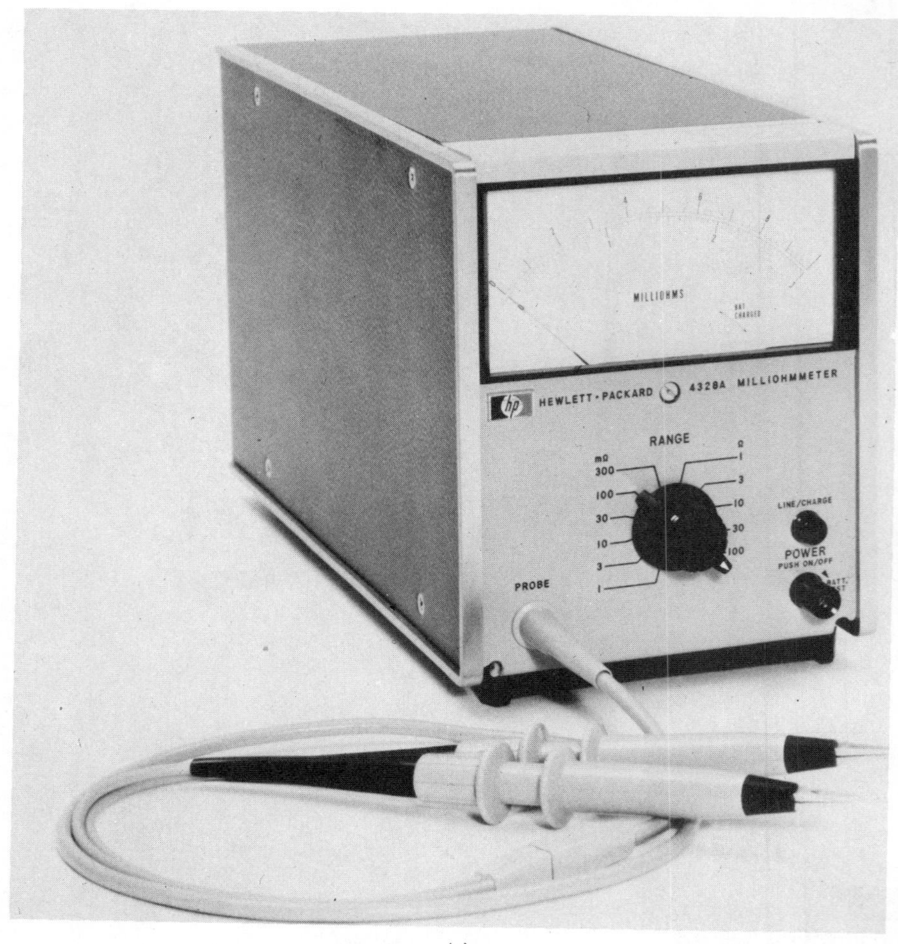

(c)

Figure 1-1. *Continued.*

1-4. ELEMENTS OF ELECTRONIC INSTRUMENT SYSTEMS

The elements that make up an electronic instrument system for providing a measurement or test indication are shown in generalized block-diagram form in Fig. 1-3.

The Transducer

The first element, the *transducer or sensing element*, is required when measuring a nonelectrical quantity, such as temperature, in order to convert the quantity to be measured (the degree of heat in this case) into a corresponding electrical output. A transducer is not required, of course, when measuring a quantity that is already in electrical form.

The Signal Modifier

The *signal modifier* is required in electronic instrument systems to make the incoming signal suitable for display on the indicating meter. In a great many cases, the incoming electrical signal is at too low a level (or the changes in a high-level signal are too small) for effective measurement unless the signal is *amplified.* In such cases, the ease with which electron devices can amplify electrical signals is revealed as an outstanding advantage of electronic instrumentation. In other

Figure 1-2. Electronic recording instruments. (Courtesy Honeywell, Inc.)

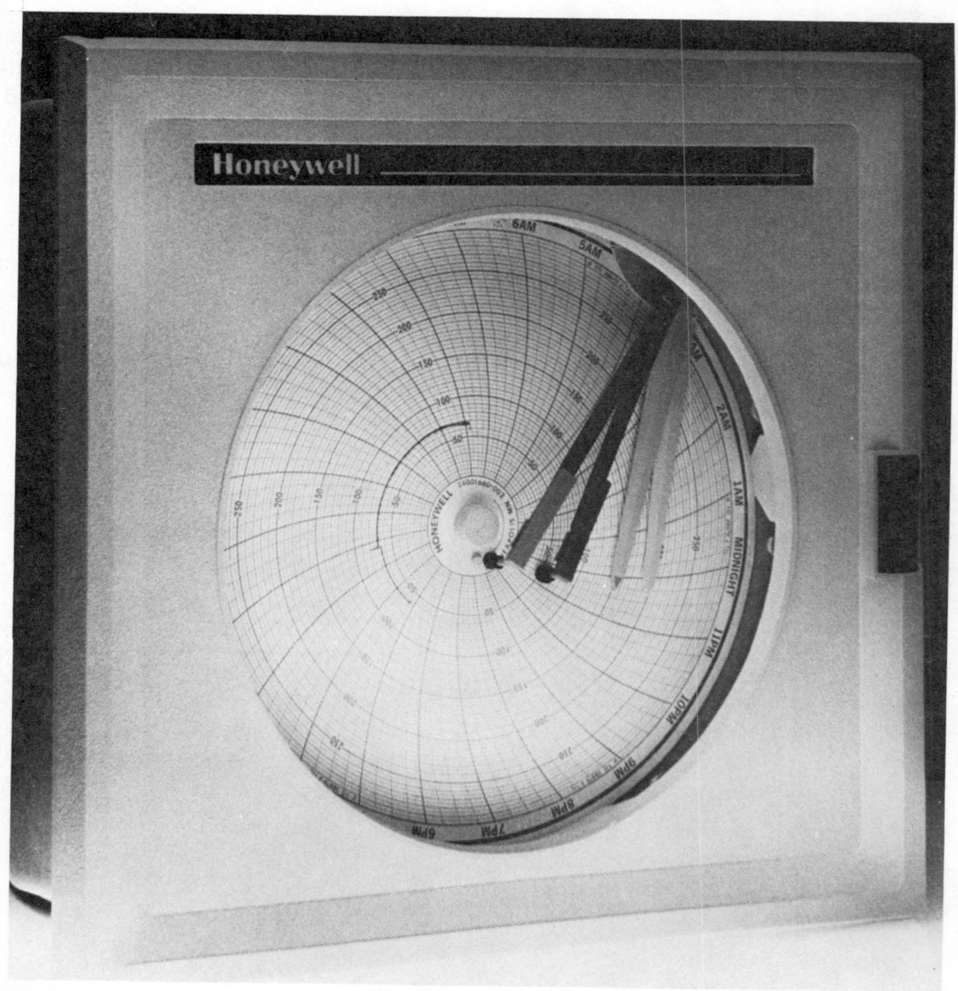

Figure 1-2. *Continued.*

instances, the electrical signal may be modified by such processes as *filtering action, rectification, and wave-shaping* to produce the desired indication.

The Indicating Meter

In the great majority of tests or measurements the result is indicated on an electrical meter of some sort, such as a voltmeter, ohmmeter, or milliammeter [the so-called multimeter or volt–ohm–milliammeter (VOM)]. Such a meter is predominantly of the pointer-and-scale (or *analog*) type, as opposed to the increasing trend toward

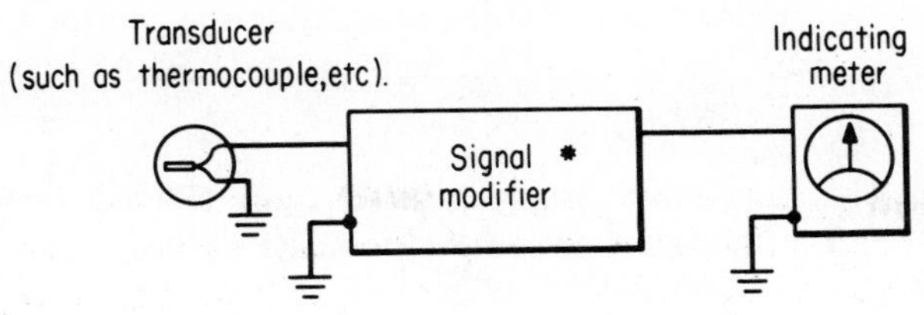

Figure 1-3. Block diagram of a generalized electronic indicating system.

numerical-reading (or *digital*) types. The digital meters are employed in those situations where speed and ease of reading become governing factors. The total variety of indicating methods, however, is quite great, including as simple a case as an indicating lamp and, at the other extreme, much more complicated analog indicators, such as pen recorders, oscilloscopes, and computer graphic displays, in addition to the digital readouts.

1-5. OVERALL VIEW OF INSTRUMENT SYSTEMS

The large variety that exists in the various forms of instrument indication provides a slight clue to the great diversity in the overall field of instrumentation systems. It is well, at this point, to take a broad overview of the instrument picture in order to appreciate the need for confining our efforts to particular segments of this very wide field. This selectivity is necessary to accomplish the double purpose of avoiding the danger of becoming too diffuse in trying to cover too wide a territory, while keeping the discussion within the useful boundary of sufficient definiteness and depth for practical understanding.

As a good start, we can divide the overall electronic-instrument field into two main areas:

1. The *mass-market entertainment field*, embracing the following popular forms:
   (a) Radio
   (b) TV
   (c) Phonograph and stereo equipment
   (d) Tape recorder devices and their associated test instruments (including newer entries, as in the automotive and camera fields)
2. The equally broad *industrial fields*, encompassing the following forms:
   (a) Commercial communication
   (b) Aerospace/telemetry activities

(c) Automatic process monitoring and control activities
(d) Versatile computer applications, both for analog simulation and for digital data processing

This condensed listing is surely quite extensive enough, without belaboring the point with such newer developments as lasers, oceanography, and chemical analysis and biomedical applications, to justify concentrating on basic representative segments while keeping a peripheral eye on the actively developing growth of the overall electronic instrument field through current literature.

The first large segment to be discussed will be the electrical (rather than electronic) indicating meter, of the volt–ohm–milliammeter type; this is basic to the electronic-instrument system. The two other main blocks, dealing with transducers and all kinds of signal modifying activities, will follow in later chapters.

2

The Basic Meter in Direct-Current Measurements

2-1. THE BASIC METER

The action of the most commonly used dc meters is based on the *fundamental principle of the motor*. The motor action is produced by the flow of a small amount of current through a moving coil, which is positioned in a permanent-magnet field. This *basic moving-coil system*, often called the *D'Arsonval galvanometer*, will be referred to as the basic meter.

A brief study of the basic meter and the various forms in which it appears as an electrical measuring instrument will help us in comparing its action with that of electronic instruments. Consequently, the basic meter will be briefly reviewed in this chapter in its dc instrument forms and in its ac forms in Chapter 3. The same basic galvanometer movement also forms the heart of a highly useful *multimeter* instrument, which is a combination volt–ohm–milliammeter (often abbreviated VOM), providing a large number of both dc and ac measurement ranges in a single instrument. This combination meter is described in later sections.

Regardless of the many different forms in which the basic moving-coil galvanometer appears, it is always essentially a *dc current meter*. The familiar details of the construction of this basic D'Arsonval dc meter are given in Fig. 2-1. The fundamental principle of motor action, illustrated in the figure, is valid for the more advanced developments, such as the 250° long-arc meter, shown in Fig.

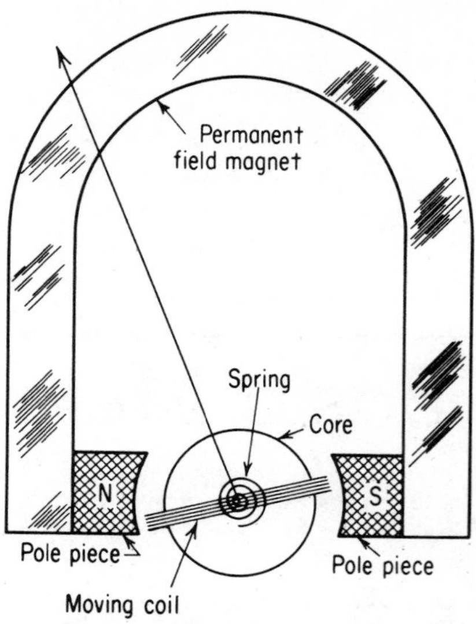

Figure 2-1. Construction of D'Arsonval moving-coil basic dc meter.

2-2, which provides a longer scale for the same-size meter body, and the "taut-band" suspension meter, shown in Fig. 2-3. Taut-band suspension allows a greater sensitivity to be obtained in portable meters than is generally available in the conventional design; a full-scale range of 2 μA, for example, is entirely practical.

Some of the different instrument forms that may be obtained by starting with the basic meter movement and adding various elements can be seen in Table 2-1 (the modifications shown apply equally well, whether the basic meter movement is in the form of a galvanometer with a zero-center scale or more generally in the form of a basic movement with the zero at the left end of the scale).

2-2. CONVERTING THE BASIC METER TO A DIRECT-CURRENT VOLTMETER

In the process of modifying the basic dc meter movement to function as a dc voltmeter, the essential point is to know the amount of current required to deflect the basic meter to full scale, a quantity sometimes called the full-scale (fs) deflection current. For an example of meter conversion, assume a value of 50 μA for the full-scale deflection current. This full-scale value will produce a voltmeter with the familiar sensitivity figure of 20,000 Ω/V, widely used for commercial dc voltmeters in electronic testing applications. The sensitivity figure is based on the fact that the full-scale deflection of 50 μA results whenever 20,000 Ω of resistance are present in the meter circuit for each volt of applied voltage. Expressed in mathe-

Long-arc (250°) meter scale: Weston

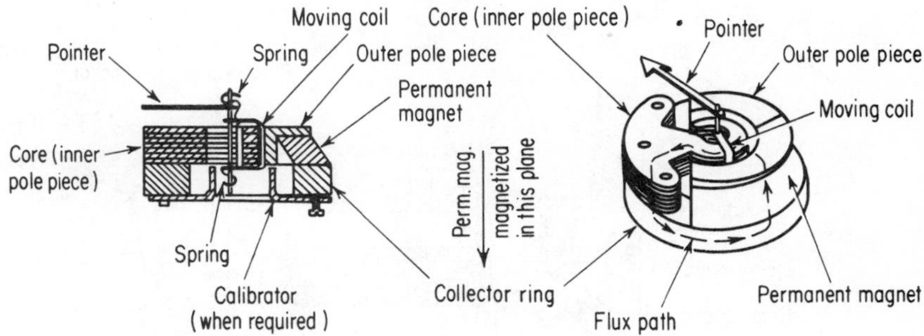

Construction detail sketch of 250°-arc meter: Westinghouse

Figure 2-2. Long-arc (250°) meter.

matical form, this is

$$I_{fs} \text{ (full-scale current)} = 50\ \mu\text{A} = \frac{1\text{ V}}{20\text{ k}\Omega}$$

or

$$\text{sensitivity (ohms per volt)} = \frac{1\text{ V}}{I_{fs}}$$

Thus, a 0–1-mA meter would have a sensitivity in ohms per volt of 1 V/1 mA = 1 kΩ (or 1000 Ω)/V.

In accordance with the sensitivity figure of 20 kΩ/V in our example, the kΩ series multiplier required for a 5-V dc range is 20 kΩ × 5 = 100 kΩ, as illustrated in Fig. 2-4 by resistor R_a. Since the resistance of the meter is of the order of 1 kΩ, it is considered negligible (within 1%) in this case, and a round value of 100 kΩ is suitable for resistance R_a on this range.

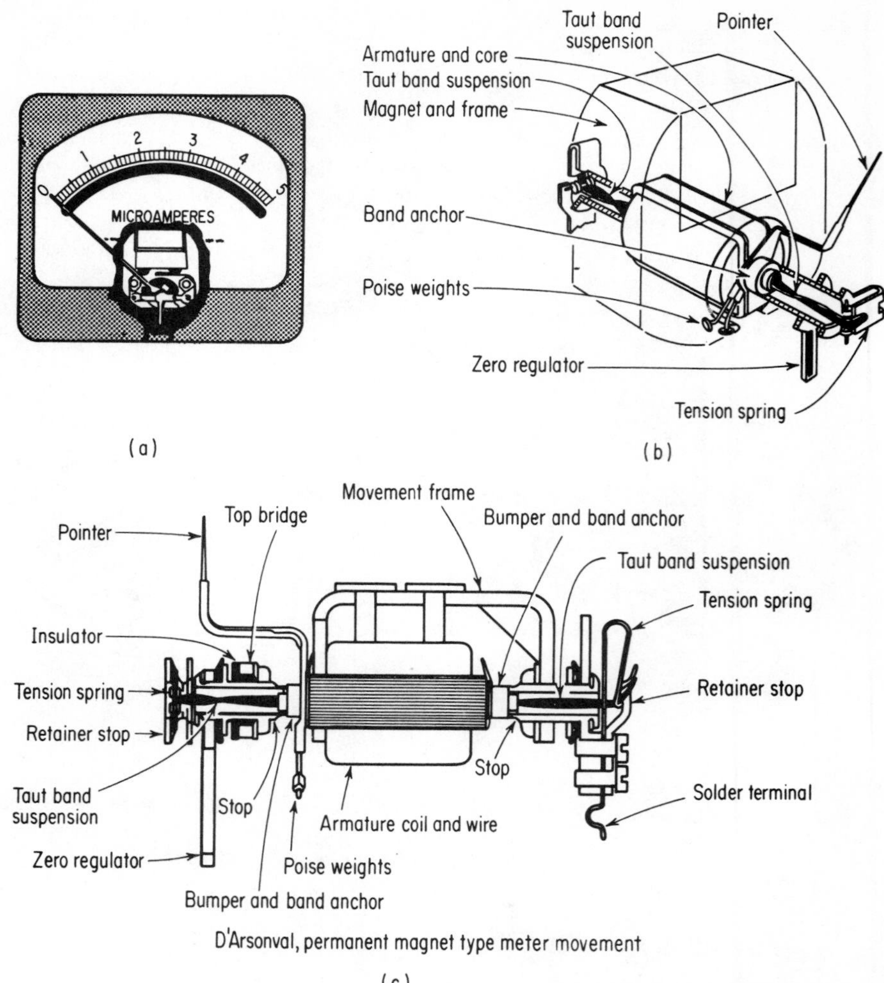

Figure 2-3. "Taut-band" suspension-type dc meter: (a) cutaway view; (b) perspective view; (c) side view. [From J. F. Rider and S. D. Prensky, *How to Use Meters*, 2nd ed. (New York: Hayden-Rider, 1960).]

If a second range is desired, a second multiplier resistance (R_b) suitable for the desired range can be selected by the switch shown in the diagram. Such switching arrangements are discussed later in a section on multirange meters.

The basic meter can also be converted to provide dc measurements of *current* or *resistance*, as shown in the previous listing of conversions. Each such conversion is discussed separately in the sections that follow.

TABLE 2-1. Variations on Basic Meter Movement

A.	Basic meter movement becomes a dc instrument measuring the following:		
	(A1) *Direct current, by adding a shunt resistor;* forming a microammeter, milliammeter, or ammeter.	(A2) *Direct voltage, by adding a multiplier resistor;* forming a millivoltmeter, voltmeter, or kilovoltmeter.	(A3) *Resistance, by adding a battery and resistive network;* forming an ohmmeter
B.	Basic meter movement becomes an ac instrument measuring the following:		
	(B1) *Alternating voltage or current, by adding a rectifier;* forming a rectifier-type meter for power and audio frequencies.	(B2) *Radio-frequency voltage or current, by adding a thermocouple;* forming a thermocouple-type meter for radio frequencies.	(B3) *Expanded-Scale by adding a thermistor in a resistive bridge network;* forming an expanded-scale (100–140 V) ac meter for power-line monitoring.

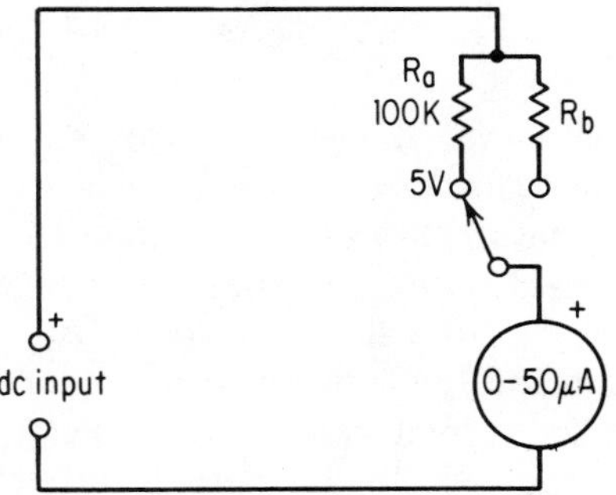

Figure 2-4. Conversion circuit for basic meter (0–50 μA) used as a 0–5-V dc voltmeter (20,000 Ω/V).

2-3. CONVERTING THE BASIC METER TO DIRECT-CURRENT RANGES

To illustrate the method for obtaining different dc ranges from a given basic meter, we take as an example a typical basic meter having a 0–1-mA full-scale range. We can arrange the instrument for higher current ranges by detouring some of the current through a shunt resistor connected across the moving coil. The value of the shunt used will determine the relative proportion of the current through the coil to the total current entering the meter-and-shunt combination. Suppose that we want to use this meter (100 Ω; 0–0.001 A full scale) as a 0–1-A meter. It will be necessary to detour 0.999 A (or 999 mA) around the coil and allow only 0.001 A (or 1 mA) to go through the meter coil. As seen in Fig. 2-5, the shunt and the meter-coil are in parallel, and therefore the voltage across both is the same. This

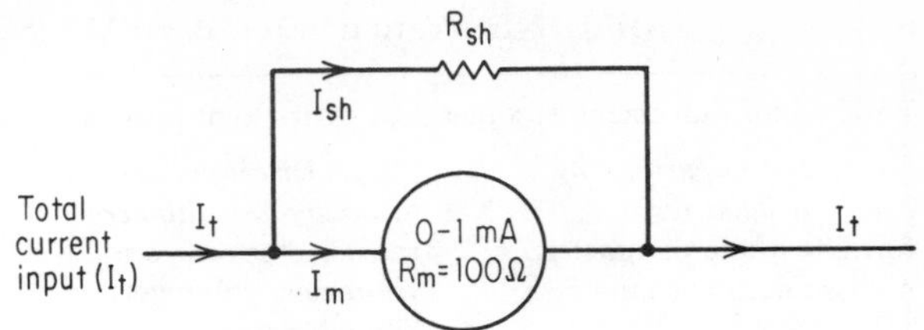

Figure 2-5. Shunt arrangement for basic meter (0–1 mA) used as a dc ammeter (0–1 A).

IR drop across the parallel combination must be 1 mA × 100 Ω, or 100 mV. The shunt must therefore have such a resistance that 999 mA will flow through it when a voltage of 100 mV (or 0.1 V) is applied across its terminals. The resistance of the shunt (R_{sh}) is

$$R_{sh} = \frac{E}{I} = \frac{0.1\text{ V}}{0.999\text{ A}} = 0.1\text{-}\Omega \text{ shunt resistance (very nearly)}$$

The result will be that for every 1000 mA in the circuit, 999 mA will go through the shunt and 1 mA through the coil; hence, the pointer will indicate full-scale deflection for the 1000-mA (or 1-A) input current. (Note that in Fig. 2-5 the solid arrows indicate direction of positive or conventional current flow.)

The value of the shunt resistance R_{sh} required for any new range can be obtained from the fundamental relation for current division, based on the fact that the smaller current in a parallel branch flows through the larger resistance. Expressed mathematically, the current through the meter I_m is given as a fraction of the total current I_t by

$$I_m = I_t\left(\frac{R_{sh}}{R_m + R_{sh}}\right)$$

In the present instance, expressing the current in amperes,

$$\frac{1}{1000} = 1\left(\frac{R_{sh}}{100 + R_{sh}}\right)$$

$$1000R_{sh} = 100 + R_{sh}$$

$$999R_{sh} = 100$$

from which

$$R_{sh} = \frac{100}{999} = 0.1\ \Omega$$

As a result, a shunt resistor of 0.1 Ω connected across the meter will change the full-scale deflection from the original 1 mA to the new value of 1 A. Other current ranges can be obtained by switching in corresponding values of shunt resistors.

2-4. MULTIPLE DIRECT-CURRENT VOLTMETER RANGES

As previously mentioned, the method for producing various voltage ranges from a basic current meter entails the addition of so-called multiplier resistors in series with the coil movement. The value of such a multiplier resistor is determined by the sensitivity figure of the voltmeter, as exemplified in the 20,000-Ω/V voltmeter derived from a basic 0–50-μA meter. Comparing this full-scale value with that of the 0–1-mA basic meter just used for the ammeter ranges, we observe that the 0–1-mA meter requires 20 times more current than the 50-μA meter, and it can therefore produce a voltmeter sensitivity of $\frac{1}{20}$ of 20,000 Ω/V, or 1000 Ω/V.

If we now start with the original unconverted 0–1-mA dc meter (Sec. 2-3) with a resistance of 100 Ω, we already have a voltmeter that registers full scale on the application of 100 mV (1 mA × 100 V), without the addition of any extra series resistance. (Such a meter, of course, would have an extremely limited use because of its low input resistance of 100 Ω, drawing an excessive current from most circuits being measured.) A more reasonable input resistance would be obtained on converting this basic 0–1-mA meter to a 0–5-V voltmeter. The total resistance (R_t) required in the circuit for a 0–5-V range would be

$$R_t = \frac{5\text{ V}}{1 \times 10^{-3}\text{ A}} = 5\text{ k}\Omega$$

Since the 100 Ω of meter resistance may be considered negligible compared with the required total of 5000 Ω, a multiplier resistance of 5 kΩ would ordinarily be used, shown as R_s in Fig. 2-6(a). (This value can be obtained more readily by multiplying the 1000-Ω/V sensitivity figure by the desired voltage range of 5 V, giving 5 kΩ as the required value for R_{s_1}.) Similarly for the 150-V range, R_{s_2} becomes 150 kΩ; and R_{s_3} is 300 kΩ for the 300-V range. In Fig. 2-6(b) the multiplier resistances (R_{s_4}, R_{s_5}, and R_{s_6}) for extending the voltage ranges are rearranged in series, as is common commercial practice. In such a series arrangement, the calculated resistance values are $R_{s_4} = 5$ kΩ, $R_{s_5} = 145$ kΩ, and $R_{s_6} = 150$ kΩ. Sometimes, when the degree of tolerance allows, the nearest standard value is used instead of the exact value; thus within a 5% tolerance, for example, a 150-kΩ resistor might be used for R_{s_5}, instead of the calculated value of 145 kΩ.

2-5. THE OHMMETER

The basic meter movement can also be used to measure resistance, if the basic meter is combined with a source of voltage and a resistive network. In the circuit diagram of Fig. 2-7(a) it will be noted that a small battery and a variable resistor are contained within the ohmmeter case. The unknown component is connected between points X–X, and the ohmic resistance is read on a scale calibrated in

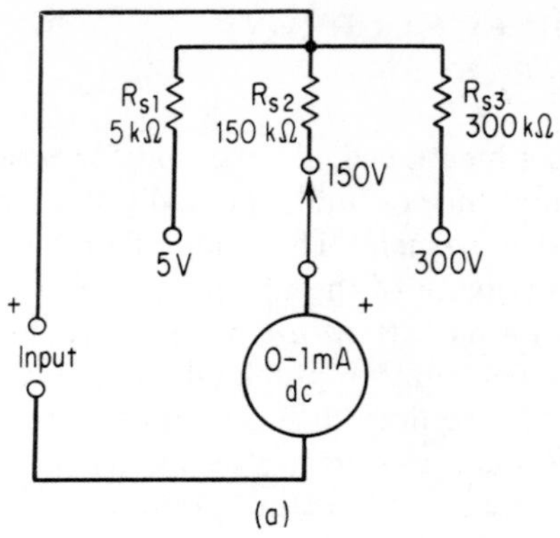

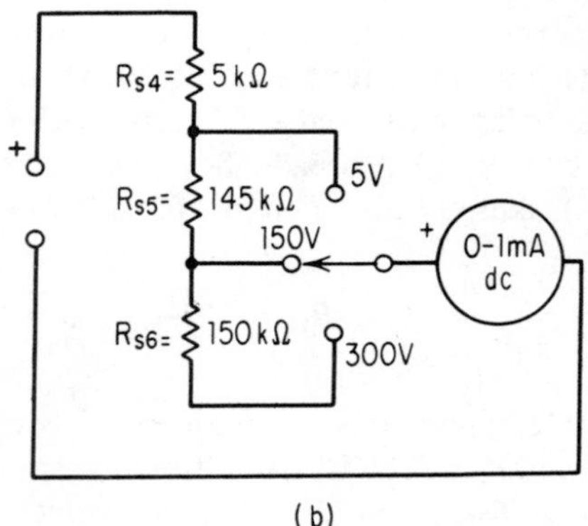

Figure 2-6. Multivolt ranges: (a) multipliers arranged individually; (b) multipliers arranged in series.

ohms. This direct-reading scale is possible because the current flow is inversely proportional to the total resistance of the circuit, with the voltage kept constant. The variable resistor compensates for the change in voltage of the battery during its life and thus keeps the effective voltage substantially constant.

To operate the instrument, the familiar procedure is to first short-circuit points X–X together, and then adjust the variable resistance (R_i) until the meter registers full-scale current (0 Ω on the ohms scale). The unknown resistance is then inserted between points X–X, and its resistance is read on the new nonlinear scale, calibrated to be direct-reading in ohms.

If we use the 0–1-mA meter as the basic meter, it will be recalled that it requires 0.001 A for full-scale deflection and has 100-Ω resistance. Let us also assume a battery potential of 3 V. Now the total resistance R_t of the circuit must be

$$R_t = \frac{E}{I} = \frac{3}{0.001} = 3000\ \Omega$$

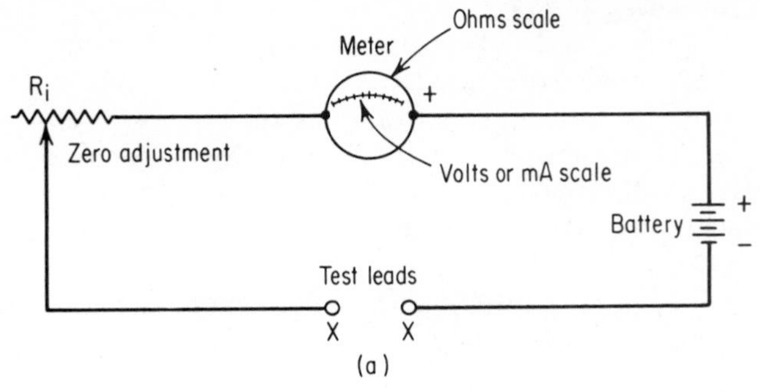

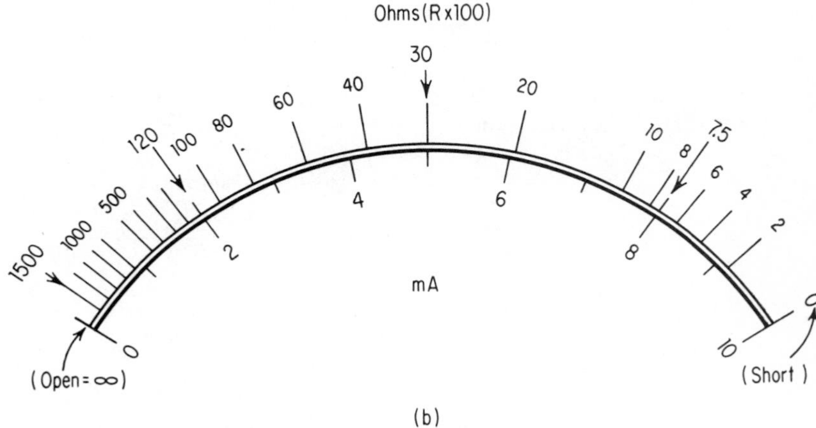

Figure 2-7. Ohmmeter developed from a basic dc meter: (a) circuit; (b) nonlinear ohms scale.

Therefore the variable resistor must be set at 2900 Ω to produce full-scale deflection, since with this value added to the 100 Ω of the meter, 3000 Ω is present when the test terminals are shorted (for the 0-Ω condition).

Now, for example, when an unknown of, say, 12,000 Ω is inserted between points X–X, the total circuit resistance will be $12{,}000 + 2900 + 100 = 15{,}000\ \Omega$. The battery is still the only voltage source, therefore the current flowing in the circuit will be

$$I = \frac{E}{R} = \frac{3}{15{,}000} = 0.0002 \text{ A, or } 0.2 \text{ mA}$$

The deflection of the meter needle accordingly will be 20% or one-fifth of full scale. This deflection would, of course, correspond to a reading of 12,000 Ω on the calibrated OHMS scale.

A few such simple calculations will readily show that the resulting ohms scale will read in a nonlinear fashion, as illustrated in Fig. 2-7(b), where some calibrated values for ohms are shown on the top scale, placed above the original milliampere scale. *Note that the half-scale reading on the OHMS scale must be equal to the internal resistor in the ohmmeter*, since the combination of the internal resistor (R_i) and an equal value for the unknown resistor (R_x) adds up to twice the original resistance. The resulting current is one-half of the original current, or 0.5 mA. Hence, the half-scale mark at 0.5 mA on the current scale corresponds to the number 30 on the ohms scale on the $R \times 100$ range of the ohmmeter (standing for 3000 Ω).

In general, the Ohm's law relation between the meter current and the amount of unknown resistance R_x is

$$I_m = \text{meter current (in amps)}$$
$$= \frac{\text{battery voltage (in volts)}}{\text{internal resistor } R_i + \text{unknown } R_x \text{ (in ohms)}}$$

This is more conveniently stated by expressing the meter current in milliamperes and each of the resistances in kilohms:

$$I_m = \text{meter current (in milliamperes)}$$
$$= \frac{\text{battery voltage (in volts)}}{R_i \text{ (in kilohms)} + \text{unknown } R_x \text{ (in kilohms)}}$$

In this case, for half-scale deflection,

$$\frac{1}{2}\text{ mA} = \frac{3\text{ V}}{3 + R_x \text{ (in k}\Omega)}$$

giving 3 kΩ for the value of R_x.

In a similar fashion, at four-fifths of full-scale deflection (0.8 mA), we find $R_x = 750\ \Omega$ (or $\frac{3}{4}$ kΩ). Since the change of 0.3 mA (from 0.8 to 0.5 of full-scale deflection) covered a difference of $2\frac{1}{4}$ kΩ and the corresponding 0.3-mA change (from 0.5 to 0.2 mA) covered a difference as great as 9 kΩ, it can be seen that the ohms scale is quite nonlinear and becomes very crowded at the high-resistance end. Considering the practical limit of readability to be one scale division out of a total of 50 divisions, this limit of $\frac{1}{50}$ scale deflection (or 0.02 mA) will set a limit for the highest practical resistance marking on the ohms scale. The corresponding value for R_x at that limit is 147 kΩ (or about 150 kΩ). It can therefore be deduced that the use of a 0–1-mA dc meter as the basis for the ohmmeter will not allow the measurement of resistors much greater than about 150 kΩ with a 3-V supply (or 75 kΩ with a 1.5-V supply).

Multirange ohmmeters (Fig. 2-8) differ considerably from the simplified single-range example given above. They generally use a 50-μA meter (20 times more sensitive than the example), along with a 1.5-V supply on the middle range (such as $R \times 100$), thereby providing a reasonably high upper limit. Then, in order to be able to use the same nonlinear scale on the other ranges, it is necessary to

Figure 2-8. WV-520B VOM. (Courtesy VIZ Mfg. Co.).

switch-in both the proper meter sensitivity and suitable supply voltage by the RANGE switch. Thus, for a lower range, such as $R \times 1$, the meter is more heavily shunted to allow greater current through the unknown; for the highest range, an extra battery is switched in to boost the supply voltage.

2-6. THE MULTIMETER

For general-purpose and service-type measurements with the nonelectronic moving-coil meter, it is general practice to combine the functions of the dc voltmeter, the ohmmeter, and the milliammeter into a single multipurpose instrument, commonly known as a *VOM, or multimeter*.

2-7. INPUT IMPEDANCE OF VOLTMETER

The impedance (ac resistance) that a voltmeter presents to the circuit in which the voltage is being measured is an important consideration. It becomes increasingly important for measurement in high-impedance circuits or in any circuit where the available current is small, because of the fact that the operation of the voltmeter requires some finite current to be drawn from the measuring circuit. Where this voltmeter current is small enough (say, less than 1% of the current originally flowing in the measuring circuit), it can be safely assumed to be negligible. There are many cases, however, where the current drawn by the voltmeter is a considerable fraction of the original current in the circuit before the voltmeter was connected; in such cases, the voltmeter reading will be in substantial error.

If a dc voltage measurement is used as an example, the significance of the voltmeter input impedance can be more clearly appreciated by comparing the two cases given in Fig. 2-9, where the same voltage is being measured, but two meters

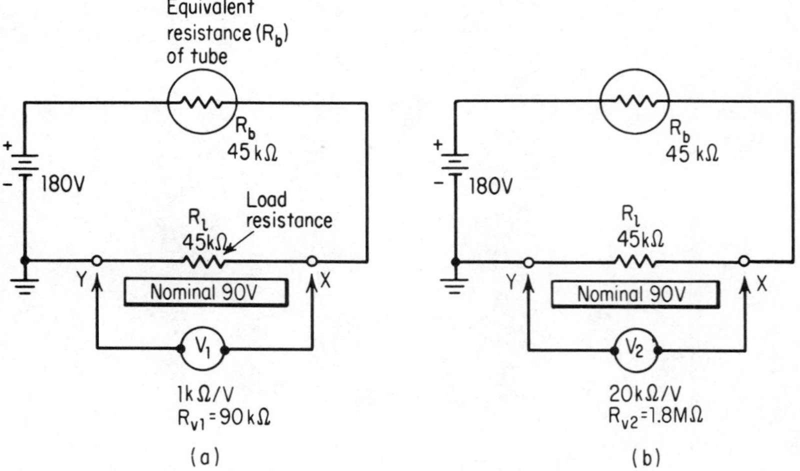

Figure 2-9. Loading effect of input resistance of voltmeter: (a) using 1000 Ω/V; (b) using 20-kΩ/V meter.

having different input resistance have been used. In Fig. 2-9(a), voltmeter V_1 at 1000 Ω/V is used to measure a nominal 90 V across a 45-kΩ load resistor R_1, which is in series with an equal tube-plate resistance R_b. Using a 90-V full-scale range for simplicity, the input resistance of the voltmeter will be 90 kΩ. As soon as the voltmeter is connected across the measuring points X and Y, the effective resistance across these points changes to the resistance of the parallel combination of $R_1 = 45$ kΩ in parallel with $R_{V_1} = 90$ kΩ, which results in a joint resistance across X–Y of 30 kΩ. Clearly, the voltmeter reading must register the voltage now existing across the joint resistance, which is

$$\begin{aligned} V_{XY} &= 180\left(\frac{R_{XY}}{R_b + R_{XY}}\right) \\ &= 180\left(\frac{30\text{ k}\Omega}{45\text{ k}\Omega + 30\text{ k}\Omega}\right) = 180\left(\frac{30}{75}\right) \\ &= 72\text{ V} \end{aligned}$$

This reading of 72 V is 18 V less than the actual value of 90 V, the true voltage that existed before the voltmeter was connected; it thus denotes a voltmeter reading error of $\frac{18}{90}$ or 20%! It should be noted that this erroneous reading is due not to any inherent malfunction of the voltmeter but rather to the impedance conditions of the circuit in which it was used.

To show that the voltmeter is operating properly within its limitations, it will be instructive to check its reading of 72 V by tracing the current distribution in the circuit under the loading conditions that exist. When the voltmeter is connected, the original current of 2 mA in the circuit changes to a new value of total current I' as follows:

$$\begin{aligned} I' &= \frac{180\text{ V}}{45\text{ k}\Omega + 30\text{ k}\Omega} = \frac{180\text{ V}}{75\text{ k}\Omega} \\ &= 2.4\text{ mA} \end{aligned}$$

The current division that takes place in the parallel circuit will give the current through the voltmeter I_{V_1} as follows:

$$\begin{aligned} I_{V_1} &= 2.4\text{ mA}\left(\frac{45\text{ k}\Omega}{45\text{ k}\Omega + 90\text{ k}\Omega}\right) = 2.4\left(\frac{45\text{ k}\Omega}{135\text{ k}\Omega}\right) \\ &= 0.8\text{ mA through the voltmeter } V_1 \end{aligned}$$

while the other 1.6 mA flows through resistor R_1. Since the meter current is 0.8 mA, the 0–1-mA meter of the voltmeter will accordingly deflect to $\frac{8}{10}$ of its full-scale reading of 90 V, or to 72 V. This checks with the IR drop across R_1, which equals 1.6 mA × 45 kΩ or 72 V, a value that also checks with the previous analysis.

When voltmeter V_2 at 20 kΩ/V (derived from a 0–50-μA basic meter) is used in Fig. 2-9(b), on its 90-V range, we find that the parallel combination now consists of $R_1 = 45$ kΩ and $R_{V_2} = 90(20\text{ k}\Omega)$ or 1800 kΩ. The parallel resistance of 45 kΩ and 1800 kΩ in this case is within 1% of the original 45 kΩ across X–Y. As a result, the voltmeter reading across the parallel combination is practically indistinguishable from 90 V (the actual reading turning out to be 89.6 V). Since the error is less than $\frac{1}{2}$% (in fact much smaller than the usual ±2% accuracy of the conventional multimeter), it can safely be considered negligible in this application.

When the two cases are compared, it can be seen that the relatively low input resistance of 1000 Ω/V for V_1 makes the voltmeter unsuitable in this application, whereas V_2 at 20 kΩ/V is quite suitable. Of course, there are any number of other applications in *low-resistance circuits* where V_1 would be entirely suitable; there are likewise many cases in *very-high-resistance circuits* where the ordinarily satisfactory input resistance of V_2 might be too low to be suitable. As a rough indication of suitability, the input resistance of the voltmeter is generally desired to be at least 20 times as great as the resistance of the circuit being measured. In those cases where a closer calculation of accuracy is desired, the circuit should be analyzed by comparing the distribution of current with and without the voltmeter connected, as was done in our example. Very often such a calculation will point out the necessity of using an *electronic voltmeter*, with its desirable high input impedance, in order to avoid a loading error.

QUESTIONS

Q2-1. Describe the action of a basic moving-coil *dc galvanometer* (center-zero type) when used in the following ac circuits:
(a) With a 60-Hz source.
(b) With a slowly—but nonperiodically varying—source, such as derived from a temperature transducer.
(c) With a 1-kHz source.

Q2-2. State the effect resulting from the following incorrect meter connections:
(a) A voltmeter is incorrectly connected in series in a position normally occupied by an ammeter.
(b) An ammeter is incorrectly connected in parallel, in a position normally occupied by a voltmeter.

Q2-3. Compared to the 1000-Ω/V sensitivity of a 0–1-mA basic dc meter used as voltmeter, what sensitivity (in ohms per volt) may be expected from a basic 0–2-μA taut-band suspension type of dc meter?

Q2-4. Describe how the construction of a taut-band suspension meter differs from the conventional D'Arsonval movement, and state an outstanding advantage and disadvantage.

Q2-5. In obtaining multiple current ranges by means of a switching arrangement for the shunt resistors, state the precautions necessary:
(a) To prevent excessive current during swtiching.
(b) To reduce errors caused by the introduction of the switch.

Q2-6. How is "zero-set" action obtained in an ohmmeter circuit, to ensure an exact zero reading for zero resistance, as the battery ages?

Q2-7. What precautions are necessary in using an ohmmeter:
(a) In checking the resistance of a low-voltage transistor on the high-OHMS range?
(b) In checking the resistance of a 0–10-μA meter on the low-OHMS range?

Q2-8. (a) Under what conditions is it *appropriate* to use an expensive voltmeter of 100 Ω/V?
(b) Under what conditions is it *not proper* to use such an instrument?

PROBLEMS[1]

P2-1. In the calibrating circuit shown in Fig. P2-1, the voltage-divider potentiometer is set to its maximum setting at point B. Disregarding the internal resistance of the meters, find the following:
(a) The maximum voltage available at point B with the switch open.
(b) The maximum voltage appearing at point B with the switch closed.
(c) The maximum current that flows through the meters when the switch is closed.

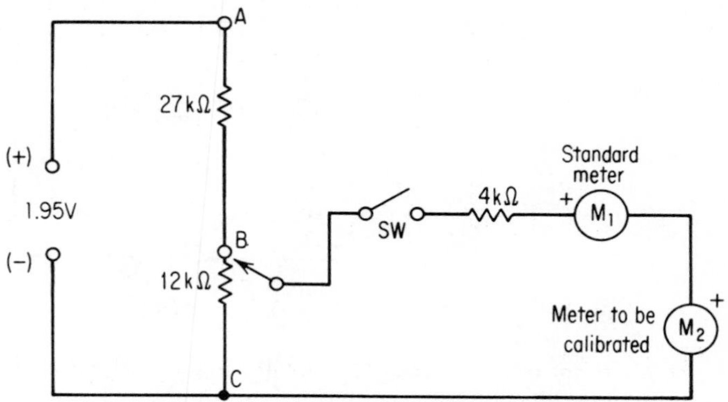

Figure P2-1

P2-2. The sensitivity of a laboratory galvanometer is sometimes expressed in terms of megohm sensitivity, indicating the resistance (in megohms) of the circuit to produce a deflection of one division (usually 1 mm), with a 1-V source. If a galvanometer, with a scale of 100 divisions of 1 mm each, has a megohm sensitivity of 20 MΩ:
(a) Express the sensitivity in microamperes per millimeter.
(b) Find the current required for full-scale deflection.

P2-3. Starting with a basic dc movement, having 200-Ω resistance and requiring 500 μA for full-scale deflection:
(a) Find the series resistance required for converting the basic meter into a 0–12-V dc voltmeter.
(b) Find the shunt resistance required for converting the basic meter into a 0–2.5-mA dc milliammeter.

[1]Answers to odd-numbered problems are given on page 446.

(c) Draw the schematic circuit arrangement providing all three ranges of 0–500 μA, 0–12 V, and 0–2.5 mA, using only one switch in the circuit.

P2-4. Using the same 0–500-μA movement with 200-Ω resistance as in Problem P2-3:
(a) Draw the schematic circuit with labeled values for an ohmmeter powered by a 1.5-V battery and employing a 1-kΩ rheostat as the zero-set control.
(b) What is the center-scale reading for this ohmmeter on its $R \times 1$ range?

P2-5. Extend the ohmmeter circuit in Problem P2-4 to include ranges for the following:
(a) $R \times 0.1$ range.
(b) $R \times 10$ range (a different battery may be used for this range).

P2-6. In the circuit of Fig. P2-6, a 6-V, 0.15-A pilot lamp is used with a dropping resistor of 40 Ω to produce the required 6 V for energizing the lamp from a 12-V source. Find the voltage reading obtained:
(a) When a voltmeter having 1000 Ω/V is used, on its 12-V range.
(b) When a voltmeter having 20 kΩ/V is used, on its 12-V range.

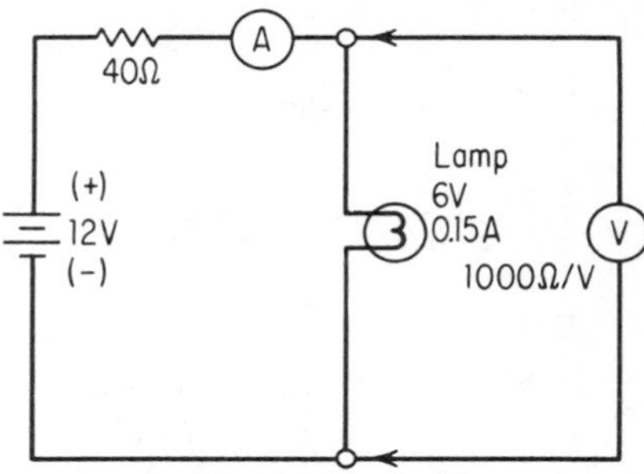

Figure P2-6

P2-7. In the circuit of Fig. P2-7, a voltmeter is used to measure the plate voltage across the tube resistance r_p.
(a) Find the reading obtained when a voltmeter having 1000 Ω/V is used, on its 360-V range, for this measurement.
(b) Find the reading across the same plate resistance r_p when a 20-kΩ/V meter (on its 360-V range) is used.

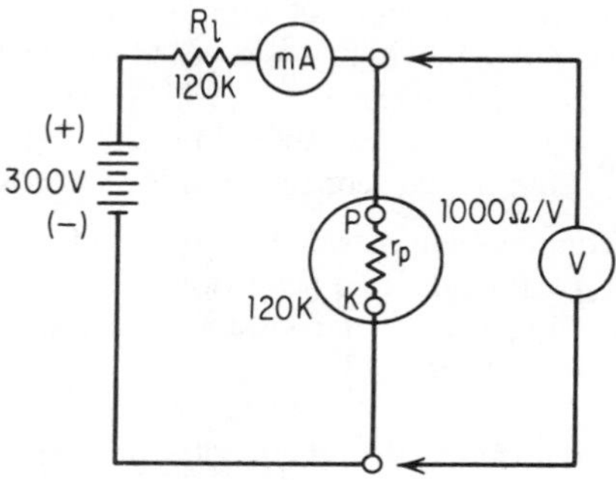

Figure P2-7

P2-8. In the series circuit of Fig. P2-8, dc voltmeter V (20-kΩ/V) is being used, on its 120-V range, to check the resistance of "black box" R. Connected in this way, it reads 60 V.

(a) Find resistance R.

(b) Find reading of voltmeter V, when used in the same position on its 300-V range.

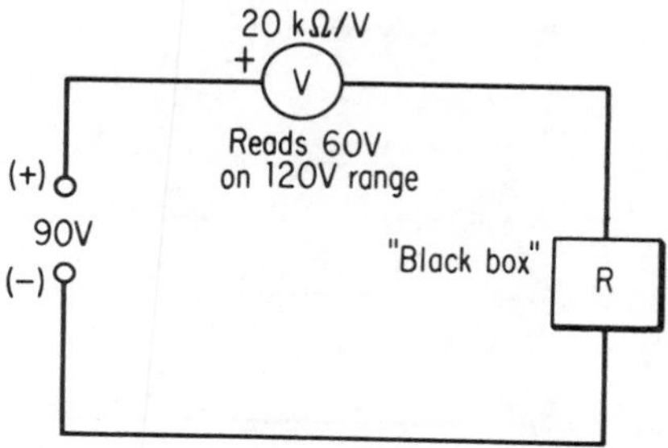

Figure P2-8

P2-9. A series combination of R_1 and R_2 is connected across a dc source, and the voltage-dividing action is being checked with a dc voltmeter having 1000 Ω/V on its 0–150-V range, as shown in the circuit of Fig. P2-9. The results obtained as the voltmeter is connected across any two terminals at a time are as follows:

$$V_{1\text{-}2} = 36 \text{ V (mA reads 0.36 mA)}$$
$$V_{2\text{-}3} = 18 \text{ V (mA reads 0.24 mA)}$$
$$V_{1\text{-}3} = 90 \text{ V (mA reads 0.80 mA)}$$

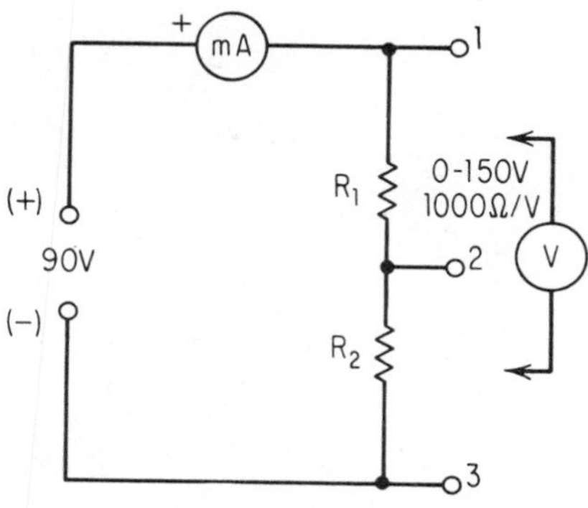

Figure P2-9

(a) Account for the fact that the two partial voltages in series do not add up to the total voltage of 90 V.

(b) Using only the information given, find the values of resistors R_1 and R_2.

P2-10. The circuit shown in Fig. P2-10 provides a dc volt–ohm–milliammeter (VOM) or dc multimeter, with voltage ranges of 0–10 V and 100 V, current ranges of 0–50 μA and 0–1.25 mA, and a single $R \times 1$ resistance range. Find the values required for the unmarked resistors R_{sh}, R_1, R_2, and R_3. (*Note:* As a voltmeter, the 50-μA meter provides a sensitivity of 20 kΩ/V.)

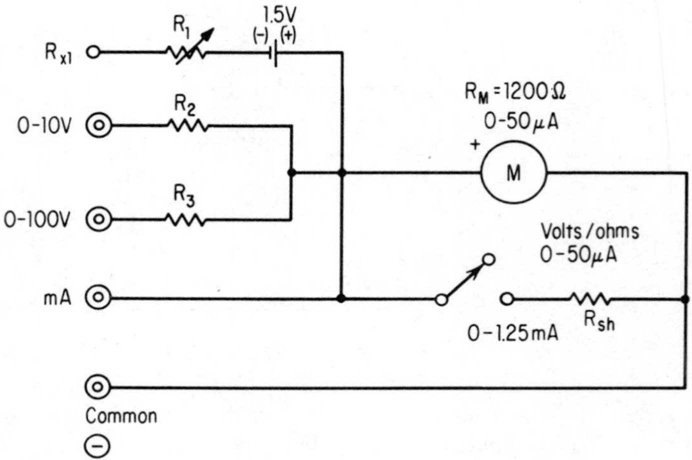

Figure P2-10. Volt–ohm–milliammeter (VOM) circuit.

3

The Basic Meter in Alternating-Current Measurements

3-1. RECTIFIER-TYPE METER

The basic dc meter also finds use in ac measurements, when a rectifier is added to the measuring circuit. Figure 3-1 gives a view of a full-wave instrument rectifier. A similar rectifier arrangement is also found as part of the AC VOLTS function in the volt-ohm-milliammeter (VOM), or multimeter test instrument (illustrated previously in Fig. 2-8). In this form, the rectifier voltmeter is widely used for general-purpose and service-type ac measurements. In comparison with other types of ac meters, the rectifier type is more widely used than either the more costly (but more accurate) dynamometer type (which is used primarily at power frequencies) or the more delicate thermal instrument types (thermocouple and hot-wire ammeters). The latter are employed mainly as transfer instruments to relate measurement at dc with those at high radio frequencies. These other ac types are discussed as separate topics later.

The simplified functional circuit of a rectifier meter is shown in Fig. 3-2; in Fig. 3-2(a) the dc meter is combined with a half-wave rectifier CR_1. If we continue using our basic dc meter as an example, we have a 0–1-mA meter, having a resistance of 100 Ω and a meter sensitivity of 1000 Ω/V. For the sake of simplicity, let the series multiplier resistor R_s be 10 kΩ, so that the circuit, without the rectifier, would be a 0–10-V dc voltmeter. (This range will allow us to consider as negligible

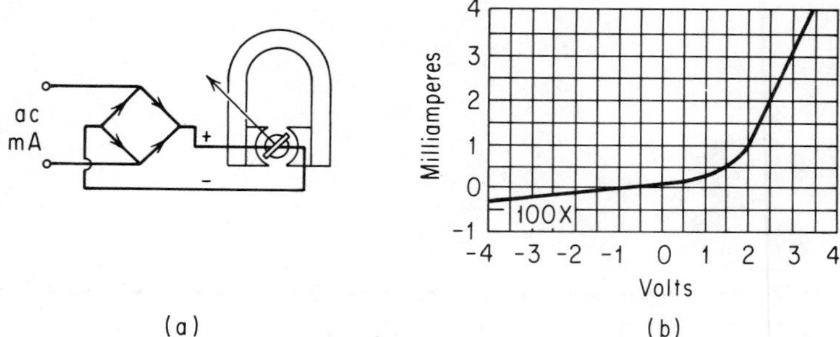

Figure 3-1. Instrument rectifier: (a) circuit when connected to a basic dc meter and mounted internally in the instrument; (b) voltampere characteristic of rectifier.

both the small resistance of the meter and the small forward resistance of the rectifier.) If a 10-V rms ac signal is now applied to the input terminals of the circuit, the input wave-form, shown at the left of the circuit, has a peak amplitude of $1.4 \times 10 = 14$ V. The output current through the meter has the half-wave rectifier form shown at the right of the diagram, and the meter deflection will be determined by the average dc of this wave-form (shown by a dashed line). The average current for the first half cycle is that produced by an equivalent dc voltage having a value of 0.636 times the peak amplitude of 14 V = 8.9 or approximately 9 V. Since there is no conduction for the second half cycle, the average over a full cycle corresponds to a *dc equivalent voltage* of half of this, which is 4.5 V, or 45% *of the rms input value* of 10 V. Stated in another way, with half-wave rectification the effective meter sensitivity on ac has been decreased to 45% of its value on dc.

Since the dc meter reads the average of the rectified output, the current through the meter for the half-wave circuit (I_m) may be expressed mathematically, using the dc equivalent of the rms input, as follows:

For half-wave rectification,

$$I_m = \frac{0.45 \times E_{\text{rms}} \text{ (volts input)}}{R_t \text{ (total resistance)}}$$

Some rectifier meters use the "three-terminal" half-wave rectifier diode circuit of Fig. 3-2(b). The three terminals encircled are ⊖, ⓒ, and ⊕. Here the horizontally arranged diode (section 1 of the instument rectifier) feeds the combination of the meter and its shunt resistor R_{sh} on the positive half cycle of input voltage; the vertically arranged diode (section 2) is nonconducting. On the negative half cycle of the input voltage, substantially all the current flow is shunted through the conducting diode (section 2), so that there is virtually no effect on the meter from the small reverse-conduction flow through the nonconducting diode section.

The presence of the shunt resistor R_{sh} across the meter allows the horizontally arranged diode to conduct more current and so operate in a more linear fashion.

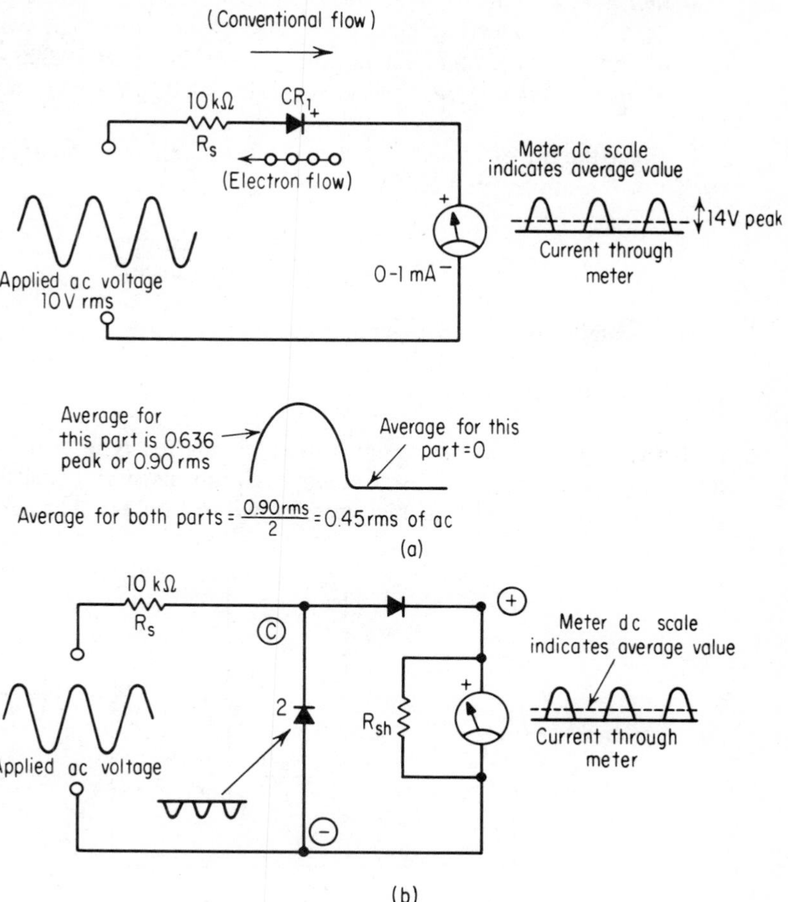

Figure 3-2. Functional circuit of rectifier-type meter: (a) simplified half-wave circuit; (b) practical half-wave circuit, using double-diode (or three-terminal) arrangement. [From J. F. Rider and S. D. Prensky, *How to Use Meters*, 2nd ed. (New York: Hayden–Rider, 1960).]

This is accounted for by the fact that the diode action has its greatest nonlinearity in the region of smallest current flow, as shown in Fig. 3-1; the action becomes much more linear after a certain minimum current flow is reached. The action of the shunt helps to provide this minimum of current and thus improves linearity. By the same token, however, the action of the shunt further reduces the sensitivity of the meter circuit. Hence, in addition to the fact that the simple half-wave rectifier circuit causes the ac sensitivity to be only 45% of the dc value, the action of the shunt acts to further reduce this sensitivity. As a result, instruments that provide 1000-Ω/V sensitivity on both dc and ac voltage measurements must initially provide a basic meter having a dc sensitivity of more than twice 1000 Ω/V. Or, seen in another way, *multimeters having a sensitivity of* 20,000 Ω/V *on dc usually have a much smaller ac sensitivity*, generally 5000 Ω/V (or, in a few cases, 10,000 Ω/V).

Since the three-terminal rectifier results in half-wave rectification, the formula for dc meter current in a half-wave arrangement:

$$I_m = \frac{0.45 \times E_{\text{rms}} \text{ (volts input)}}{R_t \text{ (total resistance)}}$$

applies equally to the three-terminal circuit.

Full-Wave Rectification

Full-wave rectifiers are sometimes used in rectifier-type ac meters, especially in laboratory instruments. The full-wave circuit usually consists of the familiar bridge arrangement of four rectifiers, shown in Fig. 3-3. The arrows represent the

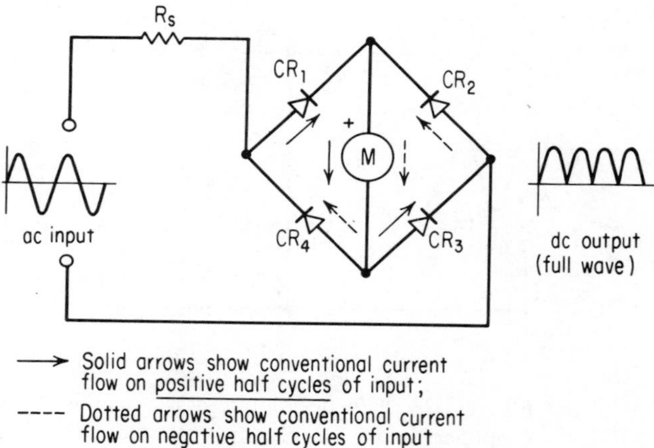

Figure 3-3. Full-wave (or bridge) rectifier circuit; the current through M is in the same direction for both the positive and negative half cycles of the ac input.

direction of positive-charge (or conventional) current flow (as elsewhere in this text, unless otherwise stated); the solid arrows in the diagram show the current direction through the two diodes CR_1 and CR_3 that conduct on the positive half cycle of the input-voltage wave-form, while the dashed arrows trace the current through the other two diodes, CR_2 and CR_4, that conduct on the negative half cycles. The resulting current through the meter is in the same direction on both half cycles, thus producing the full-wave output wave-form shown at the right of the diagram. The effective output in the full-wave case is obviously twice that of the half-wave case (to a very good approximation, where the series resistance R_s is much greater than the forward resistance of both diodes added together, as is usually the case). Hence, the equivalent dc voltage that would produce the same average current as the actual rms input voltage is accordingly twice the 45% value of the half-wave circuit, or 90% of the rms input voltage. The full-wave bridge rectifier arrangement is often used because it allows the rectifier meter to have a sensitivity value twice as great as the half-wave circuit. It also has the convenient feature that the input circuit need not provide a continuous path for dc as would be true in the elementary half-wave rectifier circuit of Fig. 3.2. This feature allows an ac circuit to be coupled to the meter through a capacitor.

3-2. COMPUTING ALTERNATING-CURRENT VOLTAGE RANGES

The calculation for the value of the series resistor on a given range of ac voltage can readily be made by interpreting the input ac voltage in terms of the *equivalent dc voltage* that produces the same average output. Thus, in the full-wave rectifier bridge arrangement of Fig. 3-3, if a full-scale range of 0–10 V rms is desired, the series resistor required has the same value it would have for an equivalent dc voltage range of 90% of this rms value, or 9 V dc equivalent. Assuming the use of a basic 0–50-μA meter (20 kΩ/V on dc volts), a full-scale deflection of 50 μA dc would be obtained from the dc equivalent of 9 V, using the following relation for meter current I_m.

For full-wave rectification:

$$I_m \text{ average dc} = \frac{E_{\text{dc}} \text{ (equivalent)}}{R_t \text{ (total)}} = \frac{0.90\, E_{\text{rms}}}{R_t}$$

or

$$0.050 \text{ mA} = \frac{9 \text{ V}}{R_t \text{ (in kilohms)}} \quad \text{and} \quad R_t = 180 \text{ k}\Omega$$

The value required for series resistor R_s is the total resistance R_t, less the resistance of the meter and the forward resistance of the two diodes. Since the forward resistance of each diode might be about 1000 Ω (under this low-current condition) and the meter resistance also about 1000 Ω, the combined resistance of 3000 Ω (or 3 kΩ) would be subtracted from total resistance R_t of 180 kΩ, to give a value of 177 kΩ for series resistor R_s (within a 2% tolerance a resistor of 180 kΩ would be satisfactory). The ac sensitivity in this case would then be 180 kΩ/10 V rms = 18 kΩ/V on ac volts.

In an actual case, the meter would probably be shunted to improve the linearity of the rectifier action, as previously shown in Fig. 3-2(b). This shunting action would lower the ac figure of 18 kΩ/V obtained in the example to perhaps 10 kΩ, or more likely, the more common figure of 5000 Ω/V on ac measurements.

When rectifier-type meters are used to make ac measurements at points that are also at a dc potential, it is important to remember that a series *blocking capacitor* is needed to isolate the ac measuring circuit from the dc potential that is present. In the general-purpose multimeter instrument such a blocking capacitor is already connected internally to the terminal marked OUTPUT. Thus, when the instrument test leads are plugged into the OUTPUT and COMMON terminals of the multimeter, the series blocking capacitor is automatically provided. Although the impedance of this capcaitor alters the ac voltage readings slightly, it does not interfere with obtaining *relative* voltage readings for comparison of ac outputs at the plate of a vacuum tube or the collector of a transistor.

3-3. FREQUENCY LIMITATIONS OF ALTERNATING-CURRENT METERS

The rectifier-type meter is used for ac measurements primarily in the power-frequency range (60 and 400 Hz) and in the audio-frequency range, with many measurements occurring at 1000 Hz. However, generally speaking, it cannot be relied upon for radio-frequency work; i.e., for frequencies much above 30 kHz. In discussing the usefulness of the *rectfier-type meter*, therefore, it should be kept in mind that the measuring circuits are assumed to apply *only to frequencies below the radio-frequency range.*

The restriction on the frequency range should not be attributed solely to the presence of the rectifier. When diodes are found combined with electronic dc amplifiers to form an *electronic voltmeter* or other electronic instruments, the meter movement is isolated from the ac portion of the circuit by the *electronic amplifier*. Here it does not contribute to frequency limitation as it does in the nonelectronic meter.

The restriction to audio frequencies and below also does not apply to the *thermocouple* and *hot-wire* meters. Even when the thermocouple is directly combined with the nonelectronic moving-coil dc meter, the combination still produces an instrument that retains its usefulness from dc all the way up into very high frequencies in the range of hundreds of megahertz. The less frequently used *hot-wire ammeter* is similarly free of these frequency limitations.

Although each has excellent frequency characteristics, neither the thermocouple-plus-dc meter instrument nor the hot-wire ammeter provides any signal amplification, and so they lack the sensitivity and flexibility possible with the electronic ac instrument. Not only do they lack sensitivity, but both nonelectronic types also suffer from their inability to withstand slight overloads without burning out, so that here again, the electronic ac instrument is more widely used.

3-4. DYNAMOMETER-TYPE ALTERNATING-CURRENT METER

The dynamometer type (also called electrodynamometer or dynamic type) depends for its action upon the relation between two sets of coils, one set fixed and the other movable. The most common form is shown in Fig. 3-4, where the moving

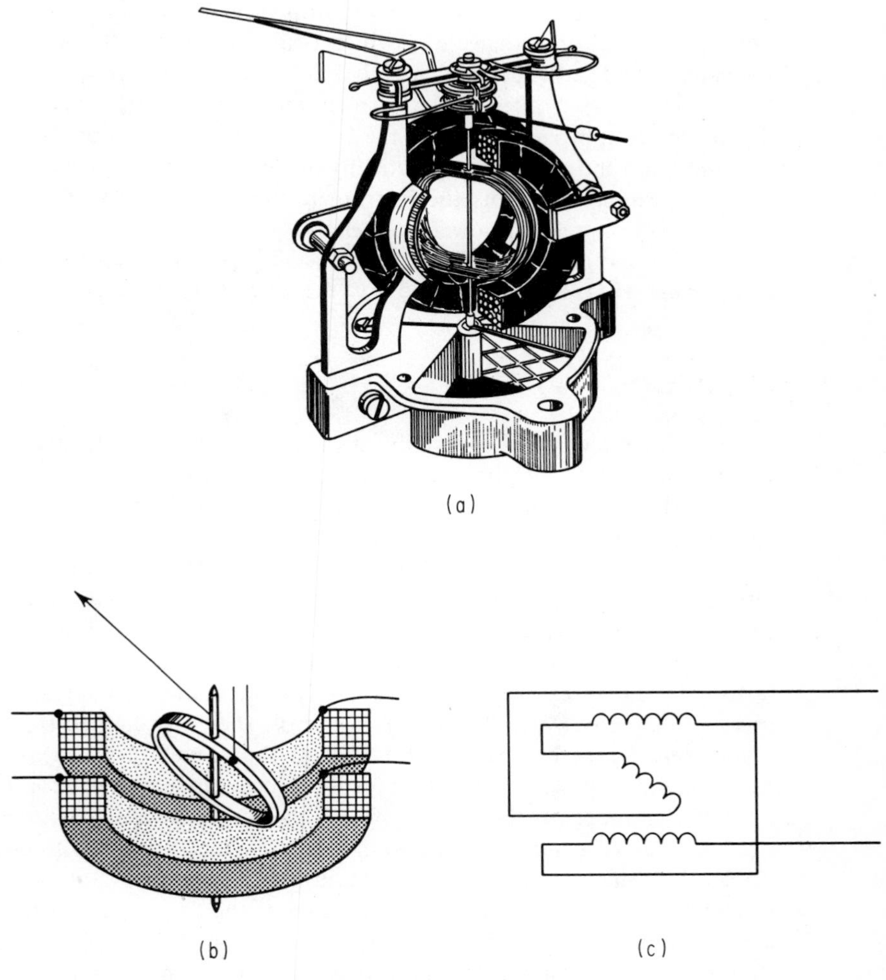

Figure 3-4. Electrodynamometer movement: (a) mechanism used in Weston wattmeters and dynamometer-type voltmeters and ammeters; (b) the moving coil rotates between the two fixed coils; (c) series connection of the two fixed coils (top and bottom) with the moving coil, as used in ammeter instruments. (Courtesy of Weston Instruments, a division of Daystrom.)

coil is a single coil, rotating between two fixed sections of the second coil. The sections of the second coil are positioned parallel to each other to provide the fixed magnetic field. The magnetic field caused by current in the moving coil interacts with this fixed field and produces a turning force, which is therefore proportional to the value of the current flowing in both sets of coils. Since the currents in these coils alternate in unison and so produce a constant torque, the instrument is suitable for ac measurements.

For *voltmeters*, the usual series multiplier will limit the current to the amount required for full-scale deflection, which is in the order of 0.1 A (or 100 mA). For *ammeters*, the fixed coils are generally built to carry 5 A, while the current through the moving coil is limited to the correct amount by shunt. When the power factor of this shunt approaches that of the moving coil, the division of current between the shunt and the coil will be practically independent of frequency, within a limited frequency range, and thus the instrument can be made to provide a fairly high accuracy at commercial power frequencies.

Ammeter ranges above 5 A are usually provided by the use of a *current transformer* (Fig. 3-5) calibrated for the sensitivity of a particular instrument.

3-5. WATTMETERS

Wattmeter instruments, especially those designed for indicating 60-Hz power consumption, generally employ the dynamometer movement.

In dc circuits, the power in watts is easily determined by obtaining readings for the voltage across the load and the current through it and then multiplying the volts times the amperes (EI) to give the product in watts. On ac, however, although both the voltmeter and ammeter read *effective* values, the product will not give the true effective power, *unless the load is entirely resistive*. When the load impedance contains an appreciable amount of inductance or capacitance—as is usually the case—the product of individual voltmeter and ammeter readings gives voltamperes (VA). The true power must then be obtained by multiplying the voltampere figure by the power factor of the circuit being measured, i.e., true power $= EI \times$ power factor (per cent). Since the power factor is not usually known, it is more practical to determine the power by a *direct-reading wattmeter*.

In the *dynamometer type of wattmeter* there are four connections in the single-phase instrument, as shown in Fig. 3-6. (Figure 3-7 shows a widely used three-phase wattmeter.) In Fig. 3-6(a), two of the connections are for the current coil, through which the load current passes in series, and the other two are for the voltage coil, which connects across the load. The current coil is physically arranged in two sections and is similar in both appearance and function to the field coil of a simple motor. The voltage coil is mounted between the two field-coil sections and is able to rotate on its axis within the limits of its spring-tension mounting. With the terminals connected, a motor torque is produced by interaction between the fields of the current and voltage coils, in a manner similar to that of

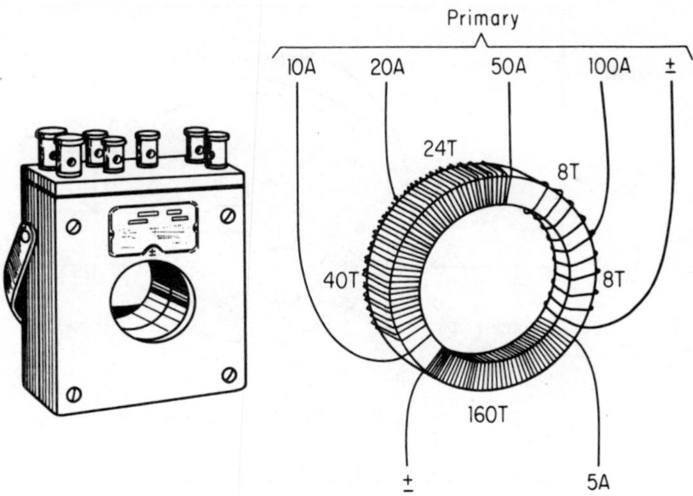

Figure 3-5. Current transformer to extend ac ranges. The secondary winding for the basic instrument range (5 A) is shown at the bottom winding, and the four extended current ranges are obtained from taps on the primary winding, shown at the top. (Courtesy of Weston Instruments, a division of Daystrom.)

the D'Arsonval galvanometer; there is, however, the important difference that the reversals of the alternating current in the wattmeter arrangement change the polarities of both the current and the voltage fields practically simultaneously. The motor torque (and therefore the wattmeter reading) is thus made proportional to the product of the instantaneous values of both the current and the voltage, and so directly indicates the power consumption of the load. The power, shown on the direct-reading dial, is the *true power* in watts, since the phase relationship also affects the reading. This can readily be realized by noting that in a purely capacitive load, as an example of a limiting case, the current would lead the voltage by 90° (or, expressed differently, the power factor would be 0%) when the current was maximum, and so, since the voltage would be zero at this time, the indication of motor torque would also read zero.

In practical cases, there will be a small power loss in the instrument itself, resulting in a slight reading error that will have to be taken into account when measuring low-power circuits. The error due to this power loss will be entirely negligible, however, in the great majority of cases where substantial power is being taken by the load. This is generally the case in the common uses of the wattmeter, as, for example, in measuring the power consumption of an ordinary receiver, which takes, say, around 50–100 W; in such cases the wattmeter loss may be neglected.

Special problems in power measurement arise when the circuit operates at

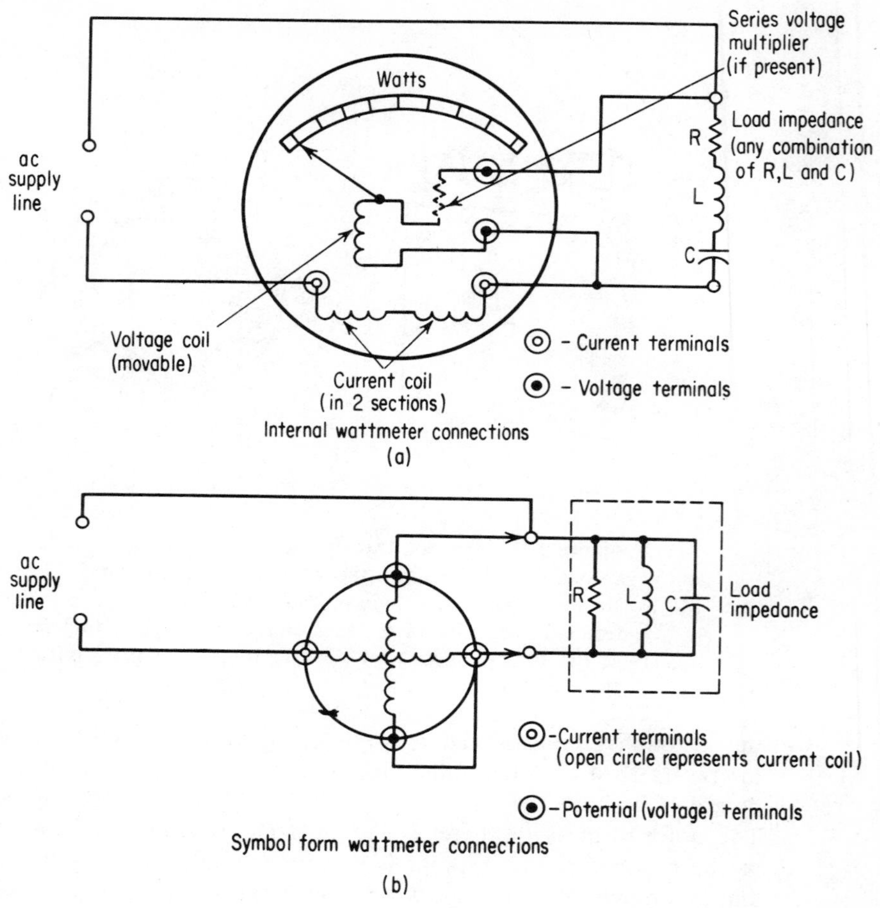

Figure 3-6. Connections of wattmeter (dynamometer type) for power measurement: (a) internal; (b) symbol form of wattmeter connections.

frequencies much higher than the 60-Hz power frequency that has been discussed in the paragraphs above. In general, it can be said that power measurement at the higher frequencies makes use of indirect methods, where the heating effect of the current in the circuit is often used as the basis for calibrating the meter in watts.

3-6. MOVING-IRON-TYPE INSTRUMENTS

Iron-vane ammeters and voltmeters depend for their operation on the repulsion that exists between two like magnetic poles. One of these like poles is craeted by the instrument coil and appears on an iron vane, fixed in its position within the coil, as shown in Fig. 3-8. The other like pole is induced on the movable iron

Figure 3-7. Electrodynamometer ac wattmeter. This model (*Weston model 432*) is calibrated in kilowatts for three-phase power.

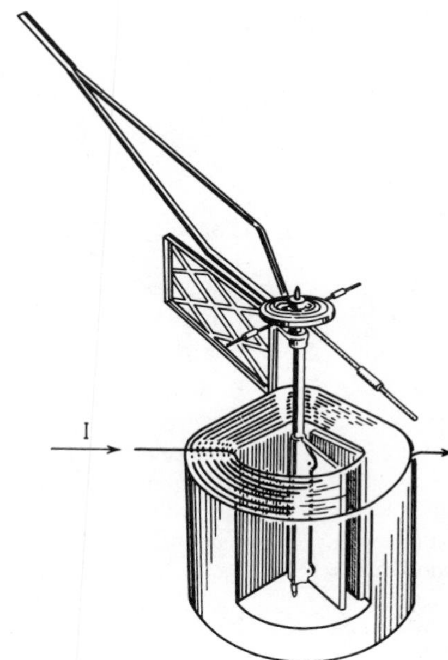

Figure 3-8. Iron-vane meter, internal view. The iron vane, of the radial type, is forced to turn within the fixed current-carrying coil by the repulsion between like poles. The aluminum vane, attached to the lower end of the pointer, acts as a damping vane in its close-fitting chamber to bring the pointer to rest quickly.

piece, or vane, which is suspended in the induction field of the coil, and to which the needle of the instrument is attached. Since the instrument is used on ac, the magnetic polarity of the coil changes with every half cycle and induces corresponding magnetic poles on both the fixed and movable iron vanes within the field coil. At any given moment, like poles will be induced on the adjacent ends of both the fixed and movable vanes, and this will cause a corresponding amount of repulsion of the movable vane, against the spring tension. The deflection of the instrument pointer, therefore, will always be in the same direction, since there is always repulsion between the like poles of the fixed and the movable vanes, even though the current in the inducing coil alternates.

The deflection of the pointer thus produced will be effectively proportional to the actual current through the instrument. It can therefore be calibrated directly in amperes or in volts. The calibration of a given instrument, however, will read correctly only on the ac frequency for which it is designed. This is so because the impedance will be different at a new frequency; therefore, the current flow for a given voltage will be affected, giving a new deflection. The moving-iron type of

TABLE 3-1. Alternating/Direct-Current Classification of Commonly Used Meter Types

Meter Type	*Suitability*	*Major Use*
D'Arsonval (moving-coil)	dc only	Most widely used meter for dc current and voltage and resistance measurements, in low- and medium-impedance circuits.
Moving-iron	dc or ac	Inexpensive type used for rough indication of large currents (such as automobile-battery charging current or corresponding voltages).
Dynamometer	dc or ac	Widely used for precise ac voltage and current measurements at power frequencies (in low-impedance circuits).
Electrostatic	dc (or ac at one frequency only)	Limited use in high-voltage measurements (at very high impedances).
Rectifier* (combined with D'Arsonval movement)	ac only	Widely used for medium-sensitivity service-type voltage measurements in medium-impedance circuits.
Electronic (VTVM, solid-state VM, or digital VM)†	dc or ac	Extensive laboratory use as voltmeter in circuits where both high-sensitivity and high-impedance requirements are combined.

*Used in the volt-ohm-milliammeter (VOM), with switch for selecting direct-current and alternating-current ranges.

†Rapidly replacing D'Arsonval and rectifier types.

instrument is usually calibrated to read *effective amperes or volts*, and is used primarily for rugged, inexpensive meters.

3-7. SUMMARY CLASSIFICATION OF ALTERNATING- AND DIRECT-CURRENT METERS

In order to show the basic dc D'Arsonval meter in its proper relation to other types of meters, both nonelectronic and electronic, a summary is given in Table 3-1, in which the various meter types are classified according to their *suitability on dc, or both*, and also according to the areas in which they find their *major use*. In the category of the electronic meter, for simplicity, only the electronic voltmeter is listed, highlighting the inherent capability of electronic instruments in general as compared with the nonelectronic type. It may be noted from this classification that the electronic instrument offers a combination of high sensitivity and high impedance, a combination that is not obtainable in the nonelectronic types of meter.

QUESTIONS

Q3-1. Using a *solid arrow for the direction of conventional current* (and a dashed arrow for the electron direction), show the correct direction for these arrows:
(a) Forward conduction in a tube diode circuit.
(b) Forward conduction in a semiconductor diode circuit.

Q3-2. Name two materials (or combination of materials) suitable for one-way conductors (or rectifiers) in the solid state, other than germanium and silicon.

Q3-3. In a diode circuit used for rectification, why is the cathode sometimes labeled plus (+) by some manufacturers?

Q3-4. Explain why the current in a dc meter used with a rectifier on an ac circuit produces a steady reading when, as a matter of fact, it is a pulsating dc current.

Q3-5. Discuss the frequency limitations of a rectifier-plus-dc-meter combination with a semiconductor diode rectifier.

Q3-6. Explain the reason for the fact that a voltmeter employing a 50-μA basic meter is rated at 20,000 Ω/V on dc measurements and only 5000 Ω/V on ac measurements.

Q3-7. Explain why the ac scale of a rectifier-voltmeter dial may have a uniform (or linear) spacing on the high-voltage ranges, but generally has a nonlinear calibration on the low-voltage ranges.

Q3-8. Explain the advantage gained by using a microammeter (rather than a milliammeter) as the basic meter in a rectifier-type voltmeter, other than the fact that lower-voltage ranges can be provided for greater sensitivity of the ac measurements.

Q3-9. Explain why both dynamometer and moving-iron types of meters may be used for either dc or ac measurements, without requiring the use of a rectifier.

Q3-10. Since an electrostatic voltmeter does not depend upon the flow of current for its operation, why is it not more widely used for sensitive voltage measurements?

Q3-11. (a) Explain why two different readings (both incorrect) are obtained if a rectifier-type meter is used to measure a dc voltage.
(b) How can you quickly determine whether a blocking capacitor is present in a rectifier type of ac instrument?

PROBLEMS

P3-1. In the *half-wave rectifier* circuit of Fig. 3-2(a):
(a) Find the value of a series resistor R_s that will allow full-scale deflection of the basic 0–1-mA meter, when 10 V (rms) is applied to the input. (Neglect meter and diode forward resistance.)
(b) Find the ohms-per-volt sensitivity of the ac voltmeter for this value of R_s.

P3-2. In the double-diode *three-terminal rectifier* circuit of Fig. 3-2(b), assume the internal resistance R_m of the meter to be 100 Ω. As in Problem P3-1:
(a) Find the value of series resistor R_s to produce full-scale deflection (1 mA), with an input of 10 V (rms). (Neglect diode forward resistance.)
(b) Find the ohms-per-volt sensitivity for this condition.

P3-3. The *half-wave arrangement* of Fig. P3-3 is often used in commercial rectifiermeter circuits to produce a more uniform scale. Assume the shunt resistor R_{sh} to be 100 Ω (the same as internal resistance R_m of the 0–1-mA basic dc meter).

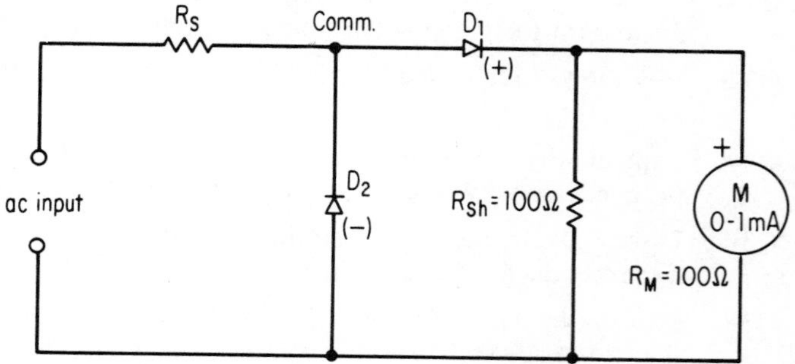

Figure P3-3. Commercial rectifier-voltmeter circuit.

(a) Find the value of resistor R_s required for full-scale deflection with an rms input of 10 V. (Disregard forward resistances of diodes D_1 and D_2.)
(b) Find the ohms-per-volt sensitivity of this commercial arrangement.
(c) If each diode D_1 and D_2 has a forward resistance of 400 Ω, what value of R_s is needed for full-scale deflection?

P3-4. Using the *full-wave arrangement* for a rectifier-type meter in Fig. 3-3:

(a) Find the value of the series resistor R_s required for full-scale deflection of the 0–1-mA basic meter, with an input of 10 V (rms); (the forward resistance of each of the diodes is 400 Ω, and the meter resistance is 100 Ω).

(b) Find the ohms-per-volt sensitivity of this full-wave rectifier arrangement.

P3-5. Under the condition where the *RC* time-constant of Fig. P3-5 is very large compared to the period of 1 cycle of a 10-V rms input—and assuming an ideal diode—sketch the input and output wave-forms of this *peak-responding voltmeter.*

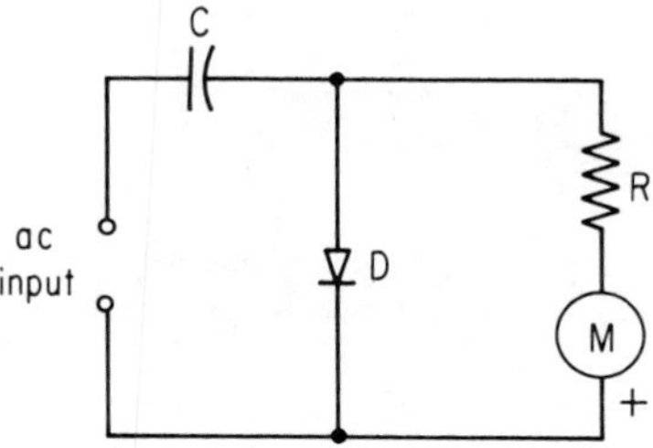

Figure P3-5. Peak-responding voltmeter circuit.

P3-6. Explain how the *half-wave voltage-doubler circuit* of Fig. P3-6 can act as a *peak-to-peak* responding ac voltmeter, when used in conjunction with a relatively large resistor *R* and a basic 0–1-mA dc meter.

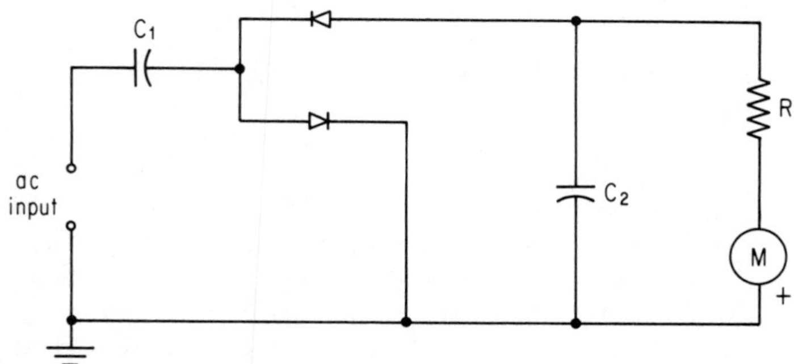

Figure P3-6. Peak-to-peak-responding voltmeter (voltage-doubler) circuit.

4

Comparison Measurement Methods

4-1. COMPARISON VERSUS DEFLECTION METHODS

The methods discussed in preceding chapters for both dc and ac measurements of current (and the voltages derived from that current) all provide an indication of the measured quantity by the *deflection of the meter pointer*. The value of an unknown quantity (say, voltage or resistance) is obtained in these deflection instruments by reading the pointer position on the scale, calibrated correspondingly in volts or ohms. The scale calibration is generally mass-produced and depends in turn on many factors not easily reproduced, such as friction, magnetic strength, and spring tensions. Hence, the *accuracy of the scale calibration* cannot be expected, in most general-purpose moving-coil meters, to be reliable to better than about 1 or 2%.

Since the accuracy of a deflection-type measurement is so greatly dependent on limitations of the scale calibration, greater accuracy can generally be achieved by using a *comparison method* that avoids this heavy reliance on the meter-scale calibration. This is especially true where a nonlinear scale, such as the one in an ohmmeter, adds further impediments to accurate readability. On the other hand, as an example of a comparison technique, the familiar Wheatstone bridge is able to measure an unknown resistance to within a small fraction of 1%, in contrast to the usual $\pm 5\%$ accuracy expected from the direct-reading deflection on an ohmmeter scale. Many other examples can be cited to show the improvement in accuracy obtainable when the indication on the meter scale shows only a null or balance point. As a general principle, greater accuracy can be obtained when the meter pointer is called upon to give only *relative, rather than direct-reading, indications*. Thus the comparison method offers a very powerful measurement technique, in freeing the measurement from dependence on precise interpretation of the amount of pointer deflection.

In distinguishing between comparison and deflection techniques, it is better not to define the terms too strictly. We can say that the *distinguishing feature of the comparison method* is the fact that the *unknown is being compared with a reference that is physically present in the measuring system*, and that the meter reading so obtained is required to give only a *relative indication* of the result of the comparison.

Comparison measurement can be divided roughly into two classes: (1) measurements where the comparison produces a *balance or null*, and (2) measurements producing *relative indications, depending on the degree of unbalance.*

Null-Indicating Instruments

The *null type of instrument*—the type of comparison instrument generally used—includes two large divisions:

1. The potentiometer-null-type instrument
2. The bridge-null-type instrument

In these types, the ability to obtain higher accuracies is the main feature that overrides the obvious disadvantages of greater inconvenience in manipulation to obtain the null and higher cost, as contrasted with deflection-type instruments.

Relative-Indication Instruments

The *relative-indication* type of comparison includes many less familiar types of measurements, of which the following might be mentioned here:

1. *Half-deflection measurements*, such as the example of the measurement of input impedance shown in Fig. 4-1(a). By varying an external resistor until the original meter indication is reduced to half, the unknown input impedance is made equal to the value of the external resistor.
2. *Equal-deflection methods*, where the same current produces equal voltage drops in known and unknown impedances; an example of a general impedance measurement by this method is given in the next section and is shown in Fig. 4-1(b).
3. *Unbalance measurements*, where departure from balance is interpreted in terms of the reference on an arbitrary scale; many "go–no go" comparators employ this type of measurement.

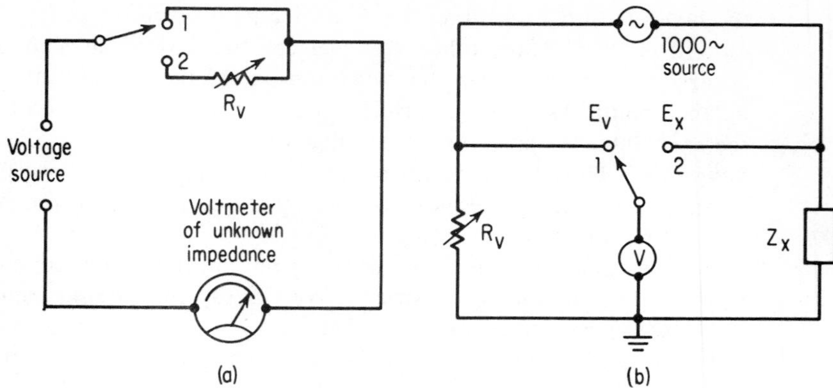

Figure 4-1. Relative-indication measurements: (a) half-deflection method for determining unknown impedance of the voltmeter; (b) equal-deflection method for comparing impedance magnitudes.

4-2. EQUAL-DEFLECTION METHOD FOR MEASURING IMPEDANCE

Before we discuss the null types of comparison instruments, we shall consider an example of the relative-indication type of measurement, using an equal-deflection form of comparison measurement for *measuring impedance*.

For impedance measurements the comparison of equal deflections is often used. The comparison of equal deflection overcomes the difficulties of the direct-reading method that arise when the scale is nonlinear, as in the case of dc resistance R. These difficulties become much greater if we are attempting to prepare a calibrated scale for *impedance* Z. The unknown impedance will generally contain a reactance (inductive X_L or capacitive X_c), having a value dependent on frequency, and very likely combined with a resistance R. Since the magnitude of the impedance Z, in ohms, follows the familiar but cumbersome relation $|Z| = \sqrt{R^2 + X^2}$, providing calibrated scales for various impedance ranges is limited to a single frequency; also, the nonlinear scale becomes quite awkward. Consequently such scales are not often provided in a direct-reading deflection type of instrument. However, it proves to be quite practical to use a comparison method, where the unknown impedance is compared with a known value at any desired frequency, thus giving an easily readable measurement of the magnitude of the unknown impedance, in ohms.

A simple comparison circuit of the equal-deflection type, for measuring impedance magnitude, is shown in Fig. 4-1(b). An ac generator supplies variable resistor R_v in series with the unknown impedance Z_x. Since the current in each series component must be the same, the voltage across each is in direct proportion to its impedance magnitude. Voltmeter V can be switched to measure the voltage drop across resistor R_v (as E_v) and the drop across impedance Z_x (as E_x). Resistor R_v is varied until the voltmeter shows $E_v = E_x$, as the voltmeter is alternately switched between them. When the two voltages are made equal, the impedance magnitude $|Z_x|$ is equal to the value R_v, which can be measured independently by ohmmeter or bridge or also, if desired, can be read off directly by a previous calibration of R_v in terms of its resistance. In this determination by comparison, it should be noted that the accuracy of the voltmeter scale does not enter. The voltmeter V must have a sufficiently high input impedance (generally 10 times as great as Z_x). Note also that in this arrangement the common point (or ground) is not connected, as it usually is, to the low side of the oscillator, but rather to the junction of R_v and Z_x, which is the same point to which the low side of the voltmeter is connected. This is done in order to avoid undesirable stray pickup that might occur if the voltmeter were operated off-ground. There is no indication of the phase angle when the magnitude of the impedance is found by this method.

4-3. COMPARISON BY NULL METHODS

The equal-deflection impedance measurement cited in the preceding section illustrates a distinction that should be made between comparison methods in general and the null method, which is a special case of the general category. If we attempted to produce a null (in a bridge arrangement) at the time that the variable resistor R_v was set equal to the magnitude of unknown impedance Z_x, we would find that the opposing voltages would not necessarily balance, even though equal in magnitude, because they would have different phase angles. In order to satisfy the requirements for a null, it would be necessary to arrange four arms of a bridge so that the components would balance in both magnitude and phase, thus producing greater complexity in the circuit and necessitating the adjustment of two interdependent parameters for obtaining a null, as is done in the typical bridge circuit. While the added complexity of the bridge arrangement is perfectly justifiable for making phase-conscious measurements, it is well to remember that the *production of a null is not a necessary factor for all cases of comparison.*

4-4. NULL-TYPE INSTRUMENTS: POTENTIOMETER AND BRIDGE TYPES COMPARED

Instruments employing a balance or null type of measurement are widely used for cases requiring a high order of precision. They fall into two large general classes: the *potentiometer type* for voltage comparisons and the *bridge type* for comparing impedance characteristics. In addition to their classical application as standards for *precise measurements* where the balance is produced manually, they also appear in the form of self-balancing instruments. The addition of the self-balancing feature offers a wide variety of instruments for industrial instrumentation to perform the functions of indicating, recording, and control of physical quantities.

Although the potentiometer and bridge types of null methods are seemingly different, it is helpful at the outset to compare the two, to bring out their basic similarity, as shown in Fig. 4-2.

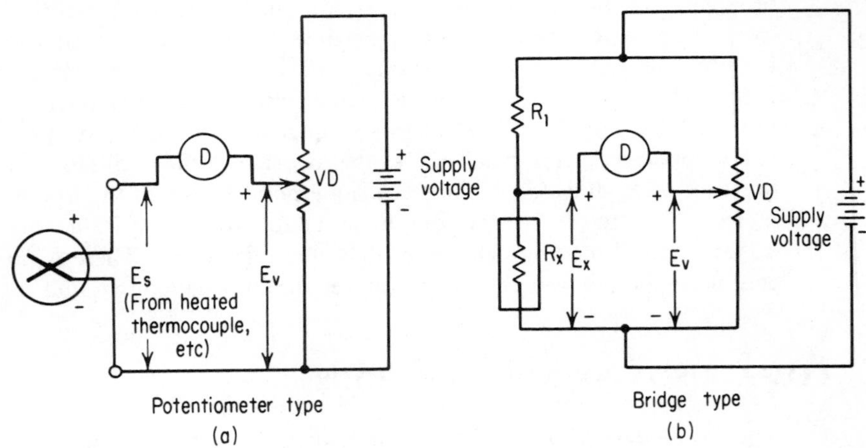

Figure 4-2. Similarity between null-type circuits: (a) potentiometer type; (b) bridge type; in both types of null circuits, zero current is produced in detector *D* at balance.

Null-Potentiometer Type

In Fig. 4-2(a) the potentiometer circuit, stripped down to its bare essentials, is seen to consist of an arrangement wherein an input signal voltage E_s is opposed by a variable voltage E_v, derived from the three-terminal voltage divider, connected across the supply voltage. When the moving arm of the voltage divider *VD* selects an amount of voltage E_v, equal to the input signal E_s, the current through the detector *D* falls to zero, producing a null reading. When the null is obtained, the value of voltage E_v can be read off from the calibrated scale of the voltage-divider *VD*, thus giving a direct-reading indication for the signal input voltage.

It will be noted that at balance, no current is drawn from the signal input source, and the potentiometer system accordingly presents an infinite impedance to the signal being measured. Combined with this characteristic of *infinite impedance*, the potentiometer type lends itself to a high order of *precise calibration* of the resistive voltage-divider. Thus, because it possesses these two highly desirable characteristics of infinite impedance and precise calibration, the potentiometer type of null instrument can serve in the laboratory as a standard for voltage calibration and in industry as a basic means for accurately monitoring voltage inputs. This is particularly true in industrial recorder instruments, where physical quantities are first converted into small electrical voltages, which are then applied to self-balancing potentiometers for indication, recording, and control, as will be shown later.

Bridge Type

The other large class of null instruments employs the bridge principle shown in Fig. 4-2(b), which can be seen to have a basic similarity to the potentiometer principle. Here the right side of Fig. 4-2(b) operates in an identical fashion to the right side of Fig. 4-2(a) in producing a variable voltage E_v. In part (b), however, the variable voltage E_v, instead of opposing a signal voltage E_s, now opposes a voltage drop E_x, which exists across the unknown element R_x. As resistance R_x assumes different values for whatever reason, a new value of voltage E_x is produced, and consequently a new value of E_v must be provided by the moving arm to oppose it and produce balance. The outstanding difference in the bridge method lies in the fact that the quantity being measured no longer is required to produce an active voltage E_s as in the previous case, but instead, it need only produce a new value of resistance R_x. Thus, as the input quantity varies, it results in a new value of E_x, derived from the action of the supply voltage across the combination of R_1 and R_x. After the new value of E_x has been balanced out, producing a new null, the same basic advantages of zero current at balance and precise calibration apply equally to the bridge method of measurement as to the potentiometer method.

In the preceding comparison, the examples cited used dc voltages in both the bridge- and potentiometer-null methods. The same principles can also be extended to ac excitation for both methods, with the substitution of values of impedance for resistance and other corresponding changes where necessary. Thus, the same principles can be applied to both dc and ac null systems, depending on the instrumentation requirements, as will be shown in later examples.

4-5. POTENTIOMETER-NULL VOLTAGE MEASUREMENT

The action in a potentiometer type of null system may be traced by the simplified operating circuit in Fig. 4-3(a). The circuit is shown in its initial balanced condition. A 50-mV signal voltage E_s in this example is the input signal derived from a thermocouple, this input being opposed by an equal voltage obtained from the variable voltage divider E_v. For the sake of simplicity, assume that the resistance of this voltage divider, in the form of a slide-wire, is 100 Ω in which 1 mA of current is caused to flow by a calibration process (to be described later). The drop across the slide-wire, then, is 100 mV, and the moving arm has been manually set at 50 Ω (indicated by 50 mV on the scale) to achieve the initial balanced condition (zero current through the zero-center galvanometer detector D). As the input voltage from the thermocouple rises, say, to 60 mV, the slide-wire arm must be manually moved upward to the 60-Ω point to achieve the new balance, at which

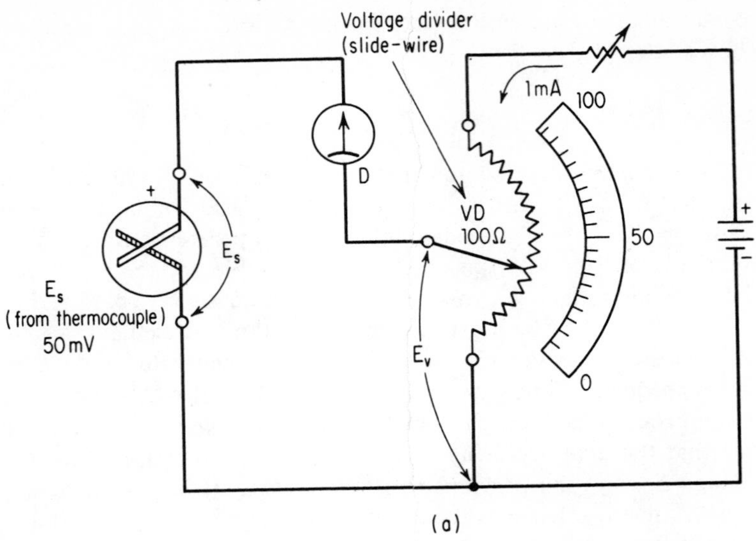

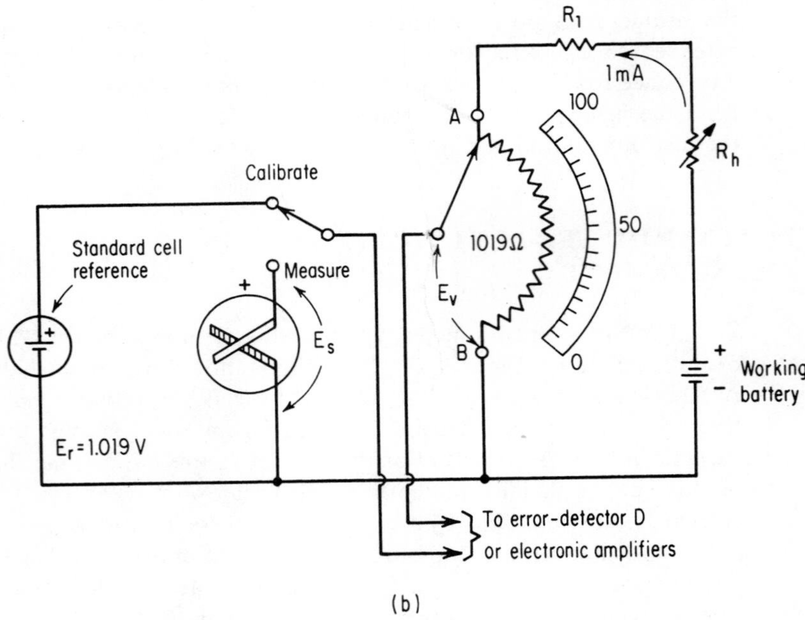

Figure 4-3. Potentiometer-null-type instrument: (a) operating circuit; (b) calibration circuit.

time the required unknown voltage can be read off directly from the calibrated scale as 60 Ω × 1 mA, or 60 mV.

The deflection obtained on the detector acting as a voltmeter in this system is directly proportional to the amount of unbalance. Thus, in the example, the effective voltage actuating the voltmeter in the unbalanced condition is the difference between the input signal $E_s = 60$ mV and the reference voltage $E_v = 50$ mV, or 10 mV. In general, then, the effective unbalance voltage driving the detector (ΔV) is equal to the change in input voltage from the balance point ΔE_s, or

$$\Delta V\,(\text{detector}) = \Delta E_s\,(\text{input})$$

It can be seen from this simple relation that the detector output voltage varies linearly with the change in input voltage. This linear relation contributes greatly to the precision of which the potentiometer-type instrument is capable, since a high degree of linearily is not at all difficult to obtain from precison resistive networks. Further study of the calibration of a representative instrument Fig. 4-3(b) will show how the unknown voltage can be read, to a high degree of precision, directly from the potentiometer dials.

4-6. CALIBRATION OF POTENTIOMETER-NULL INSTRUMENTS

To ensure that the balancing voltage will be known to a high degree of precision, the voltage across the slide-wire E_v is compared with a known reference voltage E_r furnished by a standard cell (or, in some cases, by a precisely regulated dc voltage supply).

Standard Cell Reference

The functional circuit for a calibrating arrangement using a standard cell is shown in Fig. 4-3(b). Here, the arm of the slide-wire is positioned to tap off the full voltage across the slide-wire at point A, and the switch is thrown to CAL for the calibrating position. This position connects the detector D to the standard cell, instead of to the unknown voltage. Ideally, if the current from the working battery had remained unchanged from some previous time, the voltage drop across the slide-wire would exactly equal the reference voltage, and a null would be obtained. In the more usual case, though, some slight change in the current from the working battery is to be expected, as the battery ages, or for some other reason. The current is restored to its original calibrated value by adjusting rheostat Rh to obtain a null reading against the reference voltage. The voltage drop across the slide-wire has now been made equal to the standard cell voltage and as a consequence, no current is drawn from the standard cell.

In the example shown in the calibrating circuit of Fig. 4-3(b) the standard cell is assumed to have a known value of 1.019 V, and the resistance of the slide-wire has been taken as 1019 Ω, so that a current of 1 mA would produce the required precise voltage drop of 1019 mV (or 1.019 V). The correct adjustment of the rheostat to produce the desired 1-mA current is therefore assured when the null is obtained in the calibrating position.

Zener-Diode Reference Voltage

As a substitute for the standard cell reference, a regulated voltage produced by a Zener diode is often used in newer models, particularly in industrial instruments. This method takes advantage of the voltage-regulation properties of a reverse-biased semiconductor diode that has been specially processed) to produce a stable voltage sufficiently constant to be used as a reference.

The voltampere characteristic of a typical Zener diode is shown in Fig. 4-4. The forward characteristic is identical to that of the usual semiconductor, with

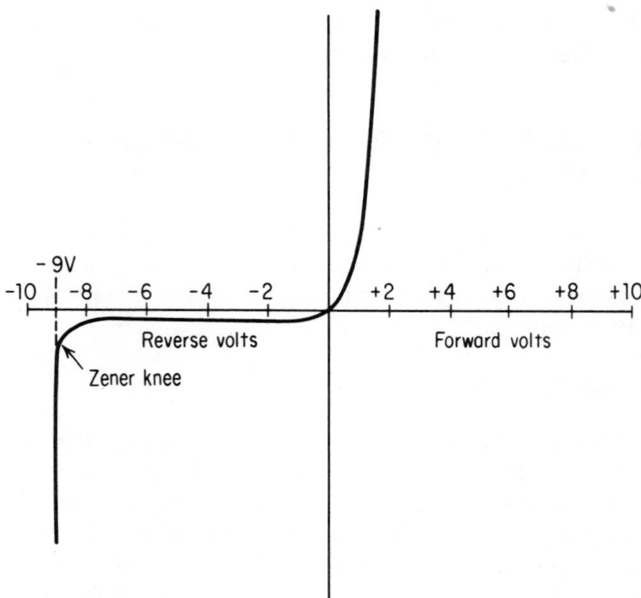

Figure 4-4. Voltampere characteristic of a typical Zener diode.

forward current increasing at an exponential rate as the forward bias gets larger and approaches 1 V. The curve is also similar at low reverse-voltage bias, showing a very small reverse current as the negative voltage is increased to the left. When the region of the Zener knee is reached and passed—in this case at negative 9 V—it is found that the reverse current abruptly increases, while the voltage drop across the diode remains very constant at 9 V. This abrupt increase is not to be confused with the voltage breakdown that occurs when the peak-inverse-voltage (PIV) rating of an ordinary diode is exceeded. Where, in the ordinary diode, the avalanche effect at voltages in excess of PIV might result in the destruction of the diode, in the specially processed Zener diode the knee occurs before the destructive voltage is reached. By the use of a series resistor to limit the potentially high Zener current to a safe value, any applied voltages greater than negative 9 V are held to the regulated 9 V within the rated operating limits of the diode.

A circuit showing the Zener diode used as a 9-V reference is shown in Fig. 4-5. Here, the input ac voltage of 12.6 V is obtained from a secondary winding of

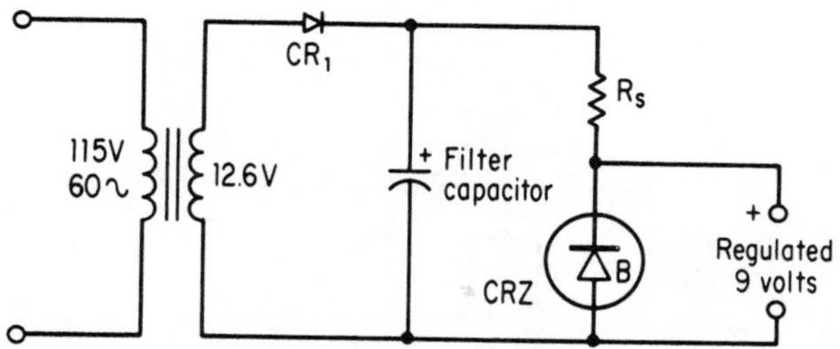

Figure 4-5. Zener-diode circuit for producing a regulated voltage suitable for reference.

the 115-V power transformer. A conventional semiconductor rectifier, in conjunction with the filter capacitor in a half-wave circuit, supplies about 17 V dc (12.6×1.4) to the series combination of the current-limiting resistor R_s and the reverse-biased Zener diode *CRZ*. The constant 9-V reference voltage is taken as an output across the Zener diode. As a reference, the Zener-diode method offers advantages over the standard cell in providing greater convenience, as well as being able to handle safely the reasonably small unbalance-currents produced during calibration, which might not be tolerated by a standard cell. Digital voltmeters, for example, might use either miniaturized standard cells or compensated Zener diodes alternately, as reference elements of satisfactory accuracy.

Self-Balancing Systems

Recent trends in commercial instrumentation practice employ the *millivolt dc potentiometer*, in conjunction with electronic amplifiers and servomotors, to achieve *self-balancing systems for recording and control.*

In keeping with the electronic-instrument viewpoint in this book, further discussion of the potentiometric system of instrumentation will concentrate on the self-balancing types employing electronic amplifiers. In the commercial instruments, the basic potentiometer-null measurement serves as an error-detection input section to an electronic amplifier, which in turn achieves the final balance by energizing a servomotor, or similar error-correction device.

4-7. BASIC DIRECT-CURRENT (WHEATSTONE) BRIDGE

The fundamental principle of the familiar four-arm bridge is shown in Fig. 4-6(a) in the form of the Wheatstone bridge. This basic circuit has retained its usefulness since the earliest days of electricity and is still vigorously producing offshoots in the large family of null instruments. As previously discussed in Sec. 4-4, the dc null in the case of the Wheatstone bridge is produced by a comparison of voltage drops in the passive arms of the bridge, as contrasted to the null produced by active voltages in the potentiometer type of null circuit.

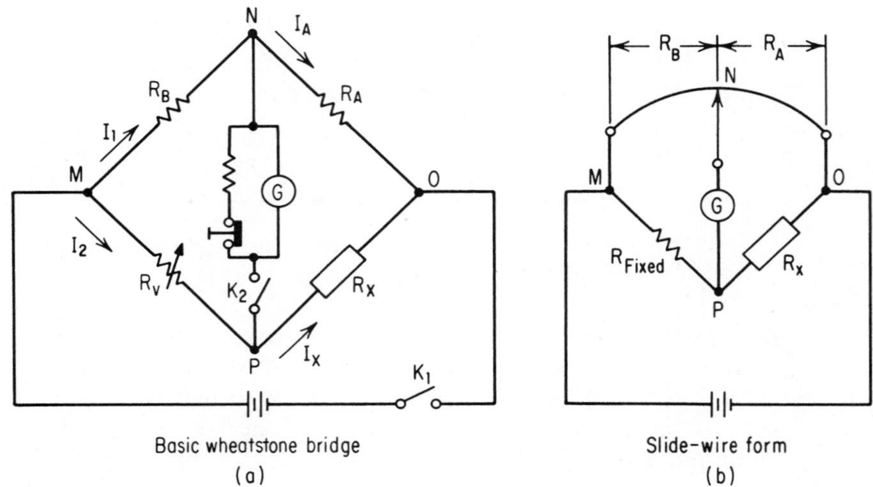

Figure 4-6. Forms of resistance bridge: (a) basic Wheatstone bridge; (b) slide-wire form.

The current and voltage relationships in Fig. 4-6(a) can be traced for simplified conditions in a very direct manner, without resorting to loop nodal equations more suitable where completely general solutions are required. To obtain the balance condition, the voltage drop from M to N caused by the upper branch current I_1 (V_{MN}), which is $I_1 R_B$ must be equal to the voltage drop from M to P caused by the lower branch current I_2 (V_{MP}), which is $I_2 R_V$;

$$I_1 R_B = I_2 R_V$$

and if the galvanometer current is zero,

$$I_1 = I_A = \frac{E}{R_B + R_A}$$

and

$$I_2 = I_x = \frac{E}{R_V + R_X}$$

Substitution in these equations gives the well-known relationship

$$R_B(R_X) = R_A(R_V)$$

or

$$\frac{R_X}{R_V} = \frac{R_A}{R_B}$$

If the ratio of A to B is known and R_V is varied to produce balance, R_X is determined from

$$R_X = \frac{R_A}{R_B} R_V$$

This balance measurement of the unknown is independent of the characteristics or calibration of the detector (galvanometer), provided only that it has sufficient sensitivity to permit the balance or null point to be found to the required degree of precision.

The slide-wire form of Wheatstone bridge is illustrated in Fig. 4-6(b). This form takes advantage of the linear relation between the length and the resistance of the slide-wire. By obtaining a continuously variable and accurately known ratio in this manner, it becomes possible to use a known fixed precision resistor for R_V, and thus avoid the more expensive precision decade resistor arrangement that would be necessary if resistor R_V were varied. The same equation applies:

$$R_X = \frac{R_A}{R_B} R_V$$

but the ratio R_A/R_B must now be calculated from the corresponding slide-wire lengths instead of being read directly as a whole-number ratio, as is usually done in the arrangement of Fig. 4-6(a). The slide-wire form is also widely used in self-balancing arrangements as a convenient method for obtaining continuous stepless variation in the balancing process, which is described in a later chapter.

4-8. SENSITIVITY OF UNBALANCED BRIDGE

A computation showing the effective voltage (and current) conditions existing for a given amount of unbalance condition is a necessary first step in determining the sensitivity of the bridge arrangement to the unbalance condition. Whether voltage or current detectors are used in a bridge, a solution to the problem is afforded by the Thévenin circuit theorem.

4-9. THÉVENIN EQUIVALENT OF UNBALANCED BRIDGE

The action in a Wheatstone bridge under various conditions of unbalance can be greatly simplified by applying Thévenin's theorem to obtain an equivalent circuit, as illustrated in Fig. 4-8. In this example equal resistors of 100 Ω each are used for the ratio arms R_B and R_A. This not only simplifies calculations, but also conforms to the equal-arm condition for greatest bridge sensitivity in a circuit, and the values used are quite similar to those encountered in practice. Consider the unknown resitsor R_X to be the resistance of a strain-gage transducer, having a nominal resistance of 100 Ω before it is stressed. Under tension, let us say that the resistance increases by an increment x. We wish to find what sensitivity a current detector (such as a galvanometer) must have in order to provide full-scale deflection on this unbalance caused by the x increase in the nominal resistance, resulting in $100 + x\Omega$ for the transducer. Using a battery of 6 V with negligible internal resistance (as is justified in most cases), we assume the circuit to be broken at the galvanometer terminals N and P, and find the Thévenin equivalent of the circuit feeding these terminals.

As seen in Fig. 4-7, the equivalent voltage E_{TH} is the voltage across NP with the galvanometer load disconnected. This equivalent voltage is the net difference between the voltage at P (to ground O) and the voltage at N (to ground O), or, letting $\Delta R = x$,

$$\begin{aligned}
E_{TH} &= V_{PO} - V_{NO} \\
&= 6\left[\frac{100 + x}{100 + (100 + x)}\right] - 6\left[\frac{100}{100 + 100}\right] = 6\left[\frac{100 + x}{200 + x} - \frac{100}{200}\right] \\
&= 6\left[\frac{200 + 2x - (200 + x)}{2(200 + x)}\right] = 6\left[\frac{x}{2(200 + x)}\right] \\
&= 6\left[\frac{x}{400 + 2x}\right] \\
&\cong 6\left[\frac{x}{400}\right]
\end{aligned}$$

when

$$x \ll 200$$

Converting this to a general form, we use E for the battery voltage of 6 V, and R for each of the 100-Ω equal-arm resistors, and ΔR for x:

$$E_{TH} = E\frac{\Delta R}{4R}$$

when

$$\Delta R \ll 2R$$

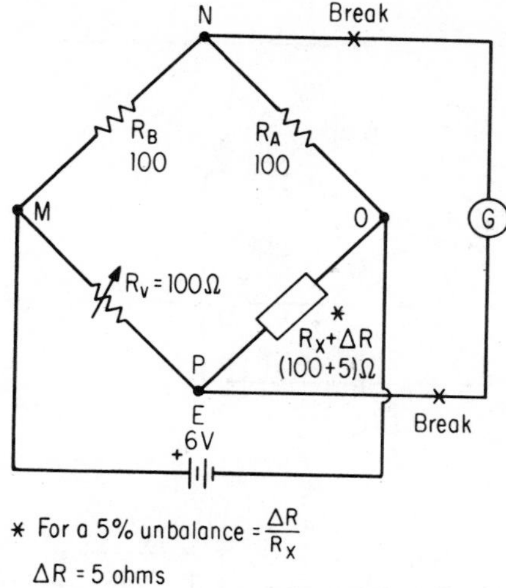

Figure 4-7. Example of an unbalanced bridge (5% unbalanced); the equivalent Thévenin circuit in Fig. 4-8 is obtained by breaking the circuit at the points shown.

Since the value of E_{TH} is being used only for finding the relative amount of galvanometer deflection when near balance, we do not require any exact figure for the galvanometer current. For an approximate value of the driving voltage E_{TH}, we are therefore justified in dropping ΔR out of the denominator if it does not exceed 10% of $2R$ (or 5% of R), as has been done in the preceding equation.

Thus, expressing the percentage increase ΔR (or x) in ohms, for a 5% increase (or less) in R, we can consider the acting voltage

$$E_{TH} \cong \frac{E}{4R}(\Delta R)$$

and for a 5% change,

$$E_{TH} = \frac{6(5)}{400} = 0.075 \text{ V or } 75 \text{ mV}$$

To ascertain the Thévenin equivalent impedance Z_{TH}, we must find, from Fig. 4-7(b), the impedance looking in from terminals NP, with the battery replaced by its zero internal resistance. This consists of the parallel combination of R_B and R_A, in series with the parallel combination of R_V and R_X; expressed in terms of R and x,

$$Z_{TH} = \frac{R}{2} + \frac{R(R+x)}{R+(R+x)}$$

and when $x \ll R$ (which generally applies to small unbalance),

$$Z_{TH} \cong R$$

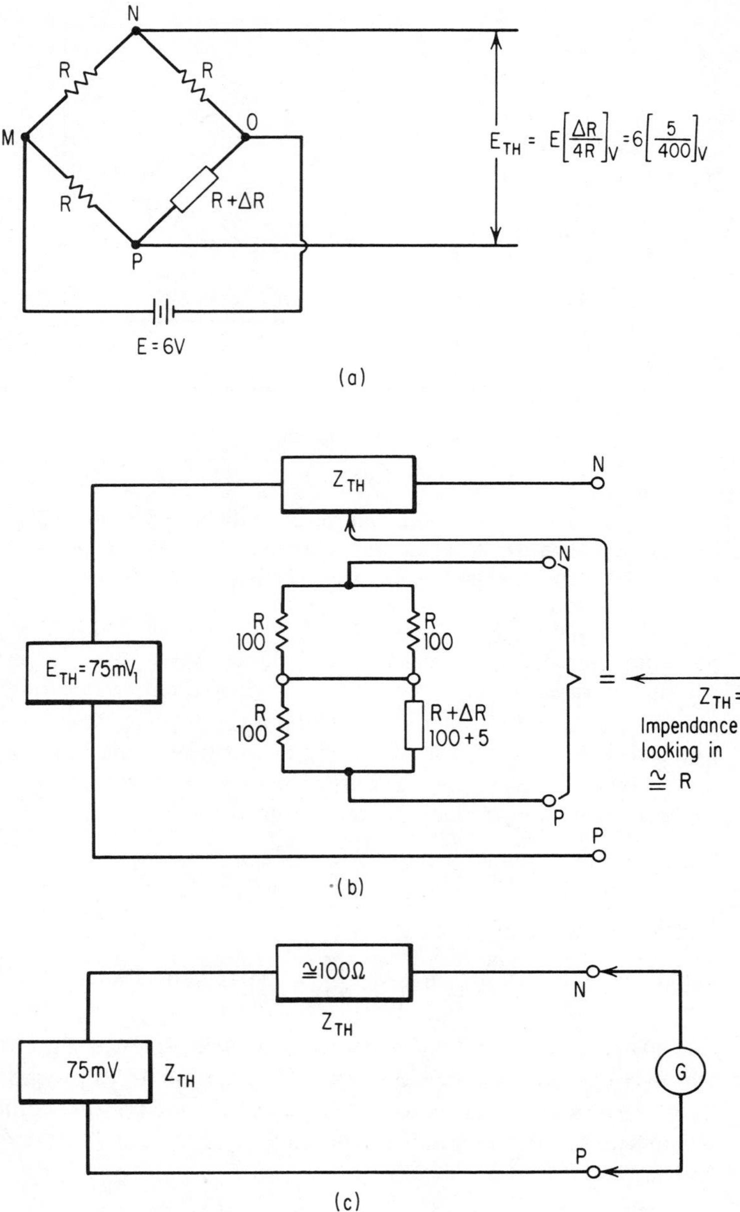

Figure 4-8. Arriving at an approximate Thévenin equivalent circuit for the unbalanced bridge of Fig. 4-7: (a) finding the Thévenin voltage E_{TH}; (b) finding the Thévenin impedance Z_{TH}; (c) the complete Thévenin equivalent circuit.

From the complete equivalent circuit of Fig. 4-8(c) we can now estimate the full-scale current required from a zero-center galvanometer detector for the 5% unbalance and 6-V battery assumed. If the galvanometer resistance could be neglected, the current for full-scale deflection to either side of zero would simply be 75 mV/100 Ω. This figure must, of course, be reduced by including the galvanometer resistance, which in this case can be taken as around 50 Ω, resulting in a current required for full-scale deflection (left or right) of 75 mV/150 Ω = $\frac{1}{2}$ mA.

We thus arrive at the type of meter detector usually found in student bridges as a 500–0–500-μA dc meter, which can be expected to deflect full scale from either side of zero for an unbalance of about 5% from the original balance value. Similar computation can now be made for any desired percentage unbalance or for other circuit changes, including the use of a voltage rather than current detector, by substituting in the simple Thévenin equivalent circuit.

4-10. ERROR DETECTION WITH UNBALANCED BRIDGE

The output of a bridge circuit is frequently used as an error-detection element in electronic systems that use a feedback arrangement to automatically correct this error, thus forming automatic self-balancing systems used in many important instrumentation applications. An example of such an error-detecting input, derived from the comparison of a thermocouple output voltage with a reference voltage, was introduced previously in Sec. 4-5. The example cited now uses the voltage output from an unbalanced bridge containing passive elements to form the error-detecting circuit, as an illustration of the numerous cases where the transducer, such as a strain gage in this case, is not a voltage generator in its own right, as was the thermocouple. In such cases, where the passive-element transducer undergoes a change in resistance, the bridge circuit converts this resistance change to a corresponding error voltage. By this means, the varying quantity being monitored produces a corresponding varying output voltage (derived from the bridge supply voltage), which is caused by the change in the resistance (or some other electrical property, such as capacity or inductance) of the passive transducer.

The strain gage, which is more fully discussed in Chapter 9, can serve at this point to emphasize how the voltage output of an unbalanced bridge can be employed for the purpose of error detection. In a typical case, we may assume that a measurement of tension is to be made involving a useful range of applied tension such that the resistance of the thin wire in the gage will vary by a maximum of 1% from its unstressed value. For the bridge circuit of Fig. 4-7, this corresponds to a change of ΔR or $x = 1\ \Omega$, in the original value of $R_X = 100\ \Omega$. Using the equivalent Thévenin voltage of Fig. 4-7 for a 4-V battery, we get

$$E_{\text{TH}} = \frac{E(x)}{4R} = \frac{4(1)}{400} = 0.01 \text{ V or } 10 \text{ mV}$$

At maximum tension, then, this 10-mV output would be applied to an electronic self-balancing instrument of the dc millivolt potentiometer type, having a range (or span) of 10 mV dc. This particular range of 0–10 mV dc is the basis of widely used general-purpose potentiometric recorders.

4-11. LINEARITY OF UNBALANCED BRIDGE OUTPUT

When considering such small unbalances as those obtained from a resistance deviation up to a maximum of about 10%, the voltage output from the bridge corresponds in a satisfactory linear fashion to the amount of unbalance caused by the resistance change. However, when larger unbalances are involved, the exact relationship for E_{TH} must be used (instead of the approximate one) as follows:

$$E_{TH} = E\left[\frac{X}{2(2R + X)}\right]$$

This indicates that a nonlinear relation exists between the percentage of resistance variation and the resulting bridge output voltage E_{TH}. The nonlinearity becomes quite pronounced for resistance deviations greater than about 10% and must accordingly be taken into account.

The relative values of bridge output voltage, resulting from increasing percentage deviation in one arm of the bridge, are plotted in Fig. 4-9. Here, the precentage resistance deviation ($X/R_X \times 100$) is plotted horizontally against the relative bridge output voltage vertically. Since the actual amount of bridge output is directly proportional to the supply voltage E, the relative output is expressed in terms of millivolts output per supply volt (mV/E). Note that for resistance deviations up to around 10%, the output is quite linear. In such cases, a simple approximation can be made for each 1% of resistance variation. Thus for a 1% resistance deviation an output of 2.4 mV per applied volt may be expected, and for a 4-V supply battery an output of 9.6 or approximately 10 mV is obtained, which agrees closely with the previous straingage calculation using the approximate formula for such a small unbalance.

A sample calculation will also show that the degree of nonlinearity differs between positive deviations of X (increasing resistance) and negative deviations (decreased resistance). Thus, for a 10% increase in the original 100-Ω R_X arm of the equal-arm bridge, X will equal 10 Ω. From the preceding equation, the output

$$E_{TH} = E\left(\frac{10}{420}\right) = 23.8 \text{ or } 24 \text{ mV}/VE.$$

Similarly, for a 10% decrease in the original resistance, X is again equal to 10 Ω,

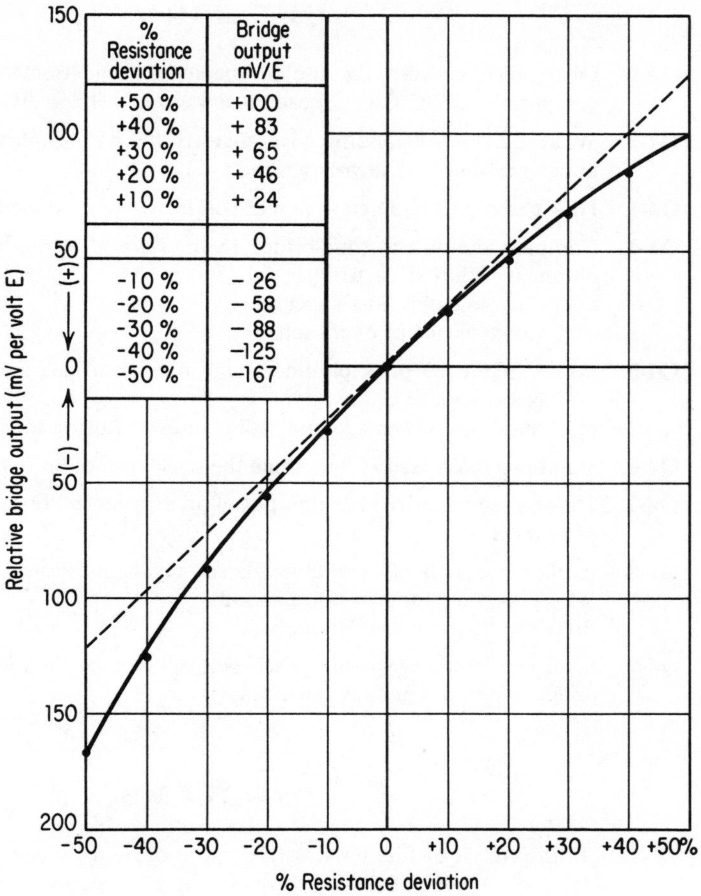

Figure 4-9. Generalized nonlinear output of an unbalanced bridge.

but the expression for E_{TH} becomes

$$E_{\text{TH}} = E\left(\frac{X}{4R - 2X}\right) \quad \text{and} \quad E_{\text{TH}} = E\left(\frac{10}{380}\right) = 26.3 \text{ or } 26 \text{ mV}/VE.$$

Plotting these values up to a 50% deviation in resistance in either direction gives the relative bridge output in the form of the generalized curve of Fig. 4-9, where the dashed straight line emphasizes the departure from linearity.

Up to this point, only the action of the dc bridge method of comparison has been investigated. Although all the points made here can also be made to hold true for an ac bridge by substituting impedances for corresponding resistances, the modifications required are sufficiently extensive to warrant separate treatment in another chapter.

QUESTIONS

Q4-1. Distinguish between the use of the term "potentiometer" when employed in comparison circuits, as opposed to its use in voltage-dividing action.

Q4-2. What are *two* outstanding advantages of the potentiometer-comparison method in determining precise voltages?

Q4-3. How can a potentiometer-comparison instrument be made to be direct-reading?

Q4-4. Compare the Wheatstone-bridge method for determining resistance, with the ohmmeter method, as to:
(a) Factors determining accuracy.
(b) Convenience of operation.

Q4-5. (a) Explain the principle underlying the use of a Zener-diode circuit in place of a standard cell.
(b) State one advantage and disadvantage of using the Zener-diode method.

Q4-6. What *two* main factors determine the sensitivity of an unbalanced bridge circuit?

Q4-7. Discuss the linearity of the output of an unbalanced bridge versus the percentage of unbalance.

Q4-8. Explain the basis of operation of a self-balancing system using:
(a) The potentiometer-null method.
(b) The bridge-null method.

Q4-9. State *two instances* where the self-balancing system is widely used in industry in preference to a manually balanced system.

PROBLEMS

P4-1. In the circuit of Fig. P4-1, with the arm of the voltage divider set at point **B**:
(a) What value of signal voltage E_s will produce a null reading of galvanometer G?

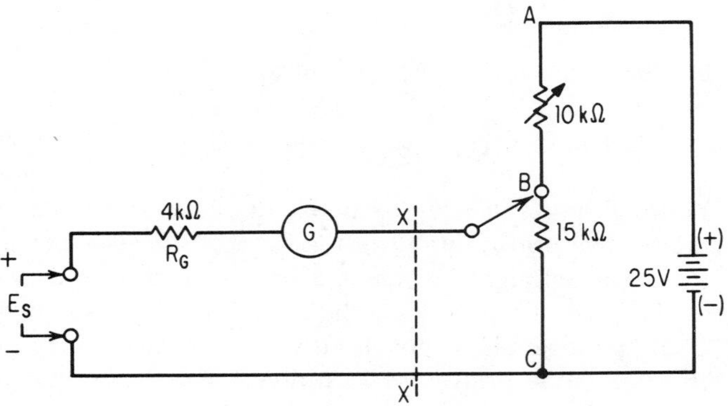

Figure P4-1. Potentiometer-null circuit.

(b) Draw the Thévenin equivalent of the circuit considered to be broken at X–X′.

(c) What current will flow through galvanometer G when the signal voltage E_s is 5 V, with the polarity shown?

P4-2. In the potentiometer circuit of Fig. P4-1, what are the limits for the signal voltage E_s in order that the current through galvanometer G should not exceed 0.5 mA (in either direction)?

P4-3. A load R_l requiring 20 mA is to be connected between points C–C′ in Fig. P4-3.

(a) Draw the Thévenin equivalent for the circuit to the left of points C–C′.

(b) Find the resistance of the load R_l.

(c) Find the voltage across the load R_l.

(d) Find the current taken from the 500-V source.

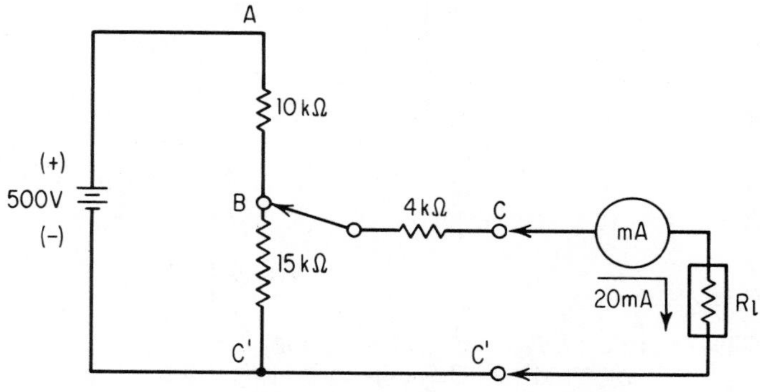

Figure P4-3

P4-4. The load resistor R_l in Fig. P4-4 requires 3 mA when connected across the points X–X′.

(a) Draw the Thévenin equivalent of the circuit to the left of points X–X′.

(b) Find the value of load resistor R_l.

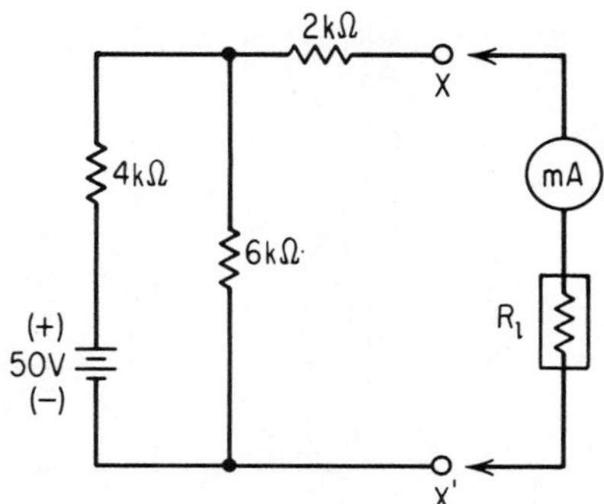

Figure P4-4

(c) Find the voltage across the 6-kΩ resistor with the load R_l connected.
(d) Find current taken from the 50-V source.

P4-5. In the Wheatstone-bridge circuit of Fig. P4-5, find the equivalent resistance between the points X–Y.

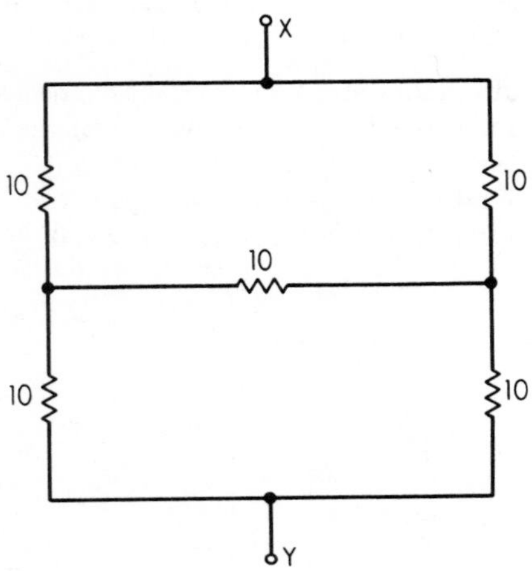

Figure P4-5

P4-6. In the *bridge arrangement of* Fig. P4-6:
(a) Find the Thévenin equivalent voltage E_{TH}, i.e., the potential of point X with respect to point Y, with an open circuit between them.
(b) If the 12-V source is replaced by a wire of zero resistance, what is the Thévenin equivalent resistance Z_{TH} between points X and Y?
(c) What is the maximum current that can flow between points X and Y?

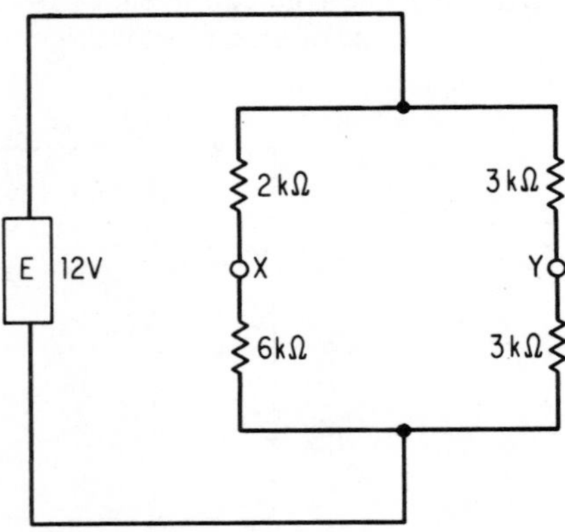

Figure P4-6

P4-7. A 3-kΩ resistor is connected as a load across the points X–Y in Fig. P4-6. Using the Thévenin equivalent of the circuit between points X and Y, find the current through the 3-kΩ resistor, when it is connected as a load.

P4-8. In the *unbalanced equal-arm Wheatstone-bridge* circuit of Fig. P4-8:

(a) Find the equivalent Thévenin voltage acting across the points X–Y (this answer may be checked by the approximate formula) for small unbalance in the equal-arm bridge $E_{TH} = E(\Delta R/4R)$.

(b) Find the *approximate current* through the 50-Ω galvanometer *G*.

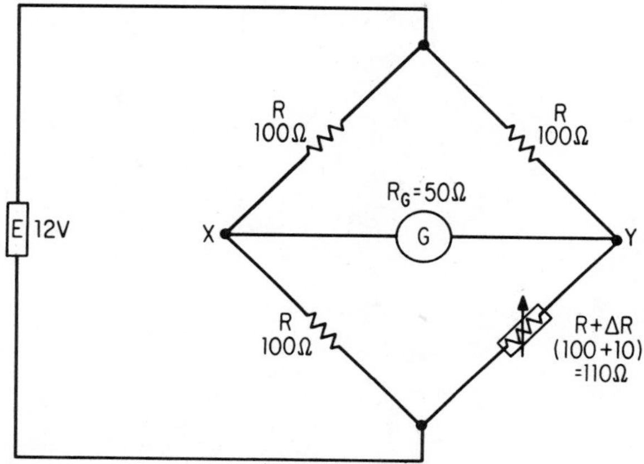

Figure P4-8

P4-9. An *unbalanced Wheatstone bridge having unequal arms* is shown in Fig. P4-9.

(a) Draw the Thévenin equivalent of this circuit, considered broken at the points X–Y.

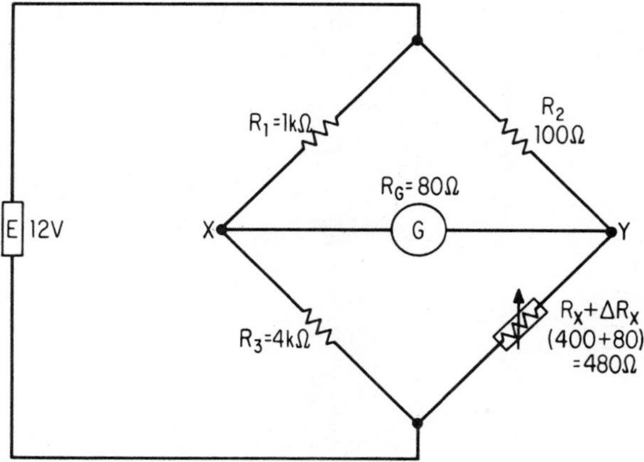

Figure P4-9

(b) Find the *approximate current* through the 80-Ω galvanometer G.

P4-10. (*Optional*) Completely solve Problem P4-7 for the current in each of the branches and for the total current from the source, using a loop analysis of the network. [See Jackson reference in Appendix B or (more advanced) E. Brenner and M. Javid, *Analysis of Electric Circuits*, 2nd ed. (New York: McGraw-Hill, 1967).]

5

Alternating-Current Bridge and Impedance Measurement Methods

5-1. THE GENERAL FORM OF ALTERNATING-CURRENT BRIDGE

The Wheatstone bridge is shown in a generalized form in Fig. 5-1 to apply to both dc and ac quantities.

If the four impedances can be so chosen that the detector current $i_D = 0$, then $i_2 = i_1$ and $i_4 = i_3$. The potentials at O and P are necessarily equal, so $i_1 Z_1 = i_3 Z_3$ and $i_1 Z_2 = i_3 Z_4$. It follows at once that the condition of balance is

$$Z_1 Z_4 = Z_2 Z_3$$

This equation among complex quantities can be shown to be equivalent to two equations among real quantities. Designating the reactive part of impedance Z as X (with the proper sign, positive for inductive and negative for capacitive), and separating the real and imaginary terms after multiplying out, we obtain two conditions for balance, as follows:

$$R_1 R_4 - X_1 X_4 = R_2 R_3 - X_2 X_3$$

and

$$R_1 X_4 - R_4 X_1 = R_2 X_3 + R_3 X_2$$

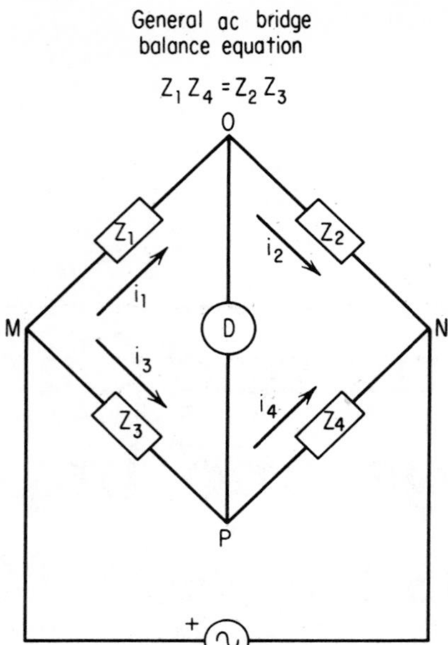

Figure 5-1. General form of ac bridge network.

These two conditions must therefore be satisfied simultaneously to balance the general bridge network. It may be noticed that the Wheatstone bridge is a special case of this network; when $X = 0$, the first specifies the familiar dc balance condition, and the other equation reduces to zero, since each term is zero if all the reactances vanish.

5-2. THE MAXWELL-BRIDGE CIRCUIT

In the circuit of the Maxwell bridge in Fig. 5-2(a) an unknown inductance Z_X, made up of L_X and R_X, is used as Z_2 in the upper branch, Z_1 being a resistor R_1. The unknown inductance L_X is compared with the parallel combination of a capacitor standard C_P and variable resistor R_P, which is used as Z_3, while Z_4 is a variable resistor R_4. This bridge is particularly suitable for measuring inductance, since ordinarily capacitances come closer to being ideal lossless standards of reactance than do the very best coils. In addition, the balance equation of the Maxwell bridge for the inductance component L_X is independent of the losses associated with the R_X of the coil, and also is independent of the frequency of measurements ω, an important point that is not achieved in many other bridge forms. The Maxwell-bridge balance equations are

$$L_X = R_1 R_4 C_P$$

$$R_X = R_1 \frac{R_4}{R_P}$$

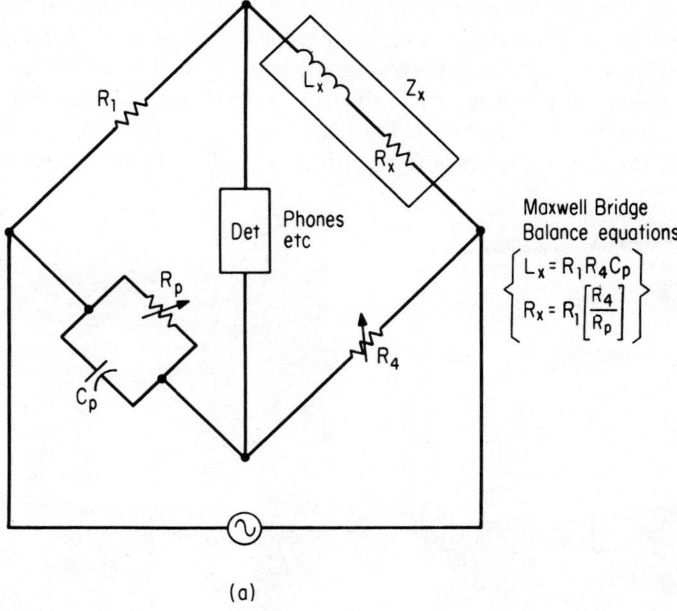

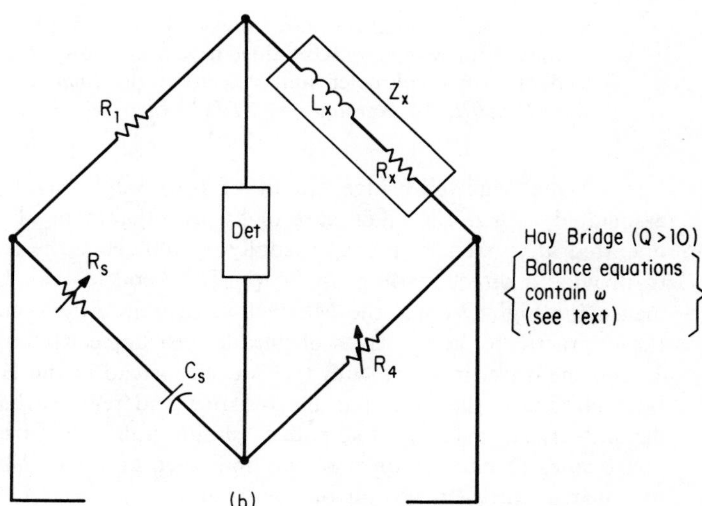

Figure 5-2. Bridge circuits of general-purpose impedance bridges: (a) Maxwell bridge and balance equations; (b) Hay-bridge arrangement, employed for impedances having a Q greater than 10.

In the ordinary arrangement of the Maxwell bridge for inductance measurements, a fixed capacitance is used as the standard for C_P, and the inductance balance is obtained by varying either resistance R_1 or R_4. In the commercial forms of general-purpose or "universal" impedance bridges, shown in Fig. 5-3, this variation is accomplished on the large main dial, calibrated directly in units of henries for inductance (and in microfarad units for capacity).

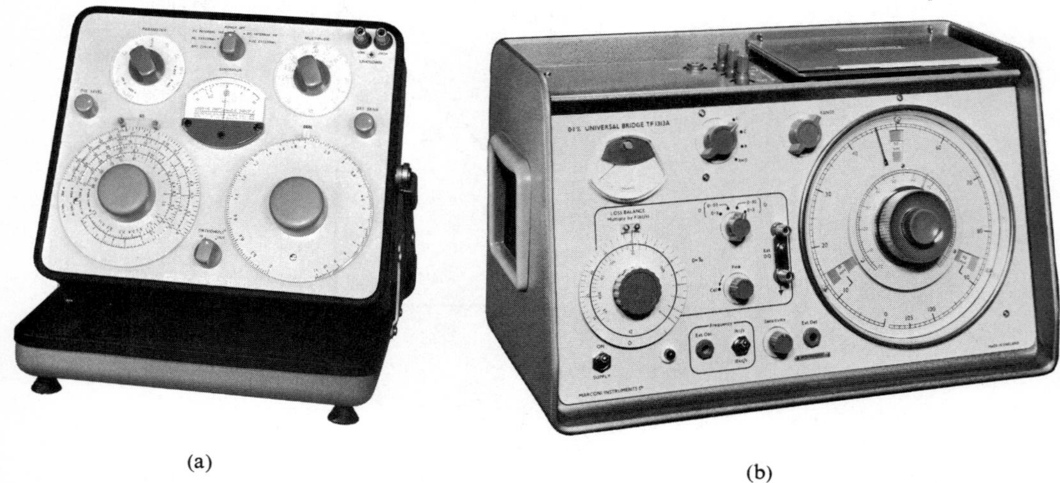

(a) (b)

Figure 5-3. General-purpose "universal impedance bridges," both types providing self-contained oscillators, power supplies, and electronic-voltmeter null detectors: (a) *General Radio model 1650B;* (b) *Marconi type TF1313A.*

In the Maxwell-bridge circuit of Fig. 5-2(a) the equivalent resistance R_X associated with L_X, is taken care of by an adjustment of R_P. When the bridge is operated at a fixed frequency (usually at 1000 Hz), the scale of R_P is calibrated to provide a direct reading of the Q of the coil ($Q = \omega L_X/R_X$). However, when measuring high-Q coils, the Maxwell-bridge circuit encounters difficulties due to the impractically large values of parallel resistance R_P required, and so the series arrangement (R_S in series with C_S) is used instead in the Hay bridge. Common to both circuits is the fact that the resistive and reactive balance points are interdependent. As a result, a so-called "sliding-null" must be obtained by successive approaches to balance on first one dial, then another. While the Maxwell bridge produces a satisfactory null on low-Q inductors, coils having a high Q (i.e., Q values greater than 10) are measured in a more satisfactory manner by the Hay circuit of Fig. 5-2(b).

5-3. THE HAY-BRIDGE CIRCUIT

The Hay-bridge circuit in Fig. 5-2(b) differs from the Maxwell circuit of Fig. 5-2(a) in using a series combination of C_S and R_S in place of the parallel combination of C_P and R_P. This arrangement avoids the difficulty of encountering excessively high values of R_P for high-Q coils, but introduces another difficulty in that the balance equations for inductance contain an expression dependent on frequency ω in the form of the quality factor Q:

$$L_X = \frac{R_1 R_4 C_S}{1 + (1/Q)^2}$$

$$R_X = \frac{R_1 R_4}{R_S}\left(\frac{1}{Q^2 + 1}\right)$$

The additional multiplier, containing the expression $1/Q^2$, nevertheless does not offer any considerable difficulty when the Q of the coil is sufficiently high (for $Q > 10$, the error is less than 1% and for $Q = 30$, it is only 0.1%). *Accordingly, the Hay circuit is generally preferred for coils of high Q, while the Maxwell circuit is preferable for coils having low Q.* In the universal type of impedance bridge, the change from the Maxwell to the Hay circuit is usually accomplished automatically when the range of Q is switched from low Q to high Q.

5-4. ALTERNATING-CURRENT NULL DETECTORS

Headphones and Rectifier-Type Meters

The contribution of electronics to the instrumentation of bridge and other ac measurements provides a great advance in the sensitivity with which it is possible to detect a null balance (or amount of unbalance) in a comparison circuit. The telephone receivers (or headphones) ordinarily used as a detector are, as a matter of fact, extremely sensitive qualitative detectors for the presence or absence of an ac signal. They are not at all as effective, however, in determining the absolute level of a signal for obtaining a clear-cut null. The presence of even a slight amount of harmonic content in the oscillator supply is enough to dull the sharpness of a null, since these harmonics generally do not balance out at the same dial setting that cancels the fundamental signal. The expedient of attenuating the harmonics by inserting a filter circuit sharply tuned to the fundamental helps a bit, but at the expense of reducing the overall sensitivity of the detector.

The problem of obtaining a clear-cut null becomes more difficult if a visual indication of output is obtained on an ordinary ac meter (such as a rectifier type), because of the inherently smaller sensitivity of the meter-and-rectifier combination.

An effective approach to a more satisfactory null is offered by the use of electronic amplification.

Electronic-Amplifier Detectors

When the level of the bridge output is amplified by an *electronic amplifier* type of detector, the sensitivity of the detector system is greatly increased and the attenuation introduced by a filter arrangement no longer constitutes a problem. Moreover, the *high input impedance* of an *electronic voltmeter* is sufficient to overcome any loading problems that might reduce sensitivity and broaden the null. The electronic detector may take the form of either the vacuum-tube (VTVM) or transistor voltmeter (TVM). As a result of the combined advantages of great sensitivity and high impedance, the electronic type of detector provides a visual indication of a definite null that can be made as clear-cut and sharp as is generally desired, under the control of the operator. The photographs in Fig. 5-3 show the use of the electronic-voltmeter type of null detector, incorporated in the impedance bridge.

Transistorized voltmeters, having a range of 1 mV full-scale, are also suitable for general-purpose visual null detection.

An advanced form of null detector consisting of a *tuned amplifier* with 1 μV sensitivity is shown in Fig. 5-4. It features a sensitive, low-noise transistor amplifier with the additional provision of continuous tuning from 20 Hz to 20 kHz, assuring a sharp null detection in spite of harmonics that may be present.

The block diagram shown in Fig. 5-4(b), indicates the use of three amplifiers: first, a preamplifier arranged for obtaining sensitivity at a minimum noise level, followed by the main feedback amplifier, which is frequency-selective, and finally a third amplifier for compression purposes, to produce a logarithmic indication. The shunt silicon-rectifier diode combinations produce limiting action to the extent that, even at the full-scale sensitivity of μV, it is possible to connect the input to a 115-V ac line without damage to the input and following transistors. Since the tuned amplifier can be switched to provide either flat or selective characteristics, the instrument may also be used as an amplifier of transducer outputs, for general use. The continuous tuning feature provides still another use as an audio-frequency wave analyzer. In all these uses, the instrument provides a possible amplifier gain of 120 dB, with an input impedance of 50 kΩ or higher (up to 1 MΩ), depending on the setting of the gain control. This high impedance is obtained by operating the first stage at an extremely low value (20 μA) of collector current.

An *oscilloscope* can be used as a null detector to take the place of the voltmeter. It can then provide a view of the output wave-form, if desired, while also functioning as a very sensitive, high-impedance null indicator.

When employing any of these electronic null detectors that incorporate an ac amplifier, care must be taken to avoid excessive pickup of hum or stray ac signals, by observing proper shielding and grounding procedures. In some cases,

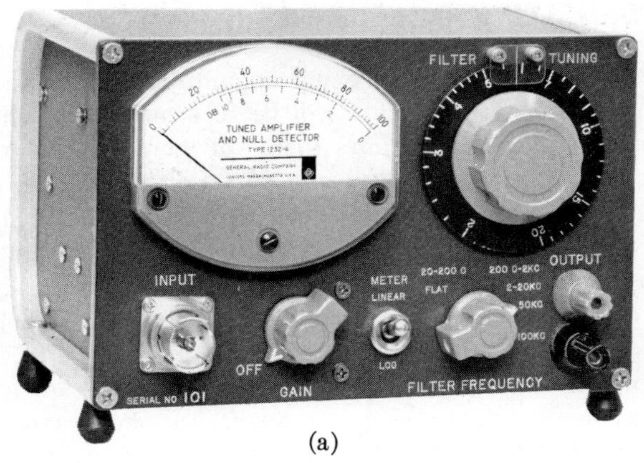

(a)

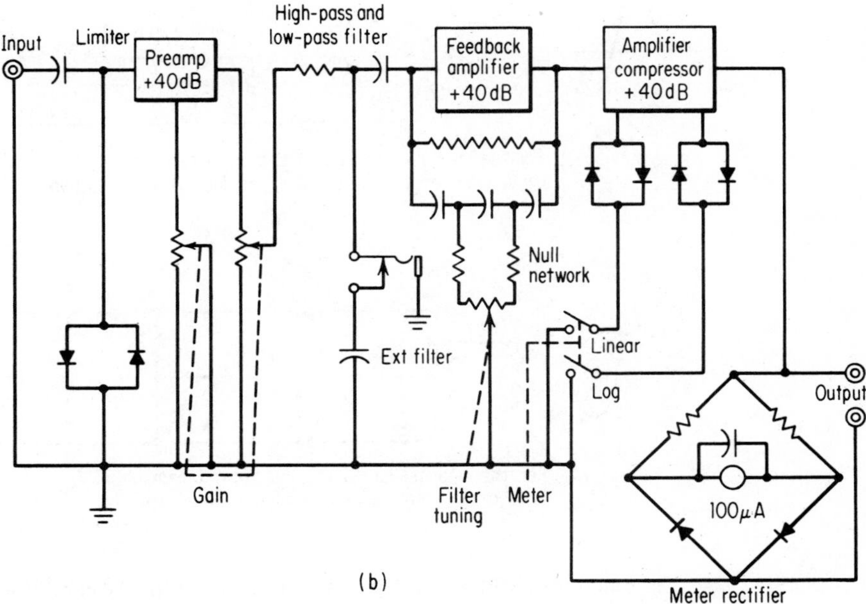

(b)

Figure 5-4. Transistor tuned-amplifier-type null detector: (a) *GR 1232-A tuned amplifier and null detector* (high sensitivity); (b) block diagram of (a).

when very high sensitivity is a major factor, the electronic null detector is powered by batteries, as is frequently the case when the transistor voltmeter is employed.

5-5. PHASE-SENSITIVE DETECTORS

When two ac voltages of equal magnitude are compared in a series opposition circuit, the resulting output will be zero only when one voltage is in phase with the other; at all other times there will be an output that will depend upon the phase relationship between the two voltages being compared. This situation applies not only to the ac bridge circuits discussed in this chapter, but also to many other cases where an input ac voltage is being compared with another voltage as a reference, as in an error-detection circuit. In all these cases, the detection of the error requires a phase-sensitive circuit arrangement.

The measurement of the absolute phase angle existing between two voltages is generally made either with an *oscilloscope* or an *electronic phase meter*. An indication of *relative phase*, however, can be obtained by a dc galvanometer, without electronic amplification, in a *dual-rectifying circuit*, shown in Fig. 5-5.

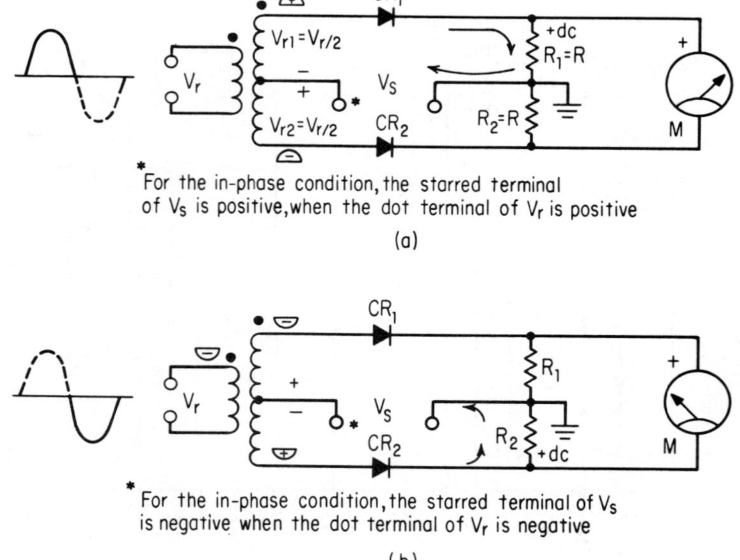

Figure 5-5. Phase-sensitive detector, shown for the condition in which the signal voltage V_s is in phase with the reference voltage V_r: (a) during the first half cycle, V_s helps cause a large deflection of meter M to the right; (b) in the second half cycle, V_s opposes V_r, causing smaller deflection to the left.

A very useful property of this circuit is the fact that the deflection of the zero-center dc galvanometer (or voltmeter) will change direction, depending on whether the two voltages V_1 and V_2 are either in phase or out of phase with each other, thus indicating not only an unbalance, but also the direction of the unbalance. In the discussion to follow, the circuit is required to be sensitive only to the extent of distinguishing in-phase from out-of-phase conditions. In this respect the circuit could be more correctly designated as a "phase-inversion-sensitive" circuit, but the shorter name is more commonly used.

In the error-detecting form in which this circuit is usually used, V_r is a reference ac voltage, generally 60 Hz, derived from a power transformer, and V_s is a signal that is substantially either in phase with V_1 (phase angle $\phi = 0$), or out of phase with V_1 (phase angle $\phi = 180°$).

When the input signal V_s is zero in Fig. 5-5, the output taken from each rectifier cathode is positive on each alternate half cycle. Thus, on the first cycle [Fig. 5-5(a)], reference voltage V_r is shown with the instantaneous polarities given in the upright semicircles, and instantaneous voltage V_{r_1} produces a positive voltage to ground across R_1, while rectifier CR_2 is nonconducting. On the second half cycle [Fig. 5-5(b)], shown with the instantaneous polarities given in the inverted semicircles, rectifier CR_1 is nonconducting and V_{r_2} this time produces a positive voltage to ground across R_2. Since these two equal and opposite voltages are averaged over the full cycle by the galvanometer, the detector reads zero over the full cycle for the no-signal input condition ($V_s = 0$).

When an input signal V_s is applied, it will either aid or oppose the reference voltage, depending upon whether it is in phase or out of phase with it. Thus, if V_s is *in phase* with V_r, the instantaneous ac voltages shown in the upright semicircles [Fig. 5-5(a)] aid each other to produce a larger R_1 dc output to ground on the first half cycle; on the second half cycle, the instantaneous ac voltages shown in the inverted semicircles [Fig. 5-5(b)] oppose each other and produce a smaller R_2 dc output to ground. The result on the galvanometer, connected as shown, is therefore to produce a deflection to the *right*, corresponding to the magnitude of the *in-phase* input signal V_s. In a similar fashion, if V_s is *180° out of phase* with respect to V_r, the galvanometer will deflect to the *left* in proportion to the magnitude of that input signal. Hence, the phase-sensitive detector has an important use in error-detecting circuits, not only in providing an output proportional to the error, but also in indicating the sense or direction of the error.

In cases where the input signal is at some phase angle other than 0° or 180° with respect to the error, the output of the phase-sensitive circuit also responds to the phase angle (ϕ). In such cases, it can be shown that the output is proportional to the cosine of this phase angle and in this case would be equal to $2V_s \cos \phi$. For use as a phase meter, the circuit is arranged to take into account both the magnitude and the phase of the input signal and usually incorporates electronic amplifiers, as discussed later.

To summarize the error-detecting portion of an instrumentation system, a comparison is made between the variable signal input voltage and a reference voltage: when the comparison results *in a balanced (or zero-output) condition, the error detector indicates a null;* when the comparison results *in an unbalanced output, the error detector responds to both the magnitude and relative phase of the error.*

QUESTIONS

Note: The questions below (and the first four problems) are in the nature of useful exercises to enable the student to *quickly estimate common impedance relations,* they are *based on the following approximate relations at a frequency of 1 kHz,* at which most commercial impedance bridges and other impedance instruments operate:

Impedance Relations (Based on $f = 1$ kHz)

Inductive (L)		*Capacitive (C)*	
Reactance $X_L = (\omega)L$ or $(2\pi f)L\ \Omega$		Reactance $X_C = \frac{1}{(\omega)C} = \frac{1}{(2\pi f)C}\ \Omega$	
1 H at 1 kHz	$X_L \cong 6000\ \Omega$	1 μF at 1 kHz	$X_C \cong 160\ \Omega$
5 H at 1 kHz 1 H at 5 kHz	$X_L = 6\ \text{k}\Omega \times 5$ or $30\ \text{k}\Omega$	5 μF at 1 kHz 1 μF at 5 kHz	$X_C = \frac{1}{5}(160)$ or $32\ \Omega$
0.2 H at 1 kHz 1 H at 200 Hz	$X_L = 6\ \text{k}\Omega/5$ or $1200\ \Omega$	0.2 μF at 1 kHz 1 μF at 200 Hz	$X_C = 5(160)$ or $800\ \Omega$

Impedance (Z) of Coil

$$Z = \sqrt{R_L^2 + X_L^2}$$

Quality Factor (Q) of Coil

$$Q = \frac{X_L}{R_L}$$

[where R_L is effective (not dc) resistance of coil]

Relations at Resonance:

X_L or $2\pi f_r L = 1/2\pi f_r C$ or X_C

and resonant frequency $f_r = 1/2\pi\sqrt{LC}$

Q5-1. Find the *inductive reactance X_L in ohms* of the following coils at the specified frequencies:

(a) $L = 2$ H at 1 kHz.
(b) $L = 1$ H at 2 kHz.
(c) $L = 2$ mH at 1 MHz.
(d) $L = 1$ mH at 2 MHz.
(e) $L = 5$ H at 400 Hz.

Q5-2. Find the *capacitive reactance X_C in ohms* of the following capacitors at the specified frequencies:

(a) $C = 4\ \mu$F at 1 kHz.

(b) $C = 1\ \mu F$ at 4 kHz.
(c) $C = 0.05\ \mu F$ at 1 kHz.
(d) $C = 0.05\ \mu F$ at 4 kHz.
(e) $C = 0.05\ \mu F$ at 400 Hz.

PROBLEMS

(See the note at the beginning of the Questions section for approximate impedance Z and Q relations for the first four problems, based on values at a frequency of 1 kHz.)

P5-1. A 10-H choke coil has a Q of 0.75. Find impedance Z at 1 kHz.

P5-2. A 6-mH voice coil has an impedance of 45 Ω at 1 kHz. Find Q of the voice coil.

P5-3. What approximate capacity C has a reactance X_C in ohms, equal to that of a 160-μH coil at 1 MHz, thus producing resonance at this frequency?

P5-4. What value of inductance L has a reactance X_L, in ohms, equal to that of a 32-pF (or 32-μμF) capacitor at 1 MHz, thus producing resonance at this frequency?

P5-5. A *generalized bridge circuit* with unequal resistive arms is shown in Fig. P5-5(a). The analytic expression for the galvanometer current I_G is given below, in terms of four arms (R_1, R_2, R_3, and R_x) and the galvanometer resistance R_G. [Note that the internal resistance of the source E (which may be ac as well as dc) is disregarded, since it is practically always low enough to be neglected.]

$$I_G = \frac{E(R_2R_3 - R_1R_x)}{R_2(R_1 + R_3)(R_G + R_3 + R_x) + R_1R_3R_x - R_2R_3^2 + R_GR_x(R_1 + R_3)}$$

When the typical values shown in the figure are substituted, the general expression reduces to the following, in terms of the variation in R_x—from which the *graph of I_G versus R_x* [shown in Fig. P5-5(b)] can be drawn:

$$I_G \text{ (in amps)} = \frac{6000 - 6R_x}{6100R_x + 500{,}000}$$

Three points have already been located on this graph, namely,

$$R_x = 1000\ \Omega \text{ (balance)},\ I_G = 0$$

$$R_x = 0\ \Omega \text{ (short circuit)},\ I_G = +12 \text{ mA}$$

$$R_x = \infty\ \Omega \text{ (open circuit)},\ I_G \cong -1 \text{ mA}$$

(a) Compute the graph by locating five more points ($R_x = 250\ \Omega$, $500\ \Omega$, $750\ \Omega$, $2\ k\Omega$, and $3\ k\Omega$).

(b) Explain the difference in the *rate of change of the slope* of this typical curve on either side of balance.

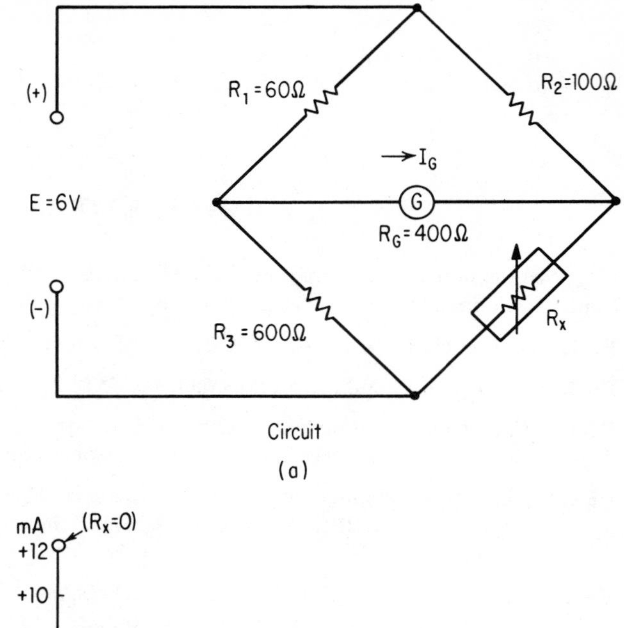

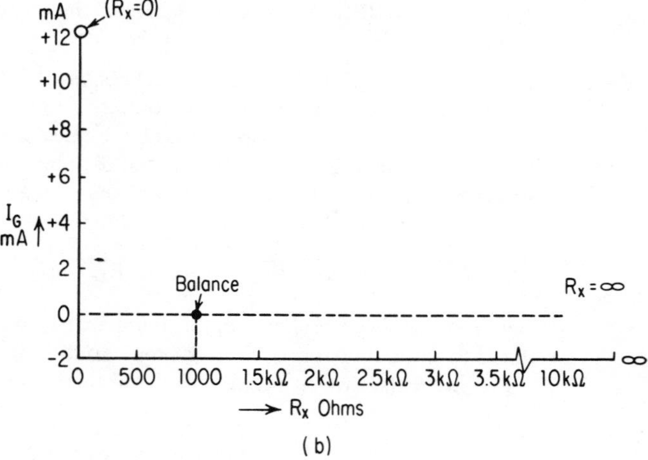

Figure P5-5. Generalized bridge: (a) circuit; (b) graph of deflection of galvanometer I_G versus unbalanced R_X.

P5-6. Simplifying the general bridge expression of Problem P5-5, after substituting the equal-arm values for $R = 100$, as shown in Fig. P5-6, the following approximate expression is obtained for the 100-Ω *equal-arm bridge*, for small deviations:

$$I_G \text{ (in amps)} \cong \frac{E(100 - R_x)}{R_x(300 + 2R_G) + 200R_G + (100^2)}$$

Note: For small deviations, assume that $R_x = R$ in the denominator of the fraction, to obtain the above simplified approximate expression.

For a 5% deviation in R_x ($R_x = 95$ or 105 Ω), show that the approximate

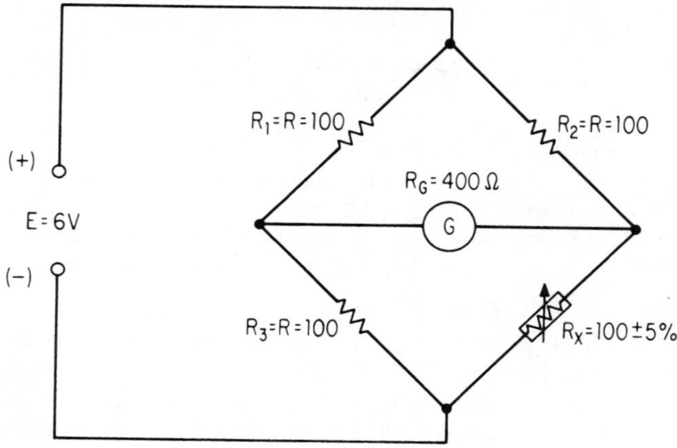

Figure P5-6. Equal-arm bridge circuit.

galvanometer current I_G, as obtained from the equation above, agrees with the solution by means of the following approximate Thévenin equivalents (as discussed in the text):

$$E_{\text{TH}} \cong E\left(\frac{\Delta R}{4R}\right) \cong E\left(\frac{100 - R_x}{400}\right) \text{ and } Z_{\text{TH}} = 100$$

P5-7. If a galvanometer resistance of 100 Ω is assumed (in addition to the assumption of equal arms of 100 Ω each), a *useful estimate of the approximate galvanometer current I_G for small unbalances* can be conveniently obtained from the further

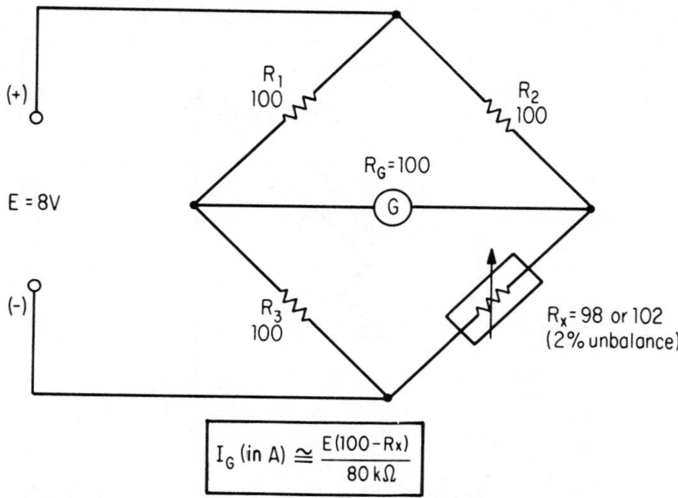

Figure P5-7. Simplified circuit for estimating galvanometer current I_G for small unbalance.

simplified expression:

$$\left.\begin{array}{l}\text{For small unbalance and}\\ \text{all } R\text{'s (except } R_x) = 100\ \Omega\end{array}\right\} I_G \text{ (in mA)} \cong \frac{E(100 - R_x)}{80\ \text{k}\Omega}$$

(This is shown in Fig. P5-7.)

(a) Estimate the galvanometer current I_G for a 2% unbalance from the equation above, using $E = 8$ V.

(b) Compare this estimate with the exact value given by either a Thévenin equivalent solution or by the general equation shown in the early part of Problem P5-5.

P5-8. If an ac source of 8 V (rms) at 60 Hz is used instead of the 8-V dc source in Fig. P5-7:

(a) What ac voltage would appear across a 100-Ω resistor substituted for galvanometer G in the figure?

(b) What type of voltmeter would be needed as a detector in this case?

P5-9. A 1000-cycle inductance-bridge circuit is shown in Fig. P5-9, arranged for a 2% unbalance (balance would be achieved by arm AB = 1000 Ω). The detector is a high-impedance ac VTVM, which measures the *unbalance voltage* between points B and D, without drawing any appreciable current.

Using the Thévenin equivalent voltage (E_{TH}) as the voltage output (E_0) at the dectector, find E_0 for an input of 15 V at 1000 cycles:

Note: Use

$$E_{TH} = E\left(\frac{Z_{AB}}{Z_{AB} + Z_{BC}} - \frac{Z_{AD}}{Z_{AD} + Z_{DC}}\right)$$

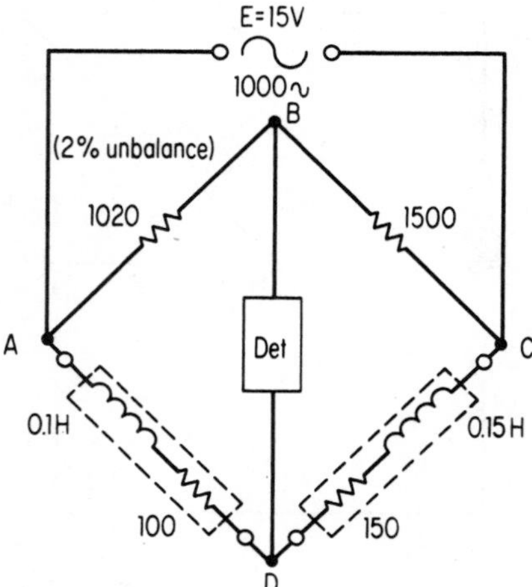

Figure P5-9. Unbalanced inductive-bridge circuit. [Adapted from M. B. Stout, *Basic Electrical Measurements*, 2nd ed. (Englewood Cliffs, N.J.: Prentice-Hall, 1960).]

P5-10. In the *direct-reading impedance circuit of* Fig. P5-10, the impedance Z consists of a 5-H choke coil (dc resistance = 100 Ω). When the calibrated variable resistor R_v is set at 50 kΩ, the reading of the voltage across R_v ($V_{C\text{-}B}$) equals that across $Z(V_{B\text{-}D})$.

(a) What is the impedance of Z?

(b) What is its approximate Q?

(c) If the voltmeter is connected across the points A–B, how would this reading $V_{A\text{-}B}$ compare with a reading obtained across the same points when a coil having the same impedance (not inductance), but a higher Q, is used?

(d) What is the relation between $V_{A\text{-}B}$ and the coil Q?

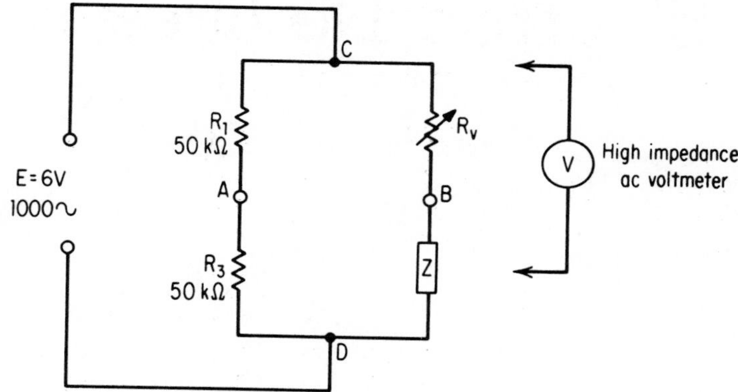

Figure P5-10. Circuit for direct-reading impedance measurement.

6

Electron Devices

6-1. NATURE OF ELECTRICAL AND ELECTRON DEVICES

When we speak of an *electronic*, rather than *electrical*, circuit, we generally look for the presence of an electron device as the main distinguishing factor in providing the electronic action. While it is true that the movement of electrons is common to both types of circuits, the point of departure lies in the relatively *more effective control of the electron movement* in the electronic types. Moreover, the circuit containing an electron device is frequently (but not always) an *active circuit*,—as contrasted to passive elements, such as resistors, coils and capacitors.

It is well, then, in discussing electronic action, to reemphasize its outstanding advantages over electrical action in two major ways: on the one hand, in its flexibility or ease of control, and, on the other hand, in its remarkable and faithful *amplification action*, running easily to voltage gains of over a million.

6-2. TYPES OF ELECTRON DEVICES

In studying the detailed operation of electron devices, it is customary to separate them into three classes: *vacuum tubes*, *gas tubes*, and *semiconductors*. For the purpose of reviewing the essential similarities in function, however, it is better to classify them as *two-element* (*diode*), *three-element* (*triode*), and *multielement* (*such as pentode*) types. There are, of course, differences in the detailed operation of an electron device—for example, a triode, depending upon whether it is a vacuum-tube triode, a gas triode (such as a Thyratron) or a semiconductor triode (such as a transistor). This chapter, however, concentrates on a fast review of the fundamental characteristics of each functional type of electronic device without going too deeply into comprehensive circuit actions, in order to avoid the hazard of not seeing the forest for the trees. Of the literally thousands of individual types of electron devices available, three types of diode are illustrated symbolically in Fig. 6-1: (a) the vacuum-tube diode; (b) the semiconductor diode; and (c) the gas diode.

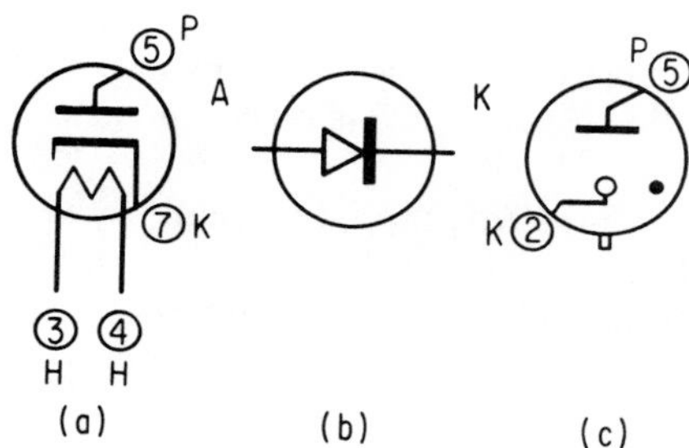

Figure 6-1. Three types of diodes: (a) vacuum-tube diode, heater type; (b) semiconductor diode; (c) gas-tube diode.

6-3. DIODE CHARACTERISTICS AND PROPERTIES

The Vacuum-Tube Diode

The characteristics of the diode as a two-element electron device are most easily understood from the examination of a vacuum-tube type arranged in a circuit to measure its characteristics, as shown in Fig. 6-2(a). The physical action in the VT diode of Fig. 6-2(a) can be visualized by imagining electrons being freed, or "boiled off", from the cathode K, as it is heated by the heater filaments H–H. These freed electrons will be attracted to the plate P, which has a positive potential E_b, obtained from the variable plate or "B_+" supply (E_{bb}). (The "A" supply to heat the cathode to operating temperature is not shown.) The only load in this circuit is the resistance of the current meter (usually less than 100 Ω), and so the test conditions yield maximum (or so-called "short-circuit") current values I_b. The voltampere characteristic of the diode is shown in Fig. 6-2(b). It will be seen that for positive values of plate voltage E_b, the curve for plate-current I_b follows the general $\frac{3}{2}$-power relation of Child's law:

$$I_b = KE_b^{3/2}$$

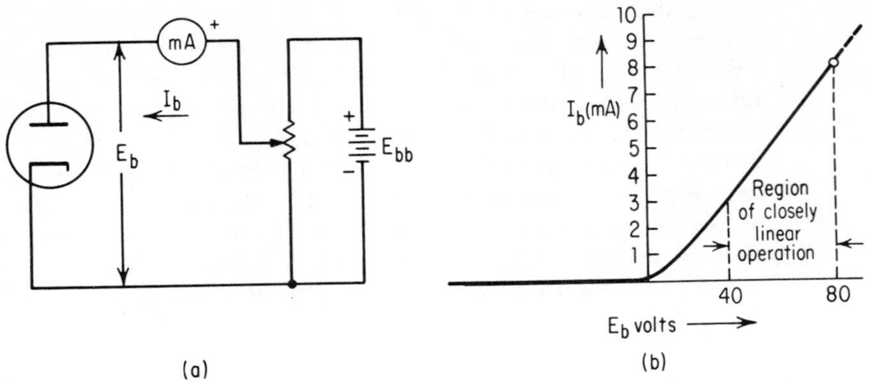

Figure 6-2. Vacuum-tube diode characteristics: (a) measuring circuit; (b) voltampere characteristic curve.

where the constant K is the infrequently used *permeance* of the diode. While this exponential relation naturally results in a nonlinear curve, it will be noted that the curve is substantially linear over its operating range portion, and from the slope of this linear portion we obtain the more commonly used value of equivalent dc resistance R_b as

$$R_b = \frac{\Delta E_b}{\Delta I_b} \qquad \text{(over the closely linear portion)}$$

Thus, for a typical single-diode portion of a dual-diode tube, such as a 6H6, the value obtained from a tube manual gives R_b as follows:

$$R_b = \frac{\Delta E_b \text{ (change from 40 to 80 V)}}{\Delta I_b \text{ (change from 3 to 8 mA)}} = \frac{40\text{ V}}{5\text{ mA}} = 8\text{ k}\Omega$$

For negative values of plate voltage, there is no measurable plate current, and the diode can be considered as an open circuit. This property of one-way conduction accounts for the great usefulness of the diode as a rectifier.

The Semiconductor Diode

The characteristic of a semiconductor germanium or silicon diode is shown in Fig. 6-3. The solid arrow, as previously interpreted in the text, indicates the conventional flow of positive charge (which is always, by definition, opposite to the electron flow that is usually indicated by dashed arrows), with the result that the semiconductor symbol can always be interpreted as an arrowhead showing the "direction of easy flow" in the conventional sense (as given by the IEEE standards for symbols and definitions).

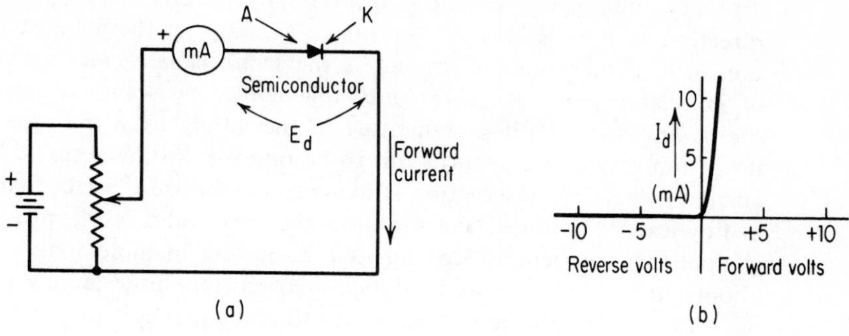

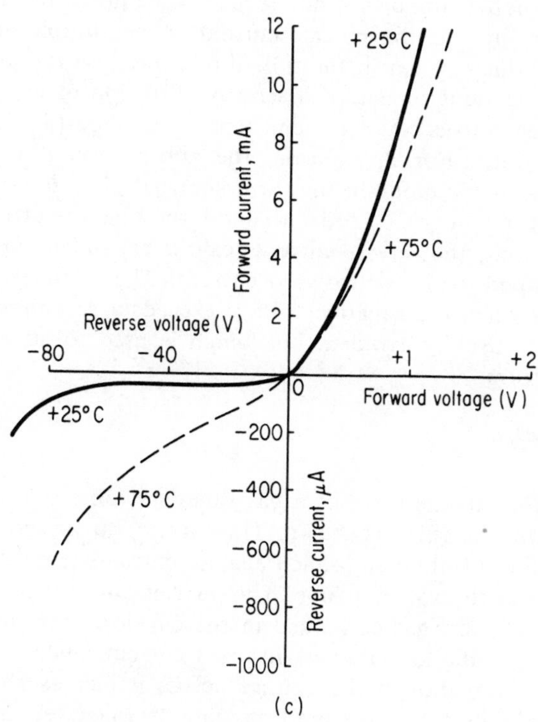

Figure 6-3. Semiconductor diode characteristics: (a) measuring circuit; (b) voltampere characteristic; (c) a more usual form of characteristic curve.

It is interesting to note that although, in an electron tube the conventional direction of flow is always opposite to the flow of the actual current carriers in the form of thermionic electrons, *in the semiconductor the conventional direction of flow can be the actual direction of flow* in all the cases *where the majority carriers are holes*, as in a p–n–p transistor. In the latter cases, the direction of electron flow (minority carriers) turns out to be opposite to the actual direction of major current flow. This observation is made to emphasize the fact that either direction of flow can be assumed and produces the same valid result, provided only that it is applied consistently. Keeping this basic fact in mind makes the "much-fuss-about-plus" arguments as to direction of current flow largely unnecessary.[1]

Continuing the comparison with the tube diode, it may be observed that the semiconductor diode requires much smaller voltages across it to produce the operating range of forward current than does the tube diode—typically 1–2 V for the semiconductor, contrasted with 40–80 V for the tube. Accordingly, the equivalent *forward resistance* in this example gives a value of around 500 Ω for the semiconductor against the 8000 Ω previously obtained for the tube.

Another important difference can be noted from the reverse-current characteristics, in that the reverse current is measurable and therefore greater in the semiconductor than in the tube diode. Because the amount of reverse current in the semiconductor is also a sensitive function of temperature, it must frequently be taken into account in semiconductor diode (and even more so in transistor) applications. For this reason, the appearance of the characteristic curve of a semiconductor diode in the form shown in Fig. 6-3(c) is more common than the graph form in (b), in order to show the reverse-current on a microampere scale. As a result, the reverse-current scale is expanded 1000 times compared with the milliampere scale for forward current. The relatively sudden increase in reverse current when the negative voltage exceeds a certain amount indicates avalanche, which is usually avoided, but which is used to advantage in the special Zener-diode form.

Gas Diode

The diode in the form of a gas tube is symbolized in Fig. 6-4 and its simplified characteristic curve is shown. The current values shown on the curve indicate the possibility of obtaining much heavier currents from the gas tube than are obtainable from the vacuum tube. The heavier currents are made possible by the ionization of inert gas contained in the envelope. An additional property possessed by the gas diode is that its forward current tends to remain constant after ionization, even though the voltage across it increases (within operating limits). In cold-cathode tubes, the inert gas ionizes whenever the anode voltage exceeds the ignition (or firing) voltage.

[1] S. D. Prensky, "Much Fuss about Plus," *Radio-Electronics*, Mar. 1956.

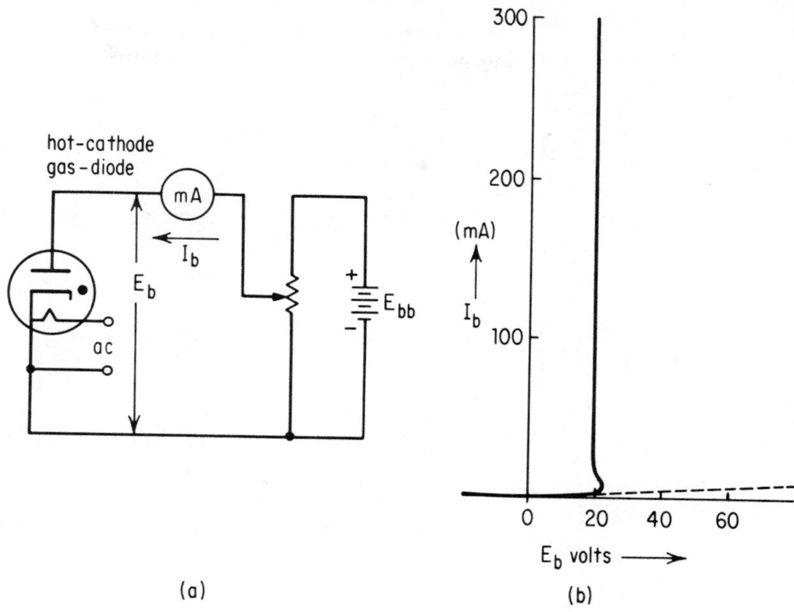

Figure 6-4. Gas-diode characteristic curve: (a) measuring circuit; (b) characteristic (solid line) compared to vacuum diode (dashed line).

6-4. DIODE APPLICATIONS

Diodes as Rectifiers in Power Supplies

The use of diodes as rectifiers may be observed in the power supply of almost any electronic instrument operated from the ac line. The familiar form of the conventional power supply is shown in Fig. 6-5. Proper operation of this circuit can be tested at the rectified dc output terminals. These terminals appear after the filter, across the equivalent load (R_1), at the points labeled B + and B −. Typical values in this application produce dc outputs in the range of around 300 V. The measurement can be made by the ordinary type of dc voltmeter, having an input impedance equal to (or greater than) 1000 Ω/V. In cases where hum trouble is present in the power supply and the rectifier is found to be good, a *low reading* on this simple voltmeter test might point to a badly shorted filter capacitor; this would load the circuit excessively by its leakage, thus causing an abnormally low voltage reading. Another very common cause of hum is traceable to a dried-out filter condenser, which has lost its filtering capacity. Such a condition is quickly checked in this circuit by the simple substitution of a good electrolytic capacitor,

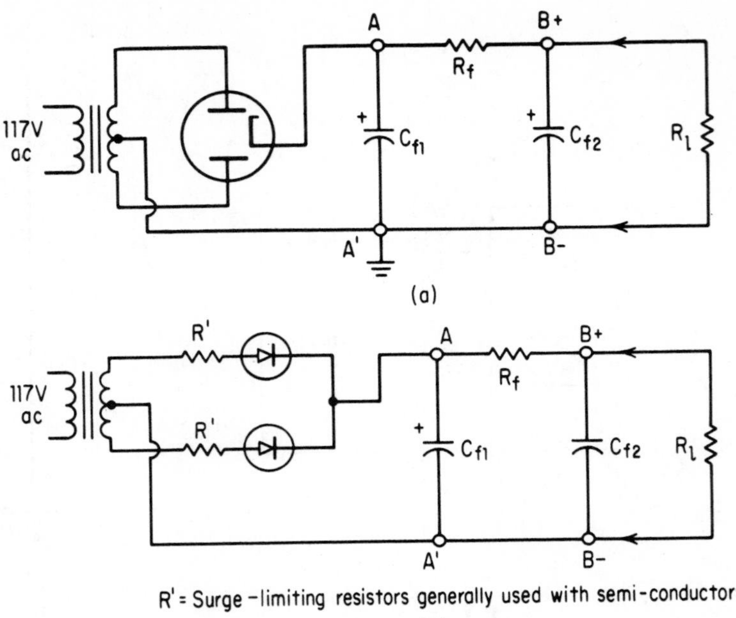

Figure 6-5. Conventional full-wave power supply: (a) vacuum tube; (b) semiconductor (silicon) type.

temporarily connected across the terminals of the suspected capacitor, after it has been previously determined by a *satisfactory voltage reading* that the filter capacitors are not shorted. If the substitution reduces the amount of the hum, it immediately points to a dried-out (or open) filter capacitor. However, for the purpose of measuring the amount of ac ripple present across the load, the ordinary multimeter (VOM) on its ac function is not at all suitable. A peak-to-peak measurement might be made, using a sensitive range of an electronic voltmeter, or preferably, the wave-form of the 120-Hz ripple would be observed and measured on an oscilloscope.

Switching and Other Diode Applications

Although the main emphasis in the previous sections has been placed on the wide use of diodes as rectifiers in power supplies, diodes are applied in many other ways. The rectifying function, for example, is the basis of *demodulation functions in detectors*. Apart from performing the function of changing ac to dc, the properties of diode conduction may also be observed in other applications, such as *switching and logic* elements in computers and *clamping* elements for establishing dc levels. Special types of diodes are also in use: microwave *mixers* for their superior high-frequency performance, reverse-biased diodes acting as *variable capacitors*, and tunnel diodes that make use of negative resistance properties for *oscillators*.

6-5. TRIODE CHARACTERISTICS AND PROPERTIES

The three-element electron device is most commonly represented as either a vacuum-tube triode or the semiconductor transistor. The vacuum-tube triode was introduced in 1903 by Lee DeForest. The triode differed from the vacuum-tube diode by the addition of a third element, the grid. This grid was another electrode which was placed between the plate and cathode of the diode. By adjusting the voltage (or potential) at the grid element, the flow of electrons from cathode to anode could be controlled. The diode could now be used as a switch, an amplifier, or an oscillator. Thus, electronics as we know it today was born.

Figure 6-6 illustrates the basic properties of the vacuum-tube triode. Here we see a simple test circuit used to measure plate current I_b and plate resistance r_p. Notice that as the grid voltage E_c becomes more negative, there is a decrease in plate current. The tube in effect becomes a variable-resistance controlling current through the load resistance R_L.

No further explanation of the vacuum diode or vacuum triode will be given in this text. While from time to time diagrams in this text will show the vacuum diode or vacuum triode, it is felt that their operation will be obvious based on this brief explanation.

6-6. TRANSISTOR CHARACTERISTICS

Transistor Advantages and Limitations

The three-electrode semiconductor device, the transistor,[2] is in many important respects similar to the tube triode, and accordingly shares the capability of the triode electron device in functions of *amplification*, *oscillation*, and *switching circuits*. Since its discovery in 1948 by the team of Bardeen and Brittain, its use has expanded rapidly, particularly in portable instruments. Two outstanding advantages of a semiconductor device are (1) its inherently greater *reliability*, due to its mechanical ruggedness and long life expectancy, and (2) its *low power requirements*, chiefly because it eliminates the necessity of heating a cathode for producing electron flow.

The Junction Transistor

Of the many types of transistors and forms of circuits using them, the *junction transistor used in a common-emitter circuit* lends itself best to a brief comparison between transistor and tube action. In emphasizing the similarities, however, it must also be kept in mind that there are fundamental differences that cannot be detailed in a brief review, but that must be left to be studied in other, more detailed works.

[2]When not qualified, the term *transistor* refers to the conventional (bipolar) transistor; otherwise, qualifying terms, such as FET for field-effect transistor (unipolar), are used.

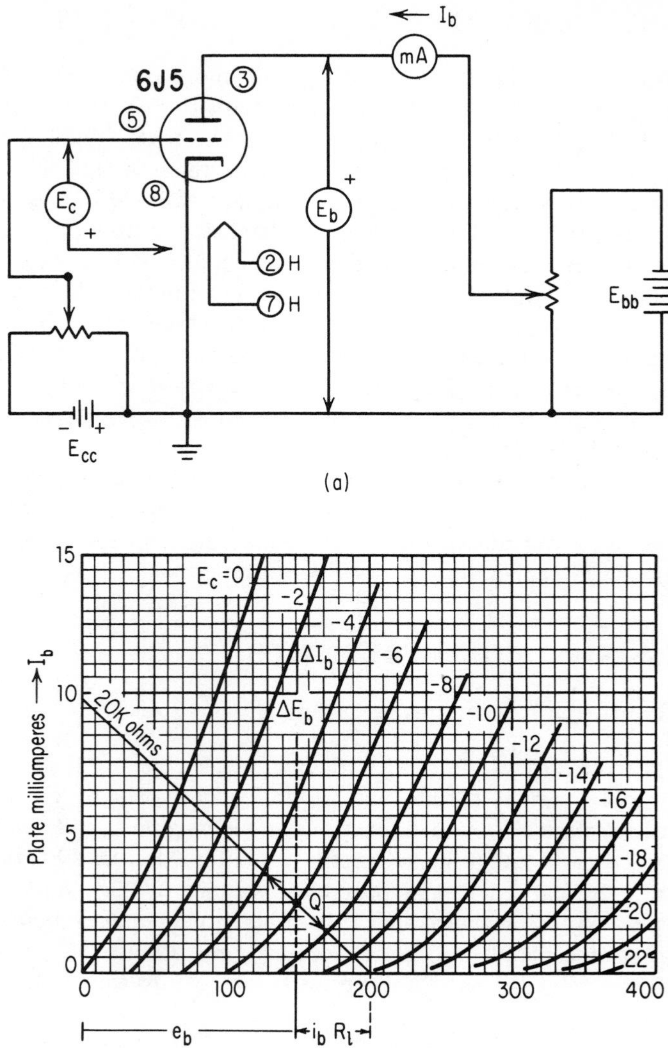

Figure 6-6. Triode characteristics: (a) measuring circuit; (b) plate characteristics.

Tube and Transistor Amplifiers Compared

For a general view of circuit action, the common-emitter circuit of an n–p–n transistor in Fig. 6-7(b) can profitably be compared to the common-cathode circuit of a vacuum-tube amplifier in Fig. 6-7(a). [The ac signal amplification with the

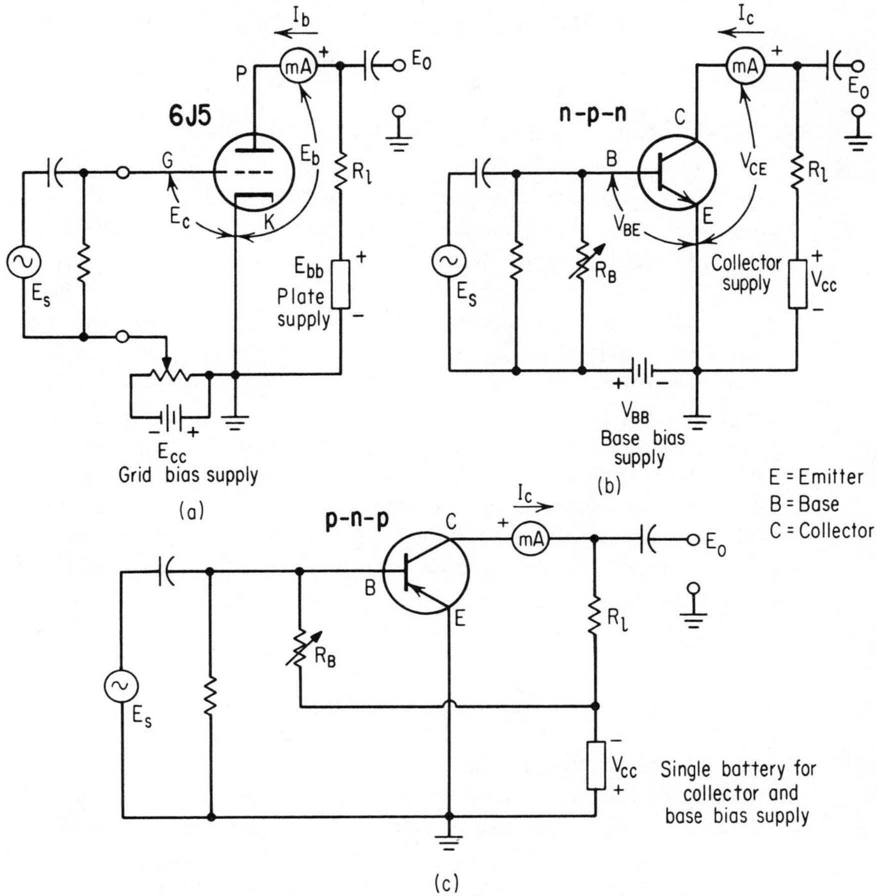

Figure 6-7. Comparing basic triode amplifiers. The vacuum-tube triode in (a) has similar battery polarities to the common-emitter *n-p-n* transistor amplifier in (b); in (c), the common-emitter single-battery amplifier using a *p-n-p* transistor shows reversal of collector and base bias voltages, compared to the *n-p-n* in (b).

p–n–p transistor in Fig. 6-7(c) is identical to the n–p–n circuit action, except for the reversal of dc polarities, to be discussed later.] If we explain the *tube* amplifier action by the essential statement that a *small change in the input grid voltage* caused a larger change in the output plate current, we can correctly explain the *transistor* action by saying that a *small change in the input base current* causes a larger change in the output collector current. In both cases there is gain that can be extracted as either predominantly voltage or predominantly current amplification; the distinction to be made is that the tube is a voltage-operated device as contrasted with the transistor, which is essentially a current-operated device.

6-7. TRANSISTOR AMPLIFIER CIRCUIT ACTION

The circuit action of the transistor can be traced graphically by the use of the collector characteristic curves in Fig. 6-8. in a manner similar to that in which the plate characteristics were used to trace tube action. With meters properly placed in the n–p–n circuit of Fig. 6-7(b) we obtain curves for the collector current I_c versus the collector-to-emitter voltage V_{CE}, with the base current I_B as the running parameter. These curves are found to be more similar to the plate characteristics of the pentode tube rather than the triode tube indicating that the equivalent resistance of the output collector circuit, r_c, of the transistor is more similar to the pentode tube r_p range of 1 MΩ than to the triode tube r_p range of around 10 kΩ. This difference, however, does not prevent us from drawing the load line as before and using the same method for finding the voltage changes in output of the transistor amplifier by projecting the load-line intersections at points P, Q, and R down to the voltage axis to obtain the resulting values for the collector-to-emitter voltage V_{CE}.

Current Gain (A_i)

A simplified example using the graphical method can be traced easily if we are satisfied with finding the output as the amount of current variation in the output collector circuit ΔI_c for a given amount of input *current* variation ΔI_B caused by the ac signal E_s in the base input circuit. For the values given in the circuit of Fig. 6-7, the load line of Fig. 6-8 is drawn between one point ($V_{CE} = V_{CC} = 12$ V, at $I_C = 0$) and the other point ($I_C = 12$ V/1.5 kΩ $= 8$ mA at $V_{CE} = 0$). If the meter showing the quiescent base current (I_B at Q) is caused to read 75 μA by proper adjustment of base bias resistor R_B, we start with the initial no-signal condition at point Q, where $I_C = 4$ mA and $V_{CE} = 6$ V.

As the input signal cuases the input base current I_b to swing ± 25 μA either side of the quiescent value of 75 μA (from point $P = 100$ μA to point $R = 50$ μA), the corresponding output change in V_{CE} is from 4 to 8 V. This results in a 4-V peak-to-peak output ΔV_{CE} for a 50 μA (p–p) input variation ΔI_B. The ac current gain A_i of the circuit is easily obtained from the approximate change in currents from P to R as

$$A_i = \frac{\Delta I_C}{\Delta I_B} = \frac{5.5 - 2.5 \text{ mA}}{100 - 50\ \mu\text{A}} = \frac{3 \text{ mA}}{50\ \mu\text{A}} = \frac{3000\ \mu\text{A}}{50\ \mu\text{A}} = 60$$

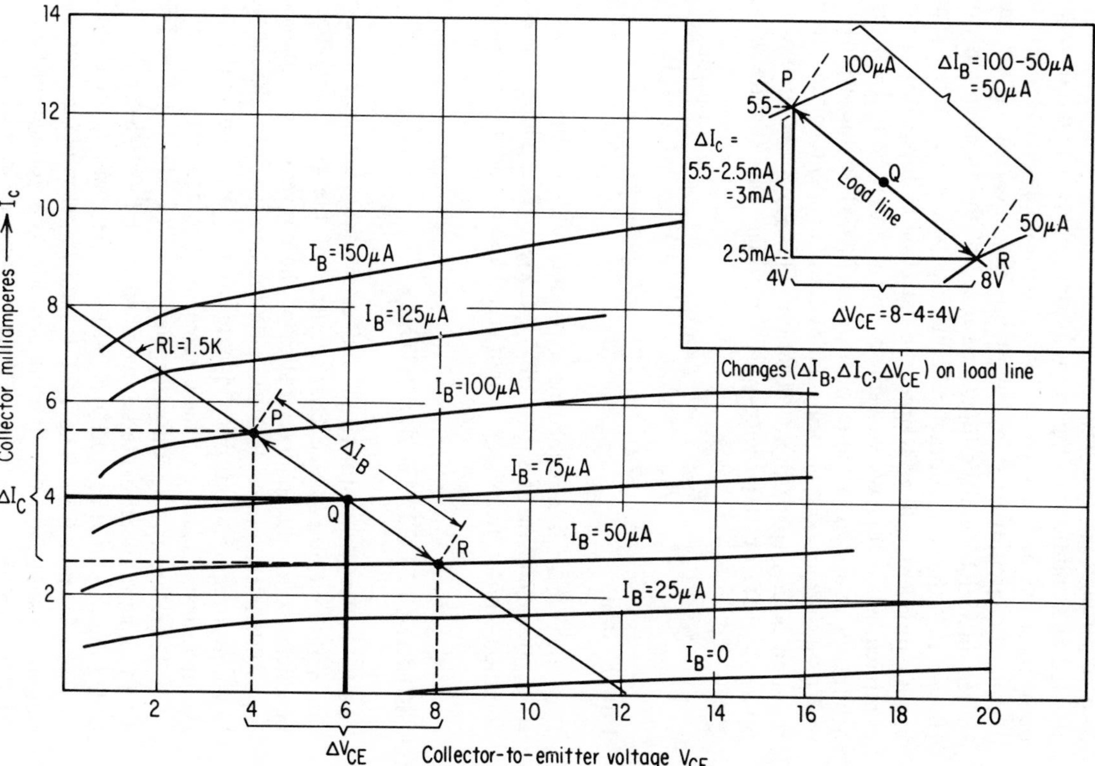

Figure 6-8. Load-line variations on transistor collector characteristics.

Voltage Gain (A_v)

The voltage gain A_v of the circuit, however, cannot be as easily obtained (except to a very rough approximation), because the input resistance of the transistor is affected by conditions in the output circuit. This inherent dc-dependent condition does not exist in the vacuum tube. For testing purposes this condition can be adequately circumvented by supplying the base with a constant-current input source, in which case the input resistance of the transistor is no longer a factor. However, in the case of practical amplifiers where the input is generally considered as an equivalent voltage source, the variation of input resistance with output is a complicating factor. We can still obtain a rough estimate of the voltage gain by arbitrarily assigning a value of around 2000 Ω to the average input resistance, and then, in our example, calculating an equivalent voltage input to produce the required 50 μA (p–p) input variation. This gives an input voltage variation of 50 μA × 2 kΩ or 100 mV (p–p), from which the estimate of voltage gain A_v is

$$A_v = \frac{E_{\text{out}}}{E_{\text{in}}} = \frac{\Delta V_{CE}}{\Delta E_s} = \frac{(8 - 4)\text{ V}}{100\text{ mV}} = \frac{4\text{ V (p–p)}}{0.1\text{ V (p–p)}} = 40$$

Power Gain

The power gain is obtained as a product of the current and voltage gains, or 40 × 60 = 2400 or about 34 dB—a fairly typical value for a single stage of a common-emitter transistor amplifier at audio and broadcast frequencies.

6-8. THE FIELD-EFFECT TRANSISTOR

The field-effect transistor (FET) makes available an electron device that possesses the high-input-impedance characteristic of the tube, together with the highly beneficial solid-state characteristics of the conventional transistor. It thus makes it possible in this single device to gain the benefits from the inherent reliability and low-power requirements of solid-state circuitry, without, at the same time, incurring the undesirable feature of the conventional transistor.

There are two broad categories of FETs: the junction FET (J-FET or just FET) and the metal-oxide semiconductor FET (MOS-FET), which is sometimes called the insulated gate FET (IG-FET). The symbols for these devices are shown in Figs. 6-9 and 6-10. The FET is said to resemble a pentode vacuum tube in operation, and the FET, like the pentode, is a voltage-operated device. Since the FET has no heater element, it does not operate at a high temperature, and since it is a semiconductor device, it lends itself to miniaturization and integrated circuits. The FET's operation depends on currents that do not cross junctions; therefore, inherent noise problems such as those in bipolar transistors are minimized.

The input circuit of the FET consists of a reverse-biased diode. The gate

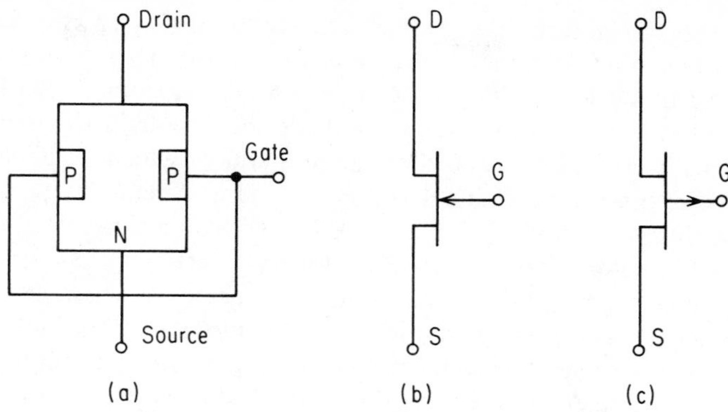

Figure 6-9. Junction FET: (a) schematic; (b) *n*-channel; (c) *p*-channel.

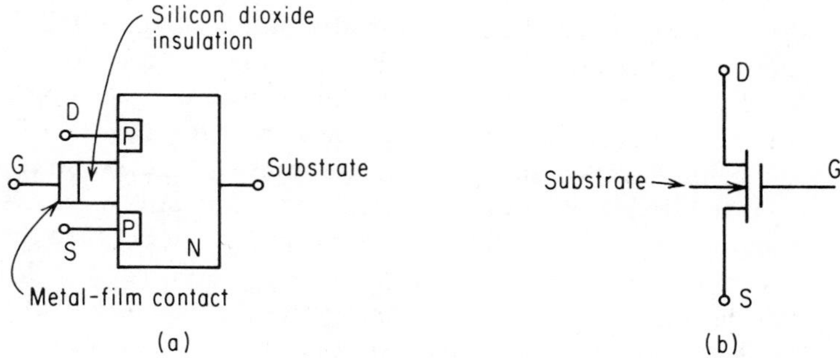

Figure 6-10. MOS-FET: (a) schematic; (b) *n*-channel.

current is on the order of 10^{-8} A in small-signal FETs, and the input resistance is on the order of 10^{8} to 10^{10} Ω. Since the gate of a MOS-FET is insulated from the channel, input resistances of 10^{15} Ω are not uncommon in MOS-FETs.

Field-effect transistors are most commonly used as RF amplifiers, mixers, choppers, small-signal amplifiers, general-purpose switches, differential amplifiers, and so on. FETs have been designed for operation from dc to UHF in switching and amplifying applications.

FET Construction

The J-FET is made of a channel (or bar) of either n-type or p-type semiconductor material. One end of the channel is called the *source* and the other end is called the *drain.* The symbol shown in Fig. 6-9(b) is for an n-channel FET, and the symbol in Fig. 6-9(c) is for a p-channel FET. The drain is analogous to the collector of a transistor or the plate of a vacuum tube, and the source is analogous to the emitter of a transistor or the cathode of a vacuum tube.

The gate of the FET is formed by diffusing two p-regions into opposite sides of the n-channel as shown in Fig. 6-9(a). These two regions are connected to a common lead. The gate lead, which is analogous to the base of a transistor or the grid of a tube, is brought outside the case by a single connection. An electric field is created between the gate and source which controls the current through the drain–source channel: hence the name "field-effect transistor." Because the operation of the FET is accomplished with an electric field, very little current flows in the gate circuit. The J-FET forms a rectifying junction (diode) between the gate and the n-channel. This junction must always be reverse-biased, so the J-FET must always operate in the depletion mode. If this junction becomes forward-biased, excessive gate current will be produced, possibly damaging the device.

The metal oxide semiconductor field-effect transistor (MOS-FET) is a device that closely resembles the J-FET; however, there are some differences. The MOS-FET is formed by diffusing two p-regions in an n-type silicon channel as shown in Fig. 6-10(b). If the MOS-FET has a p-type substrate, it is called a p-channel MOS-FET, and its symbol is shown in Fig. 6-10(c). The two p-regions diffused in the n-substrate are the drain and the source of the MOS-FET. A thin layer of silicon dioxide (SiO_2) is grown between the two p-regions to form the insulator for the gate connection. A metalic film is then deposited on the SiO_2 for the gate connection. This insulation of the gate from the substrate is why the MOS-FET is sometimes called an insulated gate transistor (IGT). Since the gate is insulated from the device, it can be operated under forward- or reverse-biased conditions. The operation of a MOS-FET under forward bias is called the enhancement mode of operation. The insulation also forms a small capacitor (1 pF) between the gate and the substrate. This capacitance gives the MOS-FET an extremely high input resistance ($10^{15}\ \Omega$). This capacitor also appears as its usual symbol on the MOS-FET symbols in Fig. 6-10(b) and (c) between the gate and the substrate.

The insulator made of silicon dioxide can be easily damaged by static voltages. Therefore, extreme care should be exercised when handling MOS-FETs because static discharges can build up on the handler's body. In the newer dual-gate MOS-FETs, gate protection is provided by instrinsic back-to-back diodes. This makes the MOS-FET less susceptible to damage by static discharges.

FET Operation

The FET operation is very similar to that of the bipolar transistor. As shown in Fig. 6-11 for a J-FET, the power source V_{DD} produces a voltage drop from the drain to the source (V_{PS}) of the FET, which causes the flow of electrons through the n-channel. As the voltage source V_0 is increased from zero to some value, the current in the n-channel decreases. This happens primarily because the depletion region of the two n-p junctions begins to spread into the n-channel and reduces the effective width of the channel; therefore, there is less current flow. The maximum amount of current flows through the n-channel when V_{GG} in Fig. 6-11 is zero. The current that flows through the n-channel is often called the drain current.

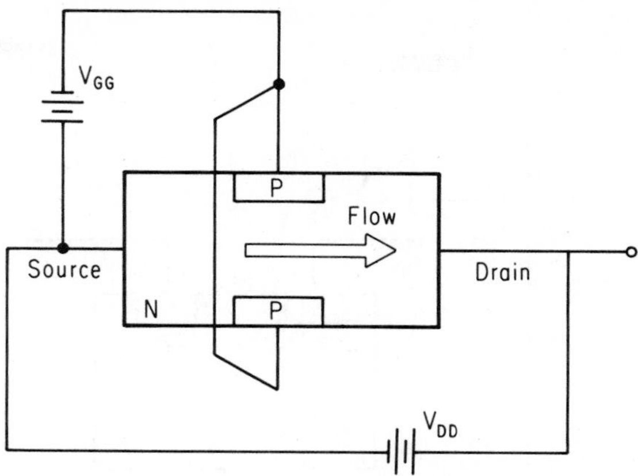

Figure 6-11. J-FET circuit operation.

Actually, the relationship given by Eq. (6-1) holds true:

$$I_S = I_D + I_G$$

The source current is equal to the sum of the drain and gate currents. But since the gate current is so small (only a few microamps or the reverse-biased leakage current associated with a reverse-biased diode), the source current is approximately equal to the drain current. The reverse-biased mode of operation for a J-FET is known as the deplection mode of operation, since the current in the n-channel can only be depleted or reduced. As we mentioned earlier, to forward-bias this junction might damage the device.

The operation of the MOS-FET is somewhat similar to the J-FET, as shown in Fig. 6-12. When the switch (S_1) is connected to V_{GG^-}, the operation, of the MOS-FET in the depletion mode is identical to the J-FETs operation in the depletion mode. However, since the gate is insulated from the n-channel by a small capacitor, the MOS-FET gate may be forward-biased. This is called operation in the enhancement mode. If the switch (S_1) is connected to V_{GG^+}, the MOS-FET is operated in the enhancement mode. The gate of the MOS-FET can either facilitate the flow of electrons from the source to the drain or hinder the flow of electrons from source to drain. It effectively widens or reduces the width of the n-channel, allowing more or fewer electrons to pass.

Circuit Configurations

The FET may be connected in three basic configurations, similar to a bipolar transistor. These are called common source, common gate, and common drain. The most frequently used connection by far is the common-source circuit. The two circuits that we discussed earlier were common-source circuits. When connected in this manner, the input is applied between the gate and the source, and the

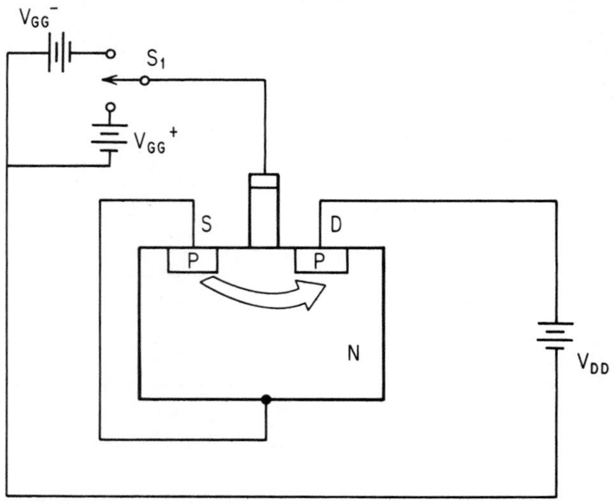

Figure 6-12. MOS-FET circuit operation.

output is taken between the drain and the source. The source is common to both the input and the output.

The circuit shown in Fig. 6-13 is a typical MOS-FET common-source amplifier circuit. The voltage gain of the FET is given as

$$A_v = \frac{\Delta V_{DS}}{\Delta V_{GS}}$$

This equation is an ac voltage-gain equation.

When connected in the configuration shown in Fig. 6-13 the output is often taken across the load resistor R_L. The value of the voltage across the load resistor V_{RL} is given in Eq. (6-2):

$$V_{RL} = I_D R_L$$

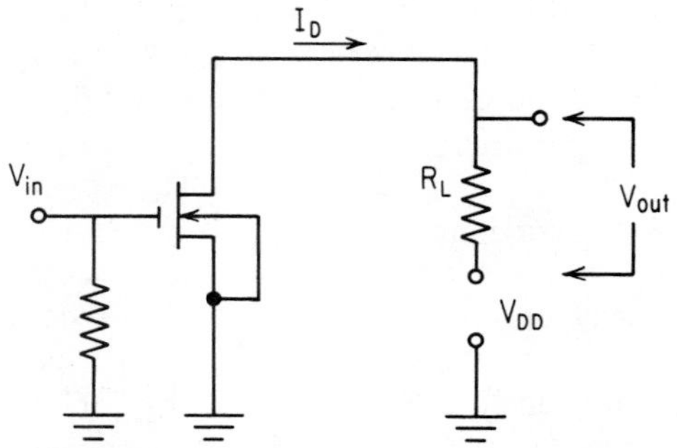

Figure 6-13. Typical MOS-FET amplifier circuit.

Of course, Ohm's law applies in this situation, since the current that flows through the load resistor is the drain current I_D. Also, if we consider the output circuit of the FET amplifier, we can write the loop voltage equation

$$V_{DD} = V_{DS} + V_{RL}$$

Substituting in the equations above, we get

$$V_{DD} = V_{DS} + I_D R_L$$

The J-FET connected as a common-source amplifier is shown in Fig. 6-14. The major difference between this circuit and the one shown in Fig. 6-13 is the source resistor R_S.

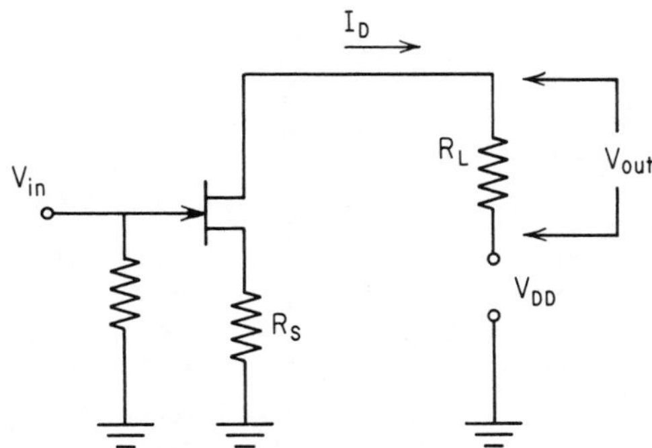

Figure 6-14. Typical J-FET amplifier circuit.

Example of FET Preamplifier

An example of a widely used input stage is found in the popular source-coupled (also called long-tailed pair) amplifier configuration, using a pair of junction FETs, as shown in Fig. 6-15. [This direct-coupled arrangement is also used extensively as a basic building block in linear integrated circuits (ICs) for the differential type of dc instrumentation and operational amplifiers, discussed in a later section.] The circuit is shown in two forms: Fig. 6-15(a) illustrates the *differential output* form, where the full output is taken between each drain terminal (neither of which is grounded); and Fig. 6-15(b), generally called a *difference amplifier*, where the single-ended output, taken between one drain terminal and ground, is less than in part (a) but has the advantage of a common ground terminal between input and output.

In Fig. 6-15(a), a high-impedance low-level signal source, such as a crystal microphone or a pH-meter probe, is fed to the gate of FET-1, while the gate of FET-2 is grounded through the 10-MΩ resistor (R_{G2}). In the no-signal condition, both gates are operating with an initial negative bias, produced by the voltage drop across the common-source resistor R_S. The input E_{in} may be either a dc or

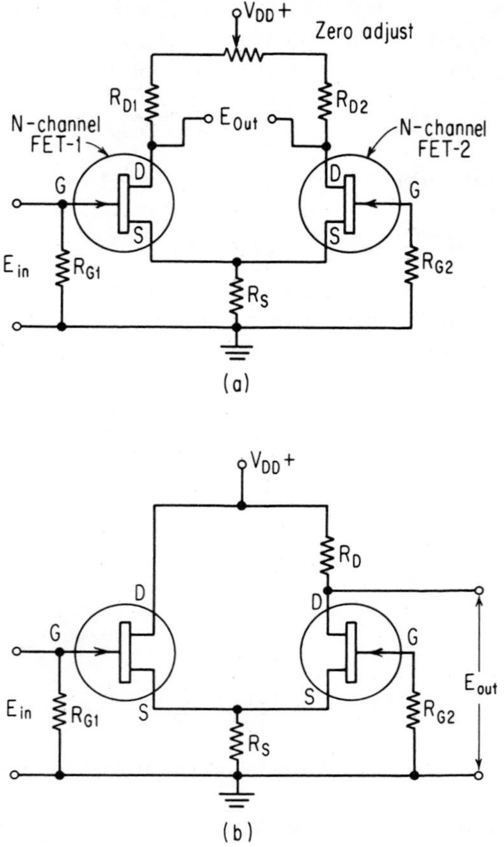

Figure 6-15. FET preamplifier common-source circuit: (a) with differential output E_{out}; (b) with single-ended output.

ac signal. When, for example, a positive dc signal is applied, the drain current in FET-1 increases, causing a larger voltage drop across source resistor R_S with positive polarity at the top making the gate of FET-2 correspondingly more negative, thus reducing the drain current FET-2. As a result of this seesaw action, the voltage at the drain of FET-1 falls, while the drain voltage of FET-2 rises. Hence, the output, taken between the two drain terminals, is the algebraic sum of the two changes in voltage level (roughly twice each voltage change). A single-ended output circuit is shown in Fig. 6-15(b).

It will be noted that the operation of this common-source amplifier is practically identical with that of the basic common-cathode differential amplifier explained in Chapter 7, as used in electronic voltmeters, except that all the advantages of the matched pair are now obtained in a solid-state version, made possible by the happy combination of FET characteristics.

6-9. ELECTRONIC OSCILLATORS

The ability of an electron device to generate sustained oscillations stems from its ability to amplify. In most cases, in an amplifying circuit capable of producing an equivalent voltage gain of three or more, one need only replace the input by a portion of the output fed back in the positive sense, to produce a self-contained oscillatory system.

LC Oscillators

A circuit containing a resonant combination of inductance and capacity (*LC*) can easily be arranged to sustain oscillations at its natural resonant frequency if a relatively slight amount of feedback is provided in a single-transistor stage. Two familiar oscillator forms are shown in Fig. 6-16, where the Hartley circuit in part

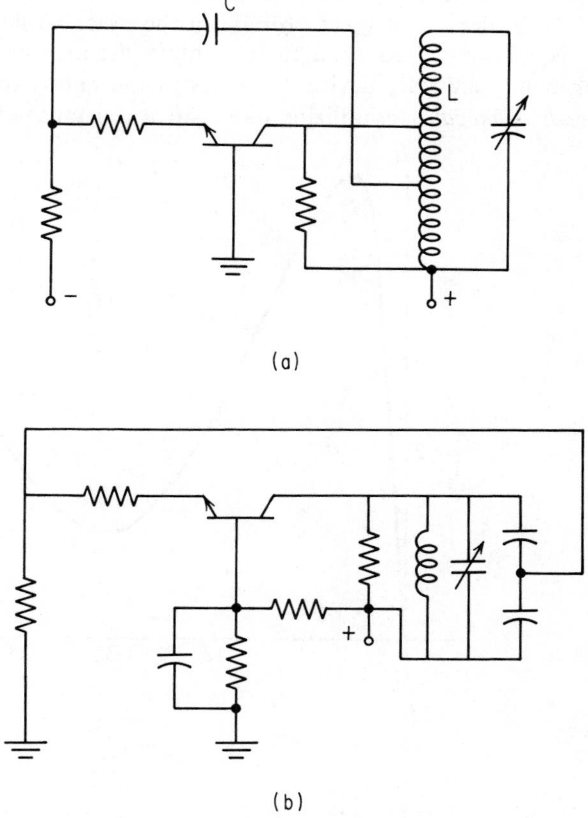

Figure 6-16. Transistor oscillator circuits: (a) Hartley circuit, shunt-fed with tapped coil; (b) Colpitts circuit, capacitive shunt-fed.

(a) illustrates the use of positive feedback by a tapped coil, and the Colpitts circuit in part (b) shows the regenerative feedback by a tapped capacitor arrangement.

Tunnel-Diode and Negative Resistance Oscillators

The effect of the positive feedback used in the *LC* oscillators can be thought of as a means of supplying energy to the tank circuit in such a way that the resulting gain in energy overcomes the resistance losses that are inherent in the oscillatory circuit. Put another way, regenerative feedback serves to introduce negative resistance action that compensates for the losses resulting from positive circuit resistance. Thus, use can be made of the negative resistance characteristic that occurs in a tetrode when the screen grid is operated at a higher potential than the plate; the circuit arrangement that applies this negative resistance to an *LC* tank circuit operates as a *dynatron* oscillator—interesting mainly from a theoretical standpoint. However, a new development in the negative resistance field, the *tunnel diode,* does hold great promise in the practical field of oscillations at very high frequencies. In addition to the simple circuitry involved in the use of this two-element solid-state device, it possesses the *ability to oscillate in the range of thousands of megahertz,* utilizing its negative resistance characteristics (Fig. 6-17).

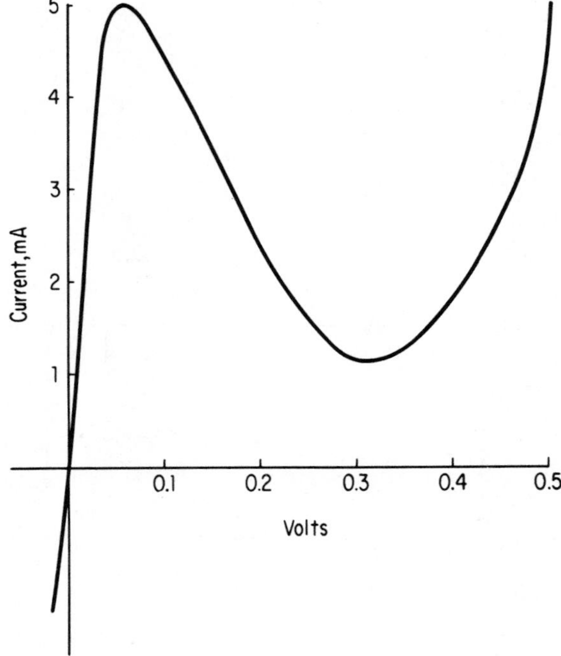

Figure 6-17. The tunnel diode (or Esaki diode) current–voltage characteristics, showing negative-resistance property.

Known also as an Esaki diode, after the Japanese scientist who first reported it in 1958; it will be noted that its curve contains a negative resistance portion, where its current decreases as the voltage across it becomes greater. This property, coupled with a high-frequency response up to one or two orders of magnitude higher than the best high-frequency transistors, points to an important potential for development in microwave high-frequency switching and low-noise applications.

RC Oscillators

For audio frequencies, where the size of the *LC* frequency-determining tank becomes too bulky, the resistance-capacitance (*RC*) type of oscillator, shown in Fig. 6-18, is preferred. The circuit, a *Wien-bridge arrangement*, requires the use

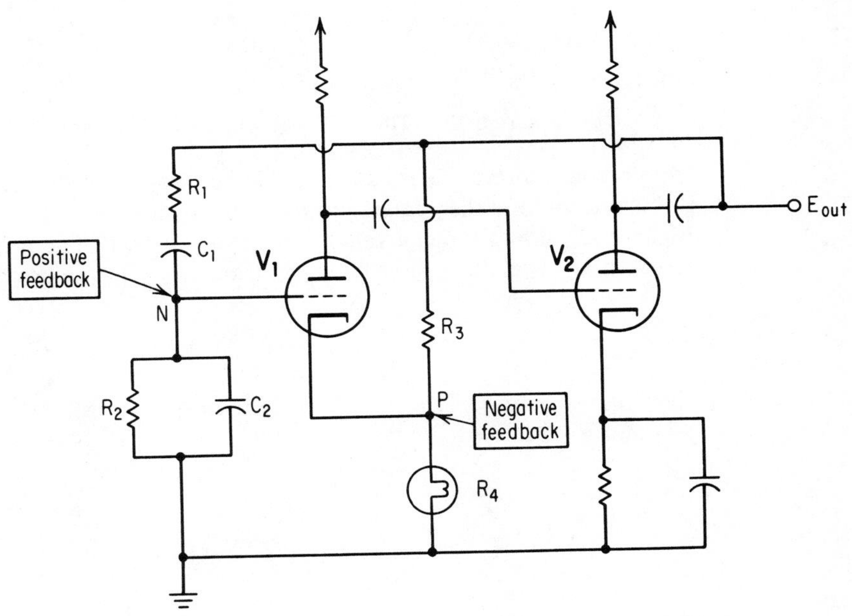

Figure 6-18. *RC* oscillator using Wien-bridge principle, showing frequency-sensitive, positive-feedback path.

of two stages. *Positive feedback* from the second to the first stage is provided by the R_1C_1 series arrangement combined with the R_2C_2 parallel arrangement. By means of an additional *negative-feedback* circuit (obtained through the connection from the junction of R_3 and R_4 to the first-stage cathode), the circuit attains a highly satisfactory degree of frequency stability; it is good for a frequency range as low as 1 Hz up to supersonic frequencies around 500 kHz, by the use of suitable range-switching circuitry.

6-10. THE UNIJUNCTION TRANSISTOR

The construction and symbol of the unijunction transistor (UJT) is shown in Fig. 6-19. The heart of the device is a lightly doped n-type silicon bar to which are attached three leads. Base 1 (B_1) and Base 2 (B_2) make ohmic connections to the bar and the emitter (E) is attached so as to form the only p-n junction.

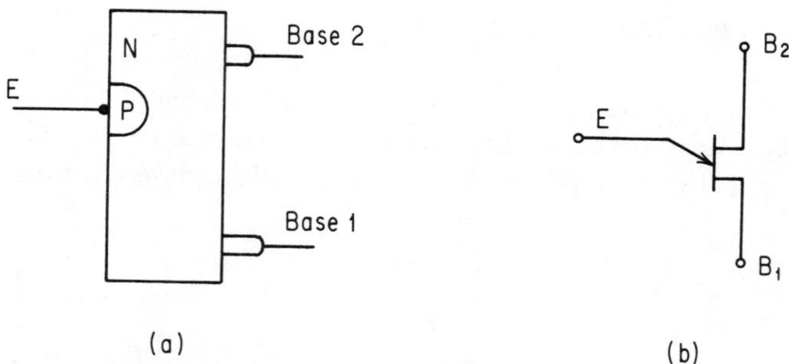

Figure 6-19. The UJT: (a) schematic; (b) symbol.

Ohmic connections are resistive (welded or soldered so that nonrectifying (equal resistance with either polarity) junctions are formed.

Figure 6-20 shows the equivalent circuit of the UJT. The n-type silicon bar is represented by two bulk resistances, R_{B1} and R_{B2}, where R_{B1} is shown as a

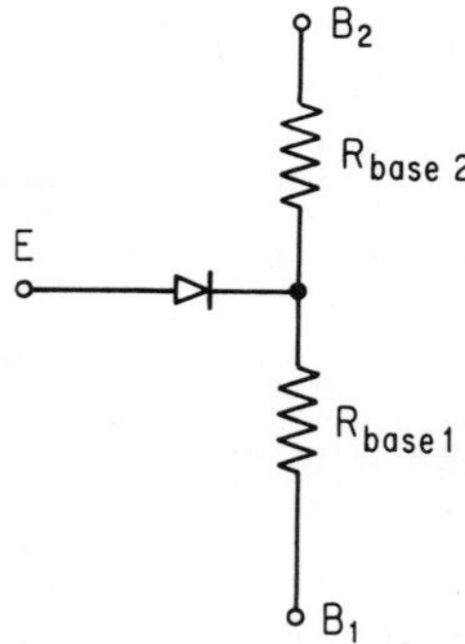

Figure 6-20. UJT equivalent circuit.

variable resistor because it changes in value when emitter current flows. Since the emitter forms a p-n junction with the bar, it forms a diode in the equivalent circuit, as shown in Fig. 6-21.

Remember that the resistance of R_{B1} is at its maximum. The resistance between B_1 and B_2 is called the *interbase resistance* R_{BB}.

$$R_{BB} = R_{B1} + R_{B2}$$

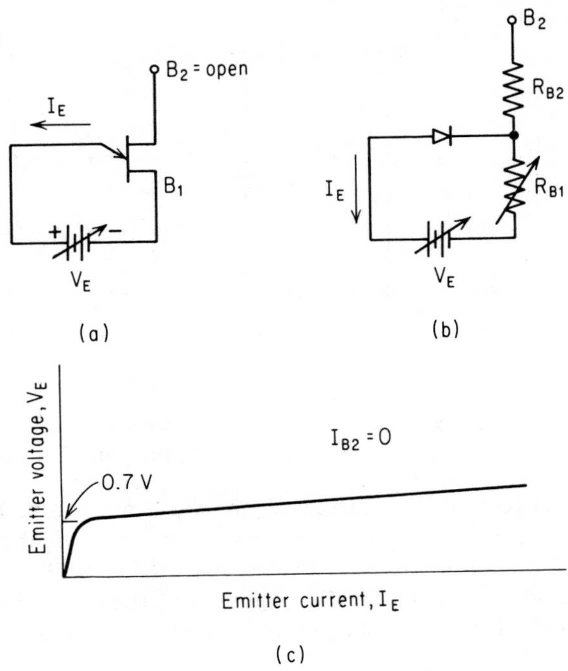

Figure 6-21. Developing the UJT static circuit: (a) circuit for developing curve; (b) equivalent circuit; (c) resulting static emitter characteristic curve with $I_{B2} = 0$.

With the emitter open, R_{B1} is larger resistance value than R_{B2}. Since voltage divides proportionally with resistance, the voltage is greater across R_{B1} than R_{B2} with emitter open. The fraction of the total resistance represented by R_{B1} or the fraction of the total voltage V_{BB} appearing at the emitter junction (between R_{B1} and R_{B2}) is called the *intrinsic standoff ratio*. The symbol for the standoff ratio is the Greek letter eta, η. And

$$R_{B1} = \eta R_{BB} \text{ or } \eta = \frac{R_{B1}}{R_{BB}}$$

UJT Characteristic Curves

Suppose that we bias the emitter of the UJT such that some emitter current flows. If only the emitter B_1 junction is biased (B_2 open), the characteristic curve is that of a diode. As the forward voltage increases, current will increase rapidly when the barrier potential (0.7 V for Si) is exceeded.

Figure 6-22 shows the circuit for generating the static emitter characteristic curve. Note that now a voltage V_{BB} is also applied between B_1 and B_2, creating a current I_{B2} as shown.

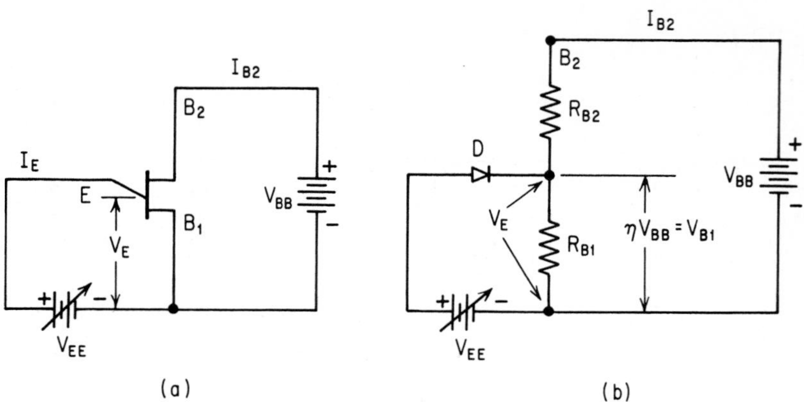

Figure 6-22. Circuit for producing static emitter characteristic curve: (a) actual circuit; (b) equivalent circuit.

As voltage V_E is increased, the diode remains reverse-biased. Because of the reverse-biased condition of the diode, emitter current I_E is low and the UJT is in the off condition (cutoff). Before appreciable emitter current can flow, V_E must exceed nV_{BB}, plus the barrier potential of the diode, V_D. This is the peak voltage, V_p, of the UJT, the voltage at which emitter current begins to flow:

$$V_p = \eta V_{BB} + V_D$$

If V_D is small compared to ηV_{BB} ($V_D \cong 0.7$), then $V_p \cong nV_{BB}$.

Figure 6-23 shows the characteristic of the UJT with voltage V_{BB} applied and emitter voltage changing from zero to the value V_p. When $V_E = 0$, I_E is

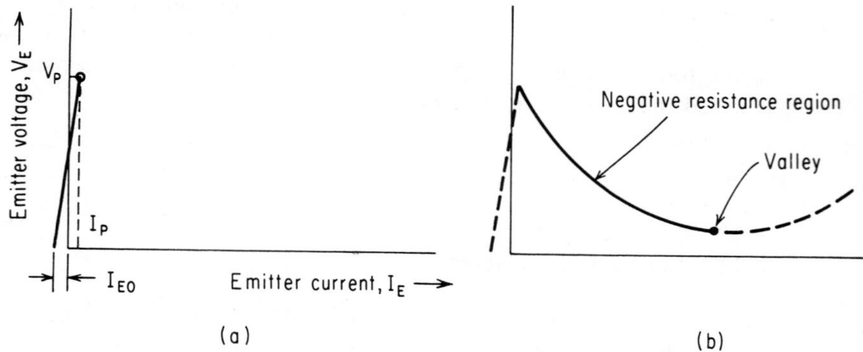

Figure 6-23. (a) Cutoff region of the static emitter characteristic curve for UJT; (b) negative resistance portion of the UJT static characteristic curve. (Courtesy of Texas Instruments, Inc.)

$-I_{E0}$. As V_E increases, $-I_{E0}$ decreases to zero. Raising V_E *still* further produces $V_p = I_E$.

Once emitter current begins to flow, things occur that are unique to the UJT (compared to a bipolar transistor). Carriers are injected between emitter and base 1, causing R_{B1} to drop drastically. As emitter current continues to increase, the resistance R_{B1} decreases, causing the voltage V_E from emitter to B_1 to decrease. Normally, when current increases, voltage increases (Ohm's law). In this case, the opposite occurs, and the region is referred to as the *negative-resistance portion* of the curve (Fig. 6-24).

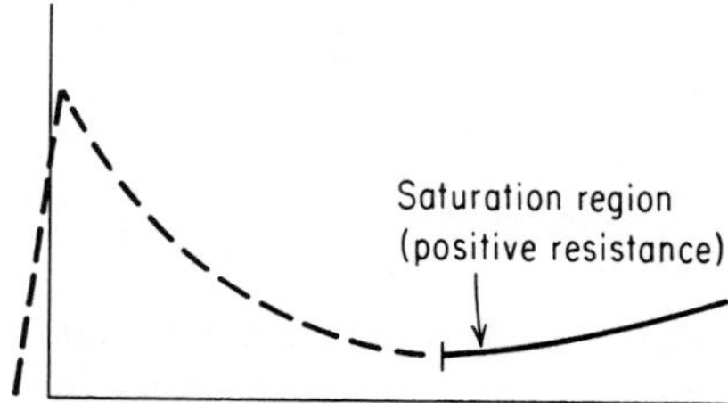

Figure 6-24. Saturation region of UJT.

When the value of R_{B1} is lowest the saturation resistance R_{sat}), voltage V_E is its lowest value. This point is referred to as the *valley* of the curve.

Raising the supply voltage V_{EE} produces emitter currents in excess of the valley current and causes a voltage drop across the bulk resistance of the emitter and base 1. As current increases, voltage increases, producing a positive saturation resistance. The saturation region is shown in Fig. 6-24.

Complete UJT static emitter characteristic curves developed thus far are shown in Fig. 6-25.

Circuit Analysis of UJT

Load lines are most helpful for analyzing and designing circuits using UJTs. As always, assume the two extreme operating conditions of the device, open and shorted. Consider the following example.

Example

Construct a load line for Fig. 6-26 using the curve of Fig. 6-27. Let $V_{BB} =$ 10 V, $V_{EE} = 5$ V, and $R_E = 100\ \Omega$.

Solution. Consider the device open:

$$I_E = 0 \quad \text{and} \quad V_E = 5\text{ V} \quad \text{(point A)}$$

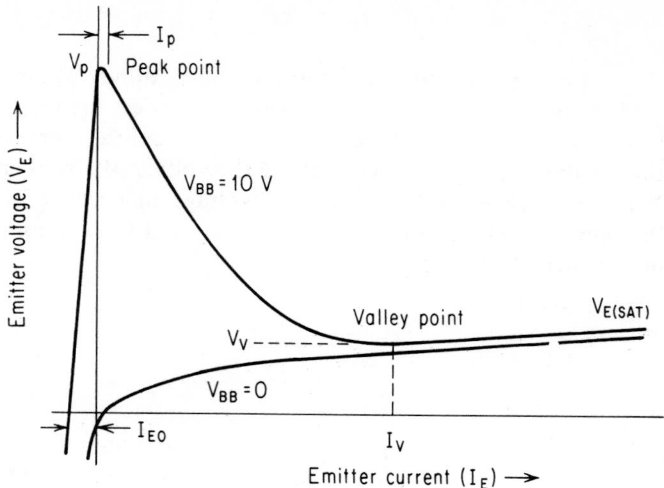

Figure 6-25. Static emitter characteristic curve of UJT. (Courtesy of Texas Instruments, Inc.)

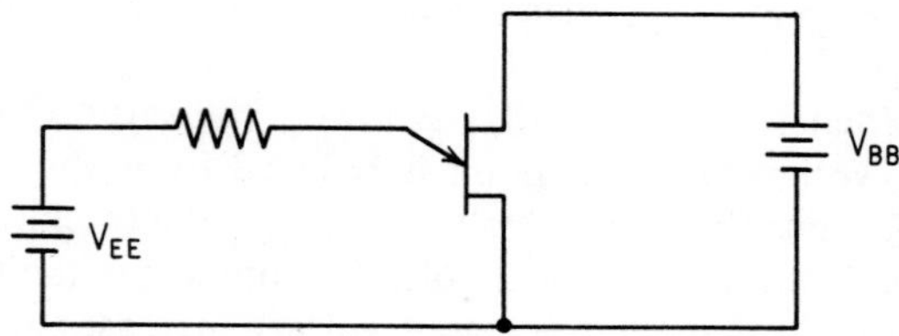

Figure 6-26. UJT circuit for example.

Consider it shorted:

$$I_E = \frac{V_E}{R_E} = \frac{5\text{ V}}{100\ \Omega} = 50\text{ mA} \quad \text{and} \quad V_E = 0 \quad \text{(point F)}$$

Connect points F and A to get load line 1 shown in Fig. 6-27.

Example

Construct a load line as in the preceding example, but use $V_{EE} = 10.0$ V and $R_E = 182\ \Omega$.

Solution. Device considered open:

$$I_E = 0 \quad \text{and} \quad V_E = 0 \quad \text{(point D)}$$

Device shorted:

$$I_E = \frac{V_E}{R_E} = \frac{10\text{ V}}{182\ \Omega} = 55\text{ mA} \quad \text{(point G)}$$

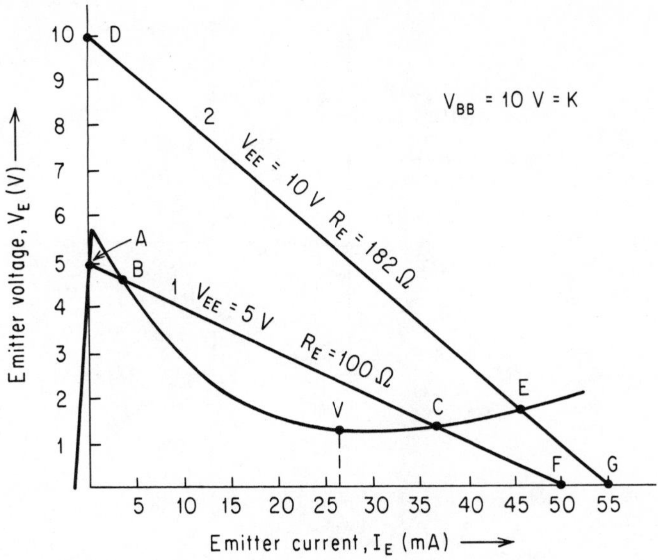

Figure 6-27. Characteristic curves with load line.

Note that load line 1 crosses the curve in three places. Point B is unstable since it is on the steepest portion of the negative-resistance part of the curves. Points A and C will be stable once either condition is reached. Because of these two stable states, load line 1 is called the *bistable circuit.* Figure 6-28 shows a

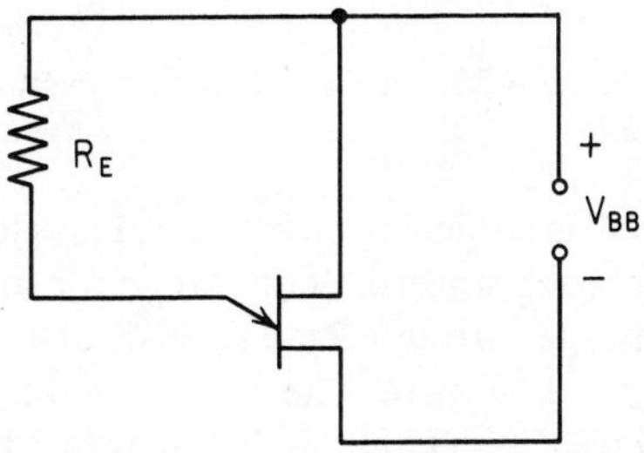

Figure 6-28. Single supply UJT circuit.

similar circuit, but one using a single supply. If V_{BB}= 10.0 V and R_E = 182 Ω, we can have load line 2. Here point E is the only stable point and we have a stable operating point. The UJT may be used to develop a circuit for any of the basic multivibrators, but perhaps its greatest use is as an astable or free-running multivibrator. Figure 6-29 shows the basic UJT astable. R_1 was added to develop a voltage at B_1. Circuit operation is as follows. S_1 is closed and the capacitor begins to charge (through R_E) at a rate determined by the RC time constant. When the

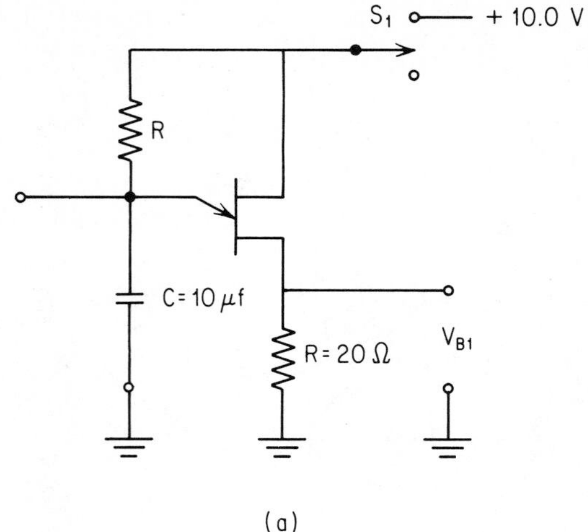

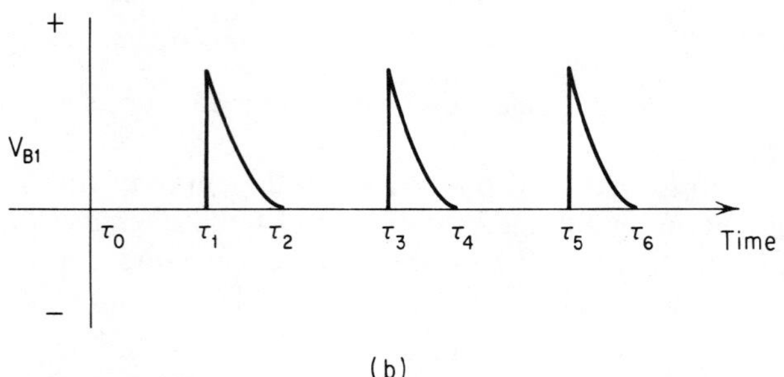

Figure 6-29. UJT relaxation oscillator: (a) circuit; (b) output.

voltage across C is sufficient to forward bias the diode portion of the UJT, internal resistance approaches zero. The capacitor now has a low-resistance path and begins to discharge through R_1. The discharge of the capacitor takes place between times t_1 and t_2 as shown. The short conduction (discharge of C) produces the sharp spike across R_1. This figure illustrates three cycles of the circuit operation. The oscillator frequency is given by

$$f = \frac{l}{R_E \ln\left(\frac{1}{1-\eta}\right)}$$

where ln signifies the natural log. The frequency may be approximated by the simple equation

$$f \cong \frac{1}{R_E C}$$

The UJT may be used in circuits to produce a time delay. One such circuit is shown in Fig. 6-30. Here we see how the UJT could be used to trigger a circuit some time after switch 1 (S_1) is closed. Since we know that period is the reciprocal of frequency, we can give the delay time T_D as

$$T_D = R_E C \ln \left(\frac{1}{1 - \eta}\right)$$

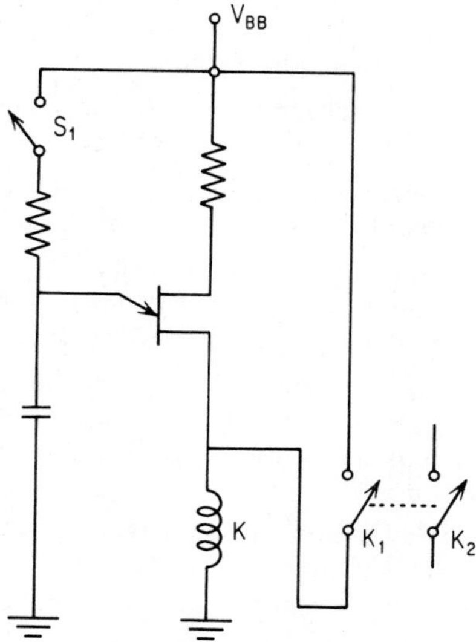

Figure 6-30. UJT time delay circuit.

where ln is again the natural log. An approximation of this equation is given by

$$T_D \cong R_E C$$

It is important to remember that R_E must be of such a value so that the load line will cross in the negative-resistance portion of the curve.

QUESTIONS

Q6-1. What conditions are necessary for thermionic emission?

Q6-2. Name at least two other means, other than heat, by which electrons are emitted from a cathode.

Q6-3. (a) Explain the difference between vacuum tubes having directly and indirectly heated cathodes.

(b) What is the reference to which the positive voltage on the plate is measured, when a directly heated cathode is used?

Q6-4. (a) What property accounts for the use of a vacuum tube as a rectifier?
(b) Explain the equivalent dc value of the output in half-wave rectification by a diode.

Q6-5. Compare the basis for one-way conduction in a tube compared with this action in a semiconductor.

Q6-6. Name two common and one less common semiconductor materials, and state where each is used.

Q6-7. Show that the resulting polarity for the dc output from a tube-diode rectifier is the same, whether the direction of electron flow or the conventional direction is used in tracing the circuit.

Q6-8. Compare a vacuum-tube and semiconductor electron device with respect to the actual flow of majority carriers.

Q6-9. (a) Name two advantages possessed by a gas diode over a vacuum-tube diode.
(b) What uses are made of these advantages?

Q6-10. Comparing tubes with transistors:
(a) State three advantages of the transistor.
(b) State two properties in which the tube is superior.

Q6-11. In transistorizing a tube circuit by the substitution of a transistor for the tube, but otherwise keeping the circuit substantially the same, what extra circuit conditions are imposed by the use of the transistor that are not present when the tube is used?

Q6-12. (a) What causes "space charge" in a vacuum tube?
(b) State its effect on the plate current of the tube.

Q6-13. (a) State two types of saturation effects in the tube diode.
(b) Which of the two saturation types is of minor importance in the ordinary operation of vacuum-tube circuits?

Q6-14. Explain the basis for considering the triode tube as an element whose resistance may be varied by the grid voltage.

Q6-15. (a) Distinguish between the dc (static) and ac (dynamic) resistance of a triode tube.
(b) What different symbols are used to represent each?
(c) What is the mathematical interpretation of the dynamic resistance of a tube?

Q6-16. (a) Does a plate resistance r_p of 10 kΩ correspond to that of a triode or a pentode?
(b) Explain its significance.

Q6-17. (a) How is a load line constructed on the plate characteristics of a tube?
(b) Explain the significance of its intersection with the family of curves representing different grid voltages E_c.

Q6-18. Discuss two ways in which the amplification obtainable in a vacuum tube is superior to that of a relay, in which a large amount of power is controlled in the relay by a small input.

Q6-19. Draw and label the conventional UJT symbol.

Q6-20. What would you expect the ohmmeter reading to be if you connected the plus lead to base 1 (B_1) and the negative lead to the emitter (E) of the UJT?

Q6-21. Define negative resistance with reference to the UJT.

Q6-22. What causes the negative resistance?

PROBLEMS

P6-1. The characteristic of a diode tube is given in the Fig. P6-1. Find the *dc plate resistance* at the plate voltage points:

(a) $E_b = 20$ V.

(b) $E_b = 50$ V.

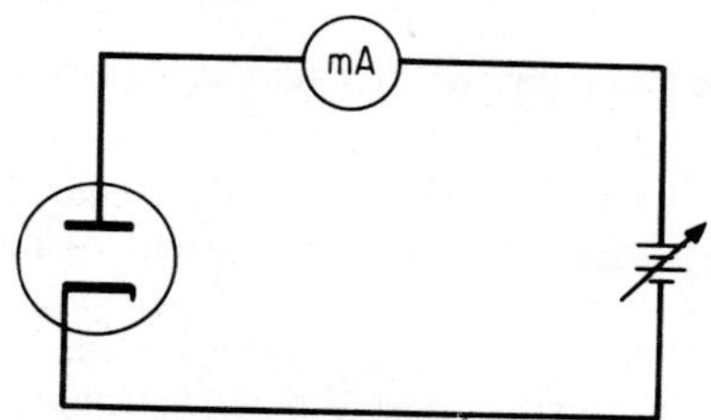

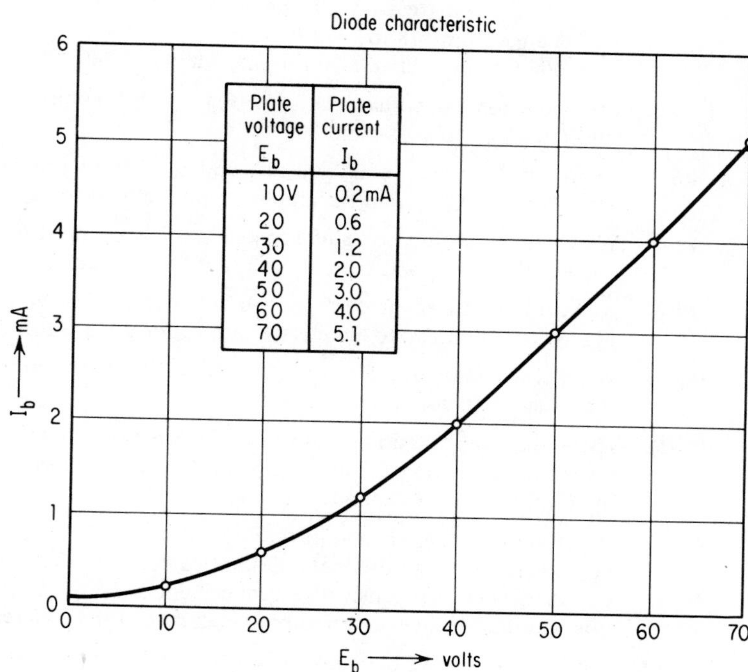

Plate voltage E_b	Plate current I_b
10V	0.2mA
20	0.6
30	1.2
40	2.0
50	3.0
60	4.0
70	5.1

Figure P6-1

(c) Find the *dynamic (ac) plate resistance* over the operating range of $E_b = 40$ to 60 V.

(d) Account for the different answers obtained in each part.

P6-2. For the same diode characteristic as in Problem P6-1, find the *dc plate resistance* at:

(a) $E_b = 10$ V.

(b) $E_b = 60$ V.

(c) Find the *dynamic (ac) plate resistance* over the operating range $E_b =$ 10–70 V.

P6-3. Over what portions of the curve is the diode characteristic of Problem P6-1:

(a) Substantially linear?

(b) Nonlinear?

(c) How does this affect the ac voltage scale of a voltmeter when the diode is used as a rectifier of the input voltage?

P6-4. Account for the fact that the diode characteristic curve of Fig. P6-1 does not start from the origin (0, 0).

P6-5. (a) Using the 6J5 triode plate characteristics [Fig. 6-6(b)], interpret the general effect on the plate current of increasing and decreasing the value of the grid bias.

(b) How does the spacing between the grid curves (whether evenly spaced or crowded closely together) influence this effect? (At a plate voltage $E_b =$ 200 V, compare a 4-V change from -8 to -4 V, with a corresponding 4-V change from -8 to -12 V.)

(c) What is the grid voltage for cutoff at $E_b = 200$ V?

P6-6. From the 6J5 plate characteristics [Fig. 6-6(b)], compare the grid voltage for cutoff at $E_b =$ 100, 200, and 300 V.

P6-7. At an operating point of $E_c = -6$ V and $E_b = 200$ V [on the 6J5 plate characteristics in Fig. 6-6(b)], find the amplification factor μ for a ± 2-V grid swing.

P6-8. When the operating point of Problem P6-7 is changed to $E_c = -10$ V, at the same $E_b = 200$ V, find the amplification factor μ for this operating point.

P6-9. At an operating point of $E_c = -6$ V and $E_b = 200$ V [for the 6J5 curves of Fig. 6-6(b)], find the ac plate resistance r_p for a plate-current swing of ± 1 mA.

P6-10. What is the value of plate conductance g_p, expressed in mhos, corresponding to the plate-resistance r_p value of 8000 Ω?

P6-11. Using the same operating point and grid swing of Problem P6-7, find the mutual conductance (or transconductance) g_m. (Check your answer against the approximate values of $\mu = 20$ and $r_p = 8$ kΩ.)

P6-12. What approximate change in the mutual conductance g_m of the 6J5 tube may be expected when an original quiescent current 10 mA is changed by shifting the operating current to a new quiescent current of 2.5 mA? (Use the chart for variation in tube parameters in the tube manual or other reference.)

P6-13. (a) Using the 6J5 plate characteristics of Fig. 6-6(b), draw a load line for a plate supply E_{bb} of 300 V and a resistive load of 20 kΩ.

(b) How does it compare with the 20-kΩ load line shown in the figure when the plate supply $E_{bb} = 200$ V?

P6-14. (a) As in Problem P6-13(b), draw a line for a 50-kΩ load resistor.
(b) How does it compare with the 20-kΩ load line shown in the figure?

P6-15. Using the 20-kΩ load line drawn on the 6J5 characteristics of Fig. 6-6(b), find the quiescent plate current I_b and plate voltage E_b at the operating point of $E_c = -2$ V.

P6-16. (a) In Problem P6-15, find the swing in plate voltage resulting from a ±2-V swing of grid voltage around the $E_c = -2$ V operating point.
(b) What is the approximate voltage gain under these conditions?

7

Electronic Voltmeters

7-1. THE ELECTRONIC VOLTMETER

The electronic type of voltmeter, whether using vacuum tubes or transistors, is one of the indispensable measuring instruments in the laboratory. Because of its electronic amplifying action, it is able to provide *highly sensitive measurements* for a wide range of signals, at the same time presenting a *very high impedance*, so that it draws a minimum amount of current from the signal source being measured. This combination of unique properties has resulted in the development of many different practical forms of the instrument. Although the vacuum-tube voltmeter (VTVM) has been replaced by solid-state electronic voltmeters for the most part, there are still a great many of them in service. Therefore, the operation of the VTVM is discussed in this chapter in addition to modern solid-state voltmeters.

7-2. TYPES OF VTVM CIRCUITS

In considering the large diversity of electronic voltmeters, it will be helpful to consider them from the standpoint of the basic types of electronic circuit action that are used to obtain the output indication. An elementary single-tube circuit will be considered first. It will illustrate, in a simple manner, how the amplifying property of the tube is used to cause deflection of the indicating meter, without requiring substantial current for this deflection from the circuit being measured (as would be the case in a moving-coil voltmeter). After the elementary single-tube circuit, the more practical version of the VTVM will be considered; a dc difference amplifier form (commonly called a "*balanced-bridge*" circuit) that is basic to most service-type VTVM instruments.

7-3. ELEMENTARY (SINGLE-TUBE) DIRECT-CURRENT VTVM

Consider, now, the simplest possible form, in which an electron-tube circuit is arranged to measure a dc input voltage, as shown in Fig. 7-1(a). When there is no input signal voltage ($E_s = 0$), the dc current meter will read a quiescent current I_{b_0} depending upon the initial operating point chosen for the tube. For the values shown, with a plate supply voltage of 180 V and a load resistor of 30 kΩ, the corresponding 30-kΩ load line will intersect the plate characteristic curves of the 6J5 at the points shown on the transfer curve of Fig. 7-1(b). This transfer curve will be seen to have a linear portion between the grid voltages E_c, from 0 to around −7 V.

If the operating point were chosen to set the fixed grid bias E_{cc} at −6 V, there would be a quiescent current of 1.5 mA through the meter any time there was no input signal E_s. It would then be possible to calibrate the current scale of the meter by marking points on the meter scale with the corresponding signal voltages E_s. Thus the 1.5 mA point would be marked 0 signal V. The current reading produced when +1 V (E_s) is applied would correspond to 2 mA (produced at an effective grid voltage E_c of −5 V), so this 2-mA point would be marked +1 V on the indicating meter. This would result in a linear scale for the input voltages marked E_s on the graph, from 0 up to +6 V. The meter would read 4.5 mA at full scale. On the other hand, the calibration would become increasingly nonlinear for negative input voltages as the tube cutoff was approached. This would require that only positive voltages be applied to the grid.

Simple as it is, this circuit will serve to illustrate the basic VTVM property by which input voltages can be measured in terms of corresponding plate-current readings. It also serves to point out the important advantage of high input impedance obtainable with the VTVM. In this example, a full-scale deflection corresponding to 6 V is obtained with an input impedance of 12 MΩ, determined by the grid resistor R_g. This results in an ohms-per-volt sensitivity of 2 MΩ/V, which is 100 times greater than the 20,000 Ω/V sensitivity of the typical VOM.

An obvious disadvantage of this simple circuit, however, is the fact that the meter shows a deflection (1.5 mA) even when the input voltage is zero. The simple solution of increasing the grid bias to cutoff (so that there will be zero output

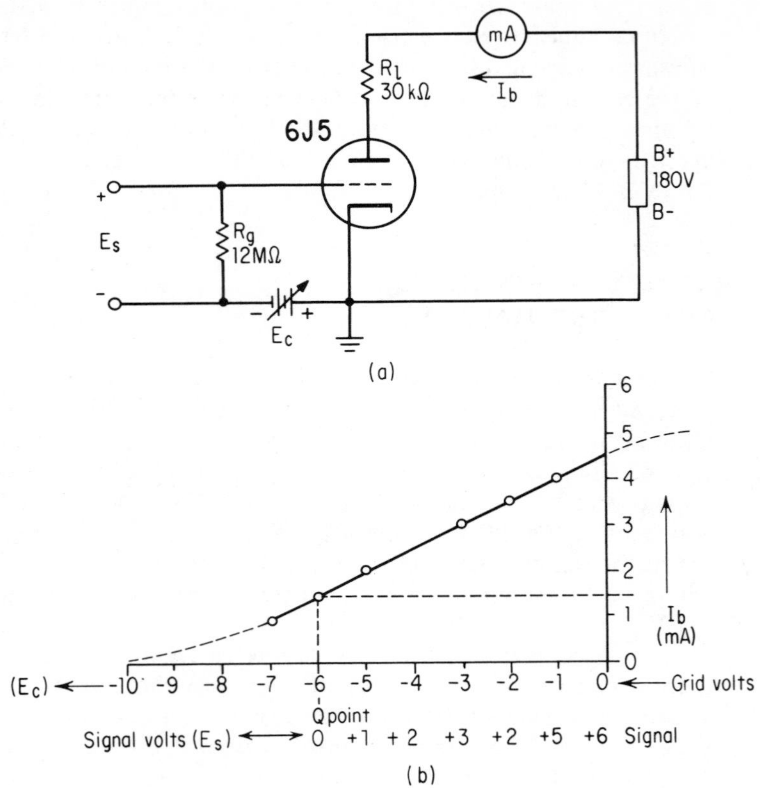

Figure 7-1. Elementary VTVM circuit: (a) circuit diagram; (b) transfer curve showing quiescent point Q, when the signal voltage E_s equals zero.

current with no voltage input) is not a feasible one here, because of the excessive nonlinearity and lack of voltage sensitivity encountered when the tube is operated close to its cutoff point. This disadvantage, however, can be effectively overcome by employing an opposing (or bucking) current to cancel out the quiescent current and thus bring the pointer to zero for the no-signal condition. This effect of quiescent-current cancellation is generally obtained by adding two resistors to form the familiar bridge-circuit arrangement, as shown in Fig. 7-2. In this way, the power supply for the tube also powers the bridge arrangement. A variable resistance is used for one of the bridge arms, thus allowing a null to be obtained for the quiescent no-signal condition. This control would accordingly be labeled as the ZERO SET knob.

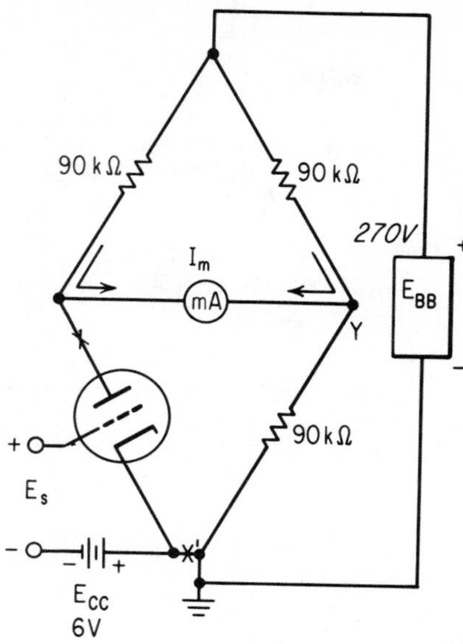

Figure 7-2. Circuit for resistor-bridge (single-tube) arrangement of elementary VTVM.

7-4. RESISTOR-BRIDGE ARRANGEMENT OF SINGLE-TUBE VTVM

An analysis of the action of the balanced resistor-bridge arrangement can best be done by the use of a Thévenin equivalent[1] circuit, as shown in Fig. 7-3. Considering the tube as the load, the circuit is broken at the load points XX′. The equivalent generator voltage V_{eq} is obtained as the open-circuit voltage across XX′. Considering the small resistance of the milliammeter as negligible compared with 90 kΩ, $V_{eq} = 270(\frac{90}{135})$ or 180 V. The equivalent series impedance Z_{eq} is the impedance seen looking into the open terminals XX′, with the 270-V battery replaced by its substantially zero resistance. This results in three 90-kΩ resistances in parallel, giving Z_{eq} as (90 kΩ) ($\frac{1}{3}$) or 30 kΩ. Having obtained this equivalent circuit, it is then an easy matter to find the plate current I_b flowing through the tube, for any given operating condition. (Note that this simplification thus far gives the tube current I_b rather than the meter current I_m; the relation between this tube current I_b and the meter current I_m is the subject of one of the problems given at the end of the chapter.)

[1]A review of the Thévenin theorem is given in Appendix A.

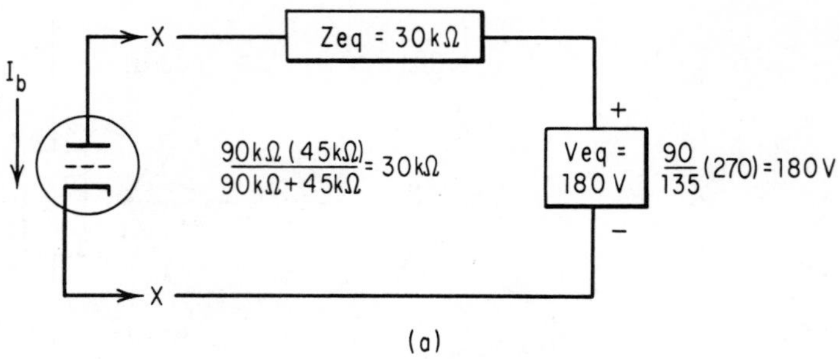

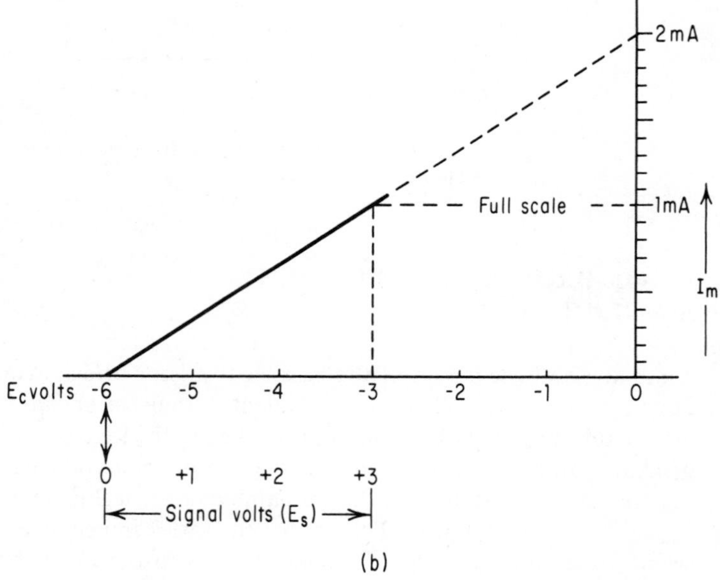

Figure 7-3. Tube action in resistor-bridge VTVM of Fig. 7-2; (a) Thévenin equivalent to find place current I_b; (b) transfer curve showing meter current I_m (1 mA full-scale) for input signal E_s from 0 to +3 V.

When we substitute a tube for the lower-right-hand resistor of the resistor-bridge arrangement, we have the popular dual-tube balanced-bridge circuit. Here, each tube performs as an active, rather than passive resistor, and a much more stable circuit results, incorporating the twin advantages of symmetry and negative feedback.

7-5. BALANCED-BRIDGE TWO-TUBE CIRCUIT (GENERAL-PURPOSE SERVICE-TYPE VTVM)

A vacuum-tube version (very similar to a field-effect transistor or FET version) of the symmetrical bridge circuit is shown in Fig. 7-4(a). It is a basic circuit, used not only in practical electronic voltmeters but also in many other amplifier applications, under the name *cathode-coupled* or differential *amplifier*. In Fig. 7-4(a) each tube V_1 and V_2 is usually one section of a 6SN7 (or 12AU7) medium-mu dual-triode tube. This arrangement may be considered as a variation of the resistance-bridge circuit previously discussed, where the effective plate-to-cathode resistance of each tube is now taking the place of each of the lower arms of a bridge circuit. Because of the symmetry of this circuit, the balance condition for zero current through the indicating meter is much more stable against unwanted variations in operating conditions than it is in the single-tube circuit. A random increase in plate voltage, for example, would cause corresponding increases in the plate currents of both tubes simultaneously and would not, therefore, necessarily upset the balance.

Another advantage of this circuit is gained from the negative current feedback introduced by the cathode resistor that supplies the coupling. The result of this negative feedback is to add additional stability to the circuit. Even through this is done at the expense of a reduction in gain, the overall amplification produced by the dual-triode arrangement is usually sufficient for the general-purpose measurements for which this instrument is intended.

The circuit action of the cathode-coupled circuit may be profitably compared with the combination of a cathode follower (for the first tube) feeding a grounded-grid amplifier (for the second tube). Such a combination can be shown on anaylsis to yield about the same amount of amplification as a single tube operated without the negative feedback and bridge-balancing features. In fact, if the output voltage of the circuit shown in Fig. 7-4(a) is considered to be $E_{o_1} - E_{o_2}$ (as it would be for a differential amplifier, where the indicating meter would draw a negligible current), and if the circuit is symmetrical ($R_{l_1} = R_{l_2}$), the analysis would yield the following exact expression for the output voltage:

$$E_{o_1} - E_{o_2} = \frac{\mu R_l}{R_l + r_p} E_s$$

This is identical with the amplification of a single tube and is independent of the value of the cathode resistor R_k. The result indicates that ample amplification is available from this stable circuit.

Under actual circuit operation it is not necessary to conform to the assumption that negligible current flows through the indicating meter. If we assume a positive input signal (say, 1.5 V), the plate current of tube 1 will increase over its quiescent value, causing a greater voltage drop across R_k. This voltage drop will make the grid of tube 2 more negative, causing the plate current of tube 2 to decrease from its quiescent value. The meter-sensitivity resistor adjustment R_m can, accordingly,

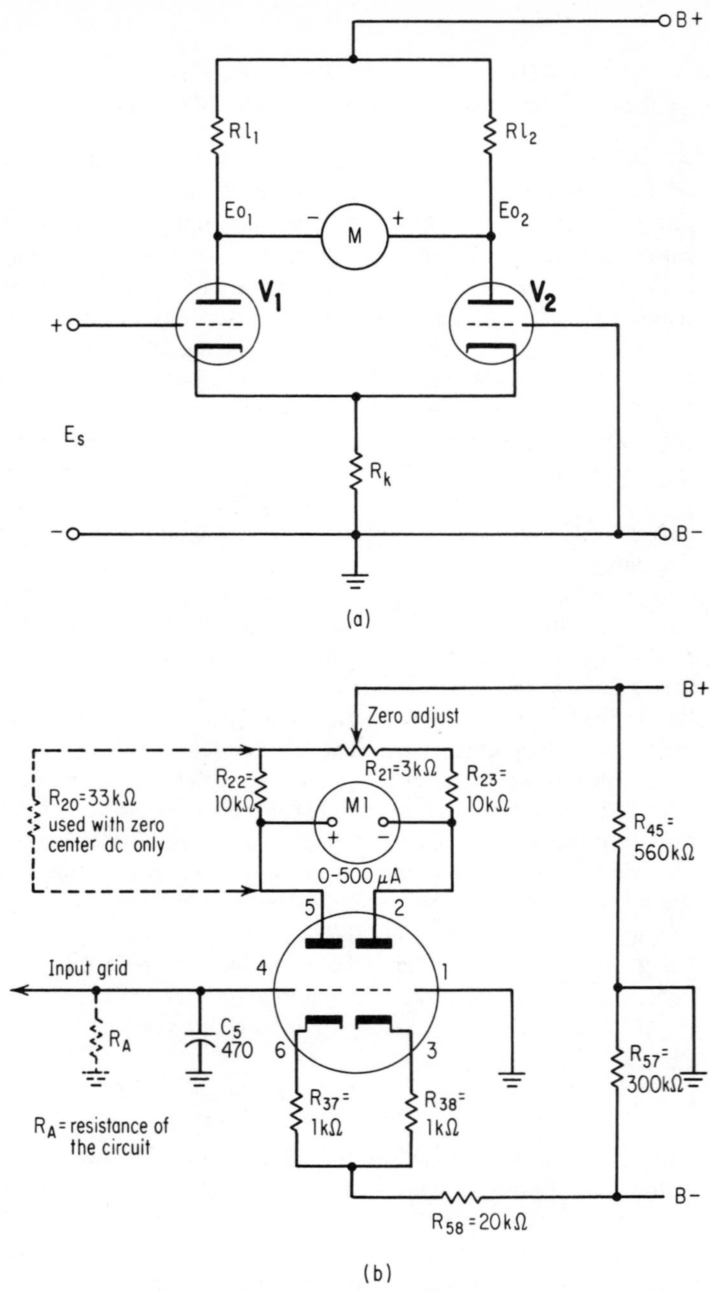

Figure 7-4. Basic two-tube bridge arrangement for dc volts: (a) functional circuit (cathode-coupled amplifier): (b) actual circuit used in *Hickok model 209C*, example of service-type VTVM. (Courtesy Hickok Electrical Instrument Company.)

Figure 7-5. A FET multimeter. (Courtesy Hickok Electrical Instrument Company.)

be proportioned to allow full-scale deflection of the indicating meter, in which case the full-scale voltage-measuring range would be 1.5 V. Since the circuit constants are chosen for operation on the linear portion of the tube characteristic, the calibration results in a linear scale for the measurement of dc volts, on the +1.5-V range in this case. Starting with this as a basic range, the measurement capability of the VTVM can then be extended in various ways to measure negative as well as positive voltage, higher dc voltage ranges, ac voltage ranges, and resistance ranges. Figure 7-5 shows an up-to-date FET multimeter.

7-6. ELECTRONIC VOLTMETERS FOR ALTERNATING CURRENT

In order to provide the high input impedance for ac measurements that the VTVM provides for dc measurements, it is necessary to place a high-impedance ac amplifier between the input and the rectifier. This was not practical for service-type VTVMs because vacuum tubes are bulky and expensive. Alternating-current VTVMs were separate instruments designed for use in the laboratory. However, with solid-state devices currently available, it is entirely practical to provide the high-impedance advantage of the electronic voltmeter for ac measurements even in a hand-held instrument.

Regardless of the type of active devices used, the rectifier-type ac electronic voltmeter generally takes the form shown in the block diagram of Fig. 7-6. The

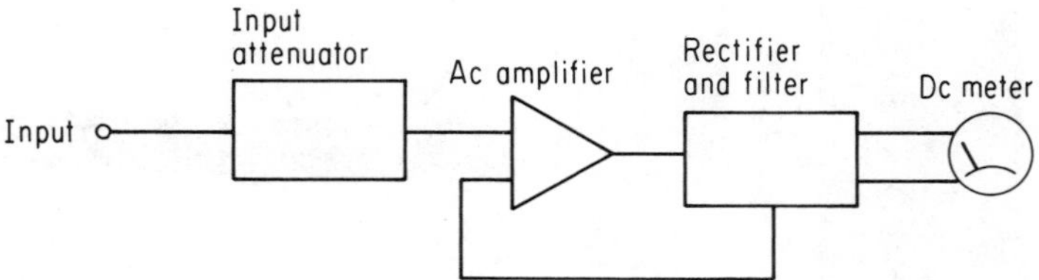

Figure 7-6. Block diagram of ac electronic voltmeter.

attenuator includes a range switch which usually selects resistors for the lower leg of a voltage divider. The amplifier provides the necessary gain to establish the sensitivity of the instrument as well as providing a high input impedance. A very-high-gain amplifier is normally employed with negative feedback to establish a stable and accurate overall gain.

The type of rectifier circuit used determines how the voltmeter will respond to nonsinusoidal wave-forms. Rectifier-type ac voltmeters will not provide accurate rms measurements of nonsinusoidal wave-forms. This problem is discussed further in the next section.

A voltmeter that measures the rms value of an ac voltage regardless of wave-form is called a true-rms voltmeter. This type of instrument measures the heating effect of the input voltage or utilizes an analog-computer type of circuit to determine the rms value directly from the definition:

$$V_{\text{rms}} = \sqrt{\frac{1}{T}\int_0^T v^2(t)\,dt}$$

The squaring and square-rooting operations can be performed using analog multiplier circuits, and the integration can be performed using an operational-amplifier integrator. The problem with the direct implementation of the equation above is that analog multipliers currently available are expensive and have very limited bandwidth. On the other hand, the heating effect of the unknown voltage can be measured using thermocouples and heating elements in a bridge arrangement which is economical and provides relatively good bandwidth.

A block diagram of a thermocouple-bridge-type true-rms voltmeter is shown in Fig. 7-7. In this type of instrument, two matched transducers are used. Each transducer consists of a resistive heating element in thermal contact with a thermocouple. The upper thermocouple generates a voltage which is a function of the heating effect of the input ac wave-form. This dc voltage is amplified and fed back

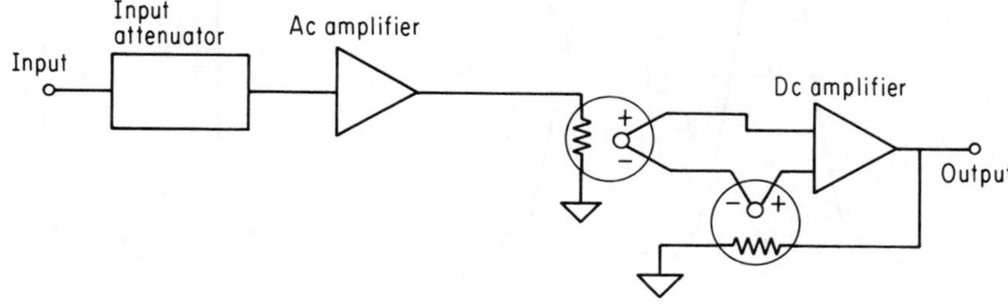

Figure 7-7. Block diagram of thermocouple-type true-rms voltmeter.

to another transducer of the same type. The thermocouple of this second transducer is connected in series with the first thermocouple, but with the opposite polarity. If the dc amplifier has a very large gain, it will insist on essentially zero input voltage. At equilibrium, the dc voltage at the output will be equal to the rms value of the ac voltage applied to the first transducer. This dc voltage can now be applied to an ordinary dc meter.

Although the output of the first thermocouple in Fig. 7-7 could be applied directly to a dc microammeter to provide an indication of rms voltage, this is not normally done. The reason for this is that thermocouples tend to have nonlinear voltage-vs.-temperature characteristics. In addition, the output of the thermocouple is affected to some extent by ambient-temperature variations. Both of these effects are canceled out by the negative-feedback arrangement in the two-transducer configuration. The only requirements for accuracy are that the dc amplifier have a very high gain and that the two transducers be well matched.

Voltmeter Measurement of Pulse Wave-Forms

Pulse wave-forms are usually observed on the oscilloscope, which effectively displays the large variety of pulses encountered, in the form of square, rectangular, sawtooth, and peaked waves. In such cases, the peak-to-peak value of the wave-form is easily obtained from the calibrated scope display. The measurement of such pulses on an ac meter, on the other hand, raises the all-important question of meter response, since meters of different response types will show widely different readings on their rms scales, for a given pulse, even though this scale is correctly calibrated for the rms voltage reading of sine waves in each case. This can best be illustrated by starting with a typical rectangular-pulse wave-form (as in Fig. 7-8) and observing the different readings obtained for this identical wave-

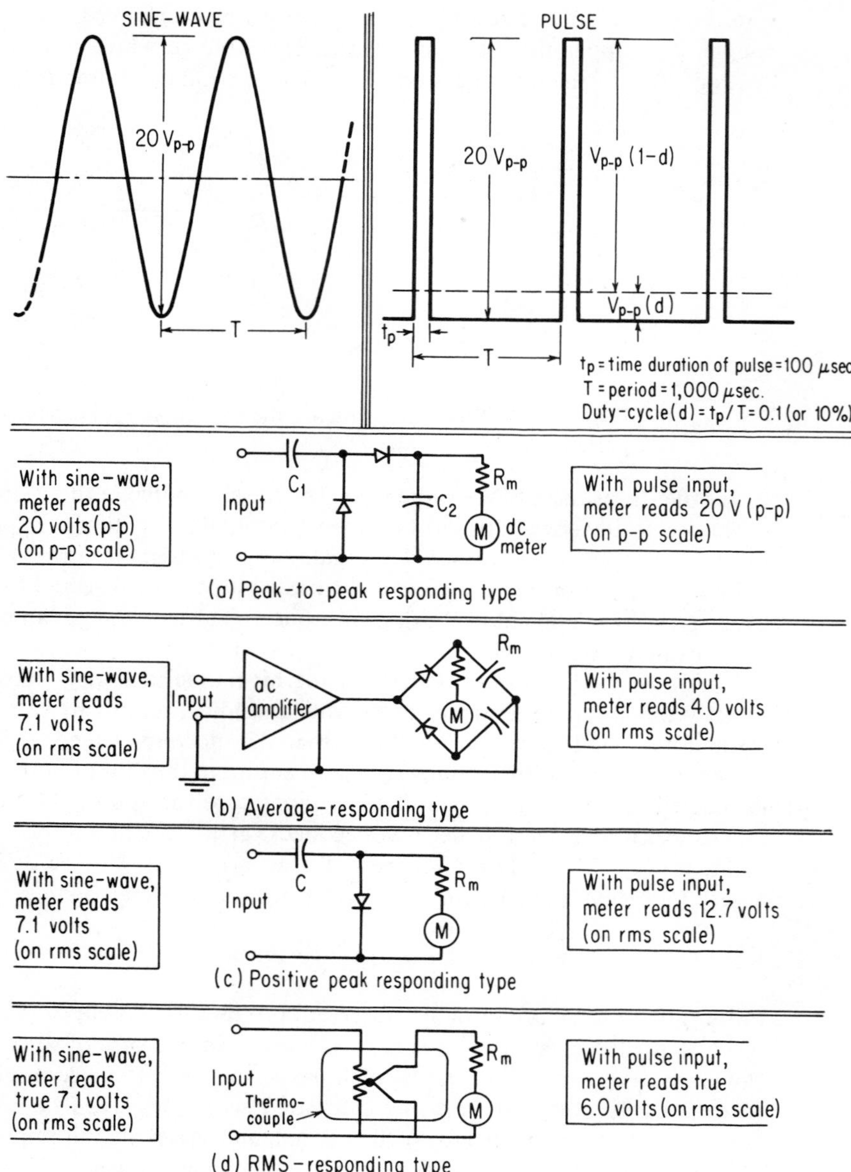

Figure 7-8. Measurement of pulse-versus-sine-wave inputs by four different types of electronic voltmeters (see the text for calculations).

form on four basic ac electronic-voltmeter types, as follows:

1. Peak-to-peak-responding voltmeter
2. Average-responding type (with full-wave rectification)
3. Single-peak (or positive-peak) responding type)
4. True-rms-responding type

The wave-form chosen for illustration is a positive 20-V pulse, having a duration of 100 μsec and a repetition rate of 1 kHz. The appearance of the 20-V pulse (V p–p), after passing through a blocking capacitor for ac coupling to the meter, is shown on the right side of Fig. 7-8 (this is how it would also appear on an oscilloscope; the dashed line for the 0-V level, which is lacking in the scope presentation, has been added and made to conform to the requirement that the average value of the ac-coupled pulse is zero, as shown by the equal areas above and below the 0-V level). The left side of the figure shows a 20-V peak-to-peak sine wave at the same frequency, for comparison purposes.

Referring to the rectangular pulse, the pulse duration of 100 μsec (t_p) is seen to occupy one-tenth of the total period (T) of 1000 μsec, so that the duty cycle, which is t_p/T, is 0.1 (or 10%), which we will designate as d.

We can now get a good idea of the different readings that will be displayed on each of the four types of ac voltmeters for the 20-V (p–p) pulse [as contrasted with the 20-V (p–p) sine wave], by calculating the response for each type as follows:

1. *On a peak-to-peak-responding voltmeter*, response to a 20-V (p–p) pulse is given by

$$\begin{aligned} \text{p–p indication} &= \text{V}_{\text{p-p}} \\ &= 20\ \text{V}_{\text{p-p}} \text{ for pulse} \end{aligned}$$
$$(\text{versus } 20\ \text{V}_{\text{p-p}} \text{ for sine wave})$$

2. *On an average-responding voltmeter* (with full-wave rectification), response to a 20-V (p–p) pulse is given by

$$\begin{aligned} \text{rms indication} &= 1.1 \text{ (average)} \\ &= 1.1[2\ \text{V}_{\text{p-p}} d(1 - d)] \\ &= 1.1[2(20)(0.1)(0.9)] \\ &= 4.0 \text{ V for rectangular pulse} \end{aligned}$$
$$(\text{versus } 7.1 \text{ V for sine wave})$$

3. *On a positive-peak-responding voltmeter*, response to a 20-V (p–p) pulse is given by

$$\begin{aligned} \text{rms indication} &= 0.707\ \text{V}_{\text{p-p}}(1 - d) \\ &= 0.707(20)(0.9) \\ &= 12.7 \text{ V for rectangular pulse} \end{aligned}$$
$$(\text{versus } 7.1 \text{ V for sine wave})$$

4. *On a true-rms-responding voltmeter*, response to a 20-V (p–p) pulse is given by

$$\begin{aligned}\text{rms indication} &= V_{p\text{-}p}\sqrt{d(1-d)}\\ &= 20\sqrt{0.1(0.9)}\\ &= 6.0 \text{ V for rectangular pulse}\end{aligned}$$

(versus 7.1 V for sine wave)

The pulse measurements are shown here to agree with the sine-wave measurements in only one of the four cases, (part 1), where the peak-to-peak measurements agree with oscilloscope display. But the spectacle of the other three ac voltmeters giving widely different results for the rms value of the same pulse (4.0, 12.7, and 6.0 V)—when each of these is perfectly correct on the pure sine wave (7.1 V)—surely deserves repeated emphasis, which may be summarized as follows:

When measuring a nonsinusoidal voltage with an ac meter:

1. *Regardless of the label on the meter face, only a true-rms-responding meter reads rms volts correctly for varied wave-forms.*
2. A peak-to-peak-responding meter (only with voltage-doubler type of circuit) reads the correct peak-to-peak value on its p–p scale for all wave-forms, as does the oscilloscope.
3. A single- (or positive-) peak-responding meter reads 0.707 times the positive-peak voltage on its rms scale (correct only for sine waves).
4. An average-responding meter reads 1.1 times the full-wave rectified average voltage on its rms scale (correct only for sine waves).

7-7. SOLID-STATE VOLTMETERS

The development of voltmeters employing dependable transistor amplifiers fulfills a long-felt need for a sensitive meter that, because of practical battery operation, is at once free of ac hum pickup and at the same time is capable of extremely high amplification. When compared with a vacuum-tube voltmeter in these respects, the transistor voltmeter has the unique advantage that the transistors do not require heater current. Also, the current drain is so small that *battery operation is entirely feasible* and, in fact, often preferable to line operation, even when ac power is readily available.

In the succeeding paragraphs, an analysis is made, first, of a solid-state high-sensitivity voltmeter designed around conventional transistors (Figs. 7-9 to 7-11), followed by an illustration (Fig. 7-12) showing how a single FET easily provides a high input impedance (whereas a much more complex double emitter-follower circuit will be required when conventional transistors are used, as in Fig. 7-10 (Further simplification using integrated circuits is discussed in Chapter 13.)

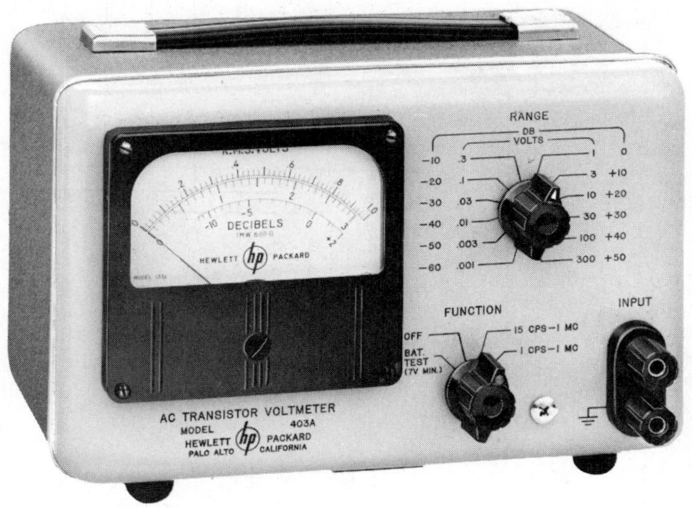

Figure 7-9. View of an ac transistor voltmeter. Compact in size ($5\frac{1}{2}$ in. high), it operates on four mercury batteries. (*Hewlett-Packard model 403-A.*)

Laboratory-Type Solid-State Voltmeter (*Hewlett-Packard Model 403A*)

By the use of the transistor amplifier, voltage sensitivity down to 1 mV full-scale is provided in this example using conventional discrete transistors.

One scheme for obtaining high input-impedance makes use of a vacuum tube operated as a cathode follower, resulting in a *hybrid tube-and-transistor type* of instrument. Another method, the one which is used in the transistor voltmeter example of Fig. 7-9, employs a *double emitter-follower circuit*, which in this case produces a 2-MΩ input impedance.

The specifications for the *Hewlett-Packard transistorized ac voltmeter, model 403A*, shown in Fig. 7-9, are listed below.

Specifications

Voltage Ranges (full-scale): 0.001 V (1 mV) to 300 V in 12 ranges.

Frequency Range: 1 Hz–1 MHz.

Operating Temperature Range: 0–50°C.

Accuracy: Within ±3% of full scale, 5 Hz–500 kHz; within ±5% of full scale at the lower and higher frequencies.

Input Impedances: 2 MΩ and 45 pF in ranges 0.001–0.1 V; 2 MΩ and 20 pF from 0.3 to 10 V; 2 MΩ and 15 pF from 30 to 300 V.

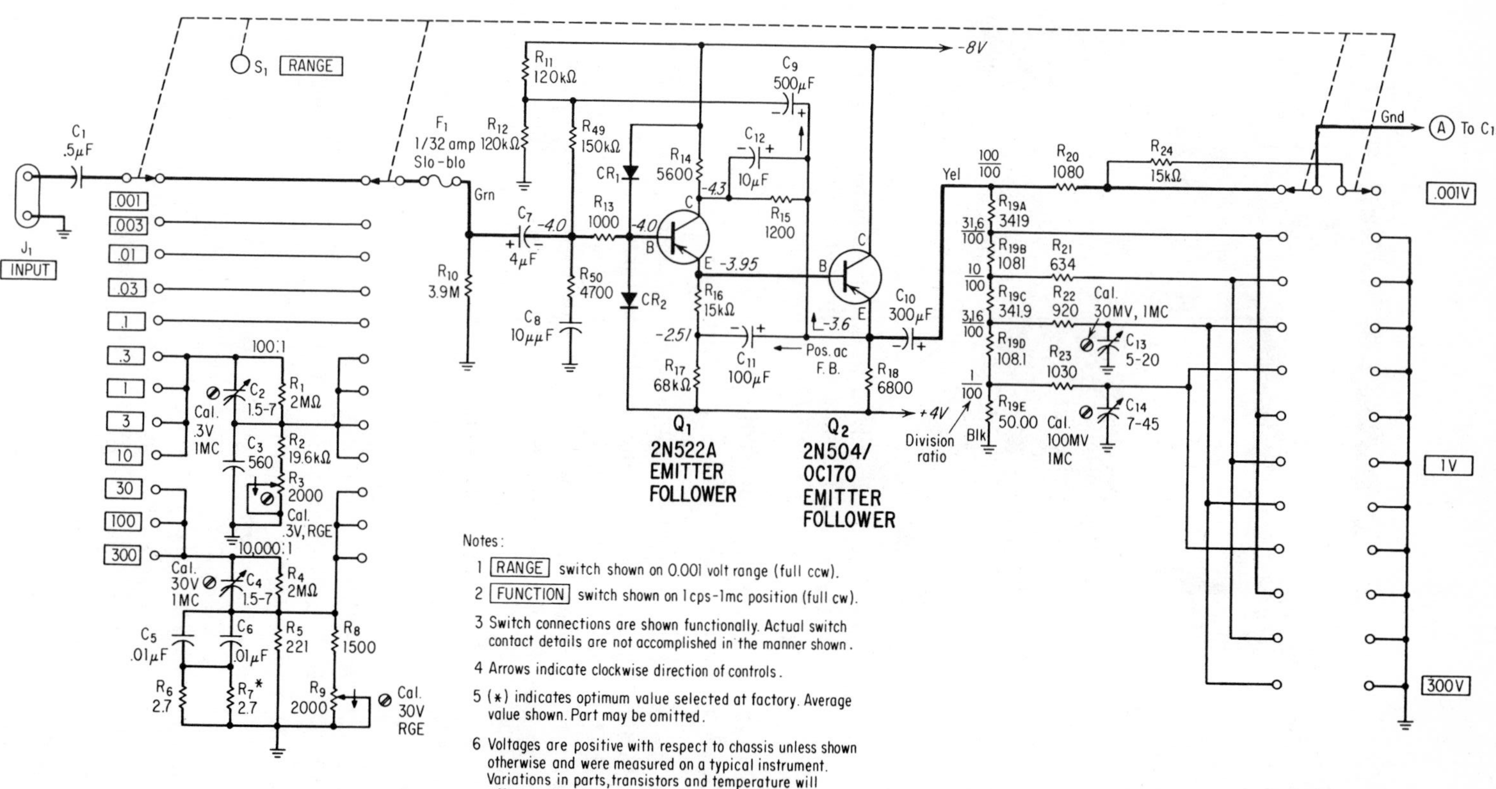

Figure 7-10. Input section of an ac transistor voltmeter; the double-emitter follower has a switched attenuator section before and after it. (Its output at point *A* connects to the ac amplifier portion.) (*Hewlett-Packard model 403-A.*)

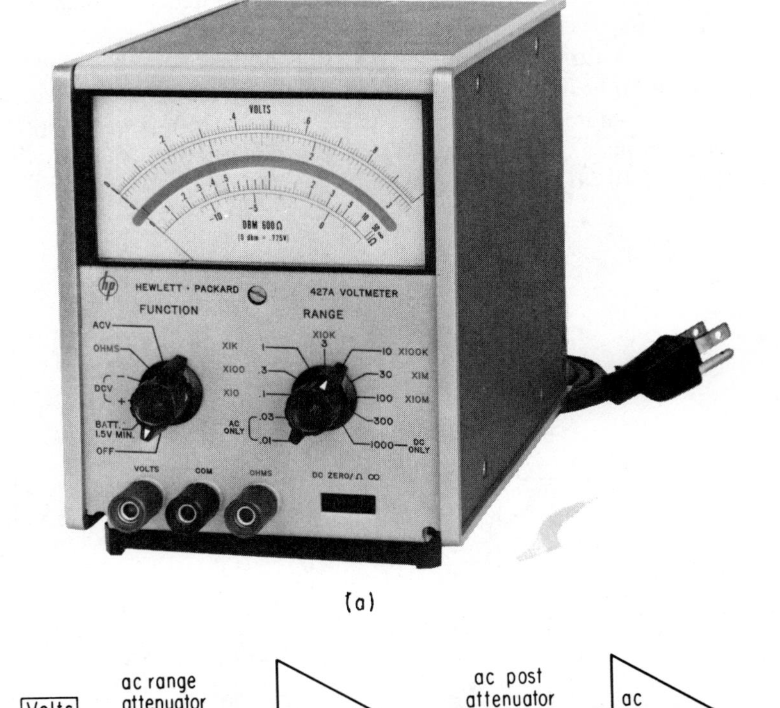

(a)

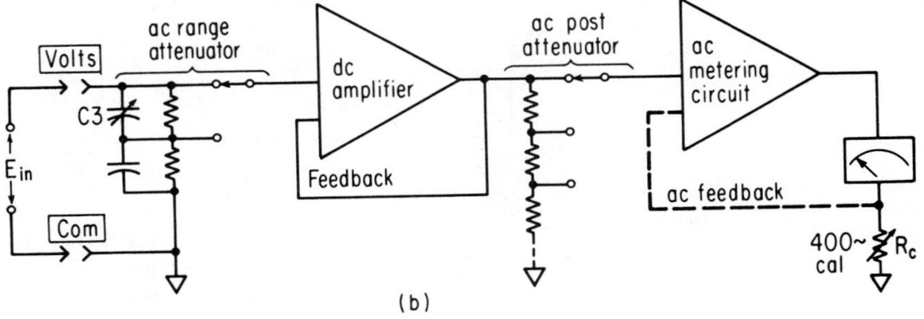

(b)

Figure 7-11. Solid-state multifunction meter, using a field-effect transistor (FET) in the input stage to provide 10-MΩ input resistance on both dc and ac voltage ranges: (a) external view; (b) block diagram of multifunction meter for ac operation. (*Hewlett-Packard model 427A.*)

Transistors: Seven.

Batteries: Four mercury batteries.

Battery Life: Approximately 400 hr.

The input circuit of the voltmeter, illustrated in Fig. 7-10, is seen to consist of a double emitter follower, which is both preceded and succeeded by the range attenuator. The output from this section feeds a five-transistor ac amplifier, which, in turn, is rectified by a bridge arrangement practically identical to the one shown for the tube version in Fig. 7-12.

As a further development in the process for obtaining the desired feature of high input impedance, *field-effect transistors* are used in the next two examples. The solid-state multifunction meter, model 427A [Fig. 7-11(a)], includes a FET as an impedance converter in its input circuit; thereby it provides an input resistance of 10 MΩ on all ranges.

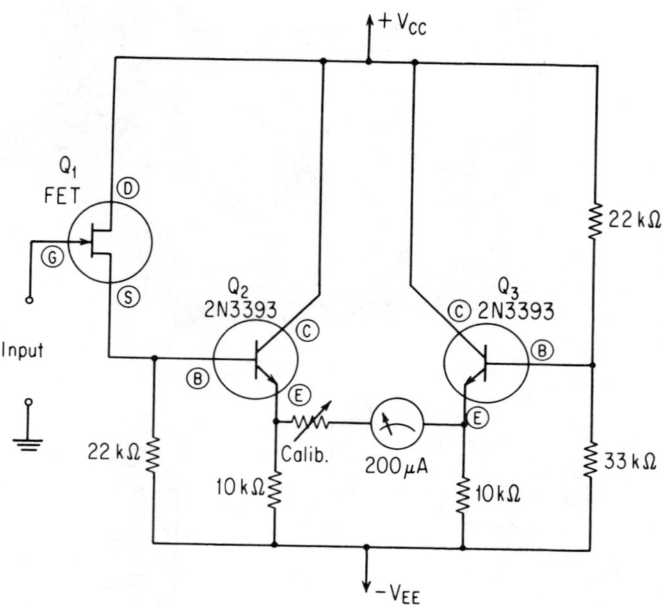

Figure 7-12. FET input circuit (in functional form); field-effect transistor Q_1 provides high input resistance (>10 MΩ) for the balanced-bridge amplifier formed by the conventional (bipolar) transistors Q_2 and Q_3. (Adapted from *Heathkit Utility Voltmeter model IM-17.*)

As the block diagram for ac operation indicates, the signal being measured sees the FET as the input impedance for both dc and ac operation, since the ac amplifier and rectifier are fed from the output of the dc amplifier. Hence, the high input resistance of 10 MΩ is seen on all dc and ac voltage ranges.

The use of a compact solid-state dc amplifier followed by the ac amplifier block, in combination with a field-effect transistor is a good example of the broad measuring capabilities (at moderate cost) that are possible with the presently available "building blocks." Condensed specifications for the high-impedance solid-state *Hewlett-Packard model 427A* used in this example follow.

Specifications

Direct-current Volts: ±100 mV–±1000 V, full-scale (9 ranges).

Alternating-current Volts: 10 mV–300 V rms, full-scale (10 ranges).

Ohms: 10 Ω center-scale to 10 MΩ center-scale (7 ranges).

Power: $22\frac{1}{2}$-V dry-cell battery; option 01: battery operation and ac line operation, selectable.

Another example of the use of the FET in an input circuit is shown in the functional circuit [Fig. 7-11(b)] of the *Heathkit utility solid-state voltmeter, model IM-17* [the photo of which was previously shown in Fig. 1-1(c)]. In this case, the unipolar FET feeds a bridge amplifier (or emitter-coupled differential amplifier) formed by a pair of conventional (bipolar) transistors.

A "do-it-yourself" version, for adding the advantages of FET input to a nonelectronic volt–ohm–milliammeter (VOM), allows the user either to use the VOM as is, for convenience, or if desired, to add on an adapter that provides both the high input impedance and increased sensitivity of a FET voltmeter.[3]

7-8. HIGH-SENSITIVITY DIRECT-CURRENT MEASUREMENTS

The advanced types of electronic voltmeters, developed for measuring only dc in the laboratory, present both higher capabilities and greater special problems than those encountered in the general-purpose (service-type) VTVM. Because the rectifying circuit is not needed, the dc measuring circuit is freed from the limitations of low input impedance that might be imposed by the rectifier and therefore is able to offer input impedances beyond the megamegohm (10^{12}) range and extending up to 10^{14} Ω—thus, for example, allowing the measurement of insulation resistances. In addition, since broadband frequency response is no longer a limitation, the gain of the dc amplifier may be made very large to produce high sensitivities resulting in full-scale sensitivities in the millivolt ranges (and with modifications, into the microvolt ranges). Such high sensitivities find applications in the measurement of extremely small bioelectric potentials, pH indications, and small voltage drops in general, resulting from minute currents as low as a small fraction of a single microampere.

When very large input impedances and high sensitivities are approached, two special problems begin to assume sufficient importance to require special attention. One is the problem of *grid current*, even though in the minute amounts, caused by gas in the tube, and the other is the problem of *drift* inherent in dc amplifier circuits, whether of the tube or transistor variety.

[3]S. D. Prensky, "FET Rejuvenates VOM," *Popular Electronics*, Nov. 1968, *Electronic Experimenter's Handbook*, 1970, Winter ed., p. 135.

In the case of grid current, a nominal value of grid current in an ordinary input tube equal to $\frac{1}{1000}$ of a microampere (10^{-9} A), flowing through an input resistance of 10 MΩ (10^7 Ω), produces a spurious voltage of only 0.01 V, a value that can well be considered negligible compared with a full-scale reading of 1 V. If, however, anything like that amount of grid current were to flow through an extremely high input resistance in the megamegohm range (10^{12} Ω), it can be readily seen that the resulting voltage would no longer be negligible. In the practical case, even a few hundredths of a volt caused by grid current would be too large to tolerate when measuring an input signal in the millivolt range. To overcome this grid-current difficulty, use is made of *electrometer-tube types*. These are special tubes processed to have grid current at least $\frac{1}{1000}$ as small as the ordinary tubes. Instruments of the electrometer type, possessing very high input impedances, will be discussed in Sec. 7-9.

The drift problem encountered in high-gain dc amplifiers arises from the fact that any slight change in dc quiescent conditions has the same effect on the dc amplifier as a change in dc input voltage. Then, as a result of random drift, the output indication would change just as if the signal input had changed, since the dc amplifier is not able to distinguish the effect of drift from the effect of a change in the signal being measured. Even though it is possible to compensate for the effect of drift at any one instant by resetting the zero, the drift problem can become quite serious when measurements are taken over a period of time, especially when the magnitude of the drift in millivolts is of the same order as the signal being measured on a millivolt range. This problem, unfortunately, cannot be met by simply applying a large amount of feedback since the effect of drift so closely imitates the effect of an input signal change. While it will be found that heavy inverse feedback is actually used in dc amplifiers, and does produce amplifiers that are highly stable against variations of some tube parameters, the amount of drift (perhaps 2 mV/hr) still remains a problem that must be overcome by periodic resetting of the zero-set adjustment.

Methods for counteracting the drift problem generally employ *balanced circuits* for some correction, or for greater correction the *modulated type of dc instrument*, where the dc input is modulated (or chopped) to produce an ac voltage proportional to the dc input. This voltage is amplified by a practical drift-free ac amplifier and then rectified to produce a dc output indication. The characteristics of this type of amplifier are capable of providing extremely sensitive dc measurements (in the microvolt range), even though they may not offer input impedances as high as the extreme megamegohm range of the electrometer type. The modulated type of dc VTVM is discussed further in a later section.

7-9. ELECTROMETER-TYPE DIRECT-CURRENT VTVM

Electrometer instruments are designed to detect minute dc currents from circuits having very high impedances in the range of millions of meg ohms (10^{12} or teraohms) such as are present in insulation testing and in the qualities of chemical and biological analyses. Hence very high input impedance and extreme stability are of prime importance in these instruments. Both tube and solid-state versions are available, since in this instance the tube has the advantage of freedom from temperature dependence, while the solid-state versions (with FET input) have the appeal of compactness and reliability.

In the case of this example of the tube version, it has previously been pointed out that it is necessary to keep grid current to a minimum, especially in instruments having a high-impedance input circuit. This need is met by employing specially designed electrometer tubes. These tubes are operated at low plate voltages (typically around 10 V) and their grid current is generally around $\mu\mu$A or pA (1×10^{-12} A), a value about one ten-thousandth as small as that present in an ordinary vacuum tube. With proper precautions, the grid-current effect of such tubes can be reduced to negligible proportions, even when the input circuit resistance is in the megamegohm range (about $10^{14}\ \Omega$ in the example to be described). This high input resistance requires special construction features: the use of suitable insulating materials (such as teflon) and specially treated surfaces.

The other special problem associated with a high-sensitivity dc amplifier is that of keeping the inherent drift down to workable levels. As previously explained, the use of heavy inverse feedback is not by itself a sufficient remedy, and the use of a chopped input is not consistent with very high input impedance. Effective methods in electrometer instruments consist of using *balanced circuits* (two tubes per stage) throughout the amplifier or closely regulating all the supply voltages. Each method requires working to very close tolerances, the balanced system calling for very close tube matching, and the regulated supply system requiring highly stabilized supplies, not only for plate and screen voltages, but for the filament supply as well. The example to be described next uses the latter system of close regulation of supply voltages.

7-10. REPRESENTATIVE EXAMPLE OF ELECTROMETER

Electrometer Circuit

The fundamental circuit for voltage measurement is shown in Fig. 7-13. The amplifier (shown in functional form) is a three-stage, direct-coupled type. The combined circuit produces a cathode-follower output in resistor R_B, which determines the reading of the indicating meter (0–5 mA). The effective transconductance of the circuit is the product of the transconductance of the third stage and the voltage gain of the first two stages. The result is a transconductance in the millions of micromhos. The input voltage is consequently duplicated within a few microvolts across cathode resistor R_B with excellent linearity, even on the most sensitive 30-mV range. The panel meter reads the current change in the cathode resistor R_B. Different values of R_B are selected for each range, allowing the meter to be calibrated directly in millivolts. The comparatively heavy current through the

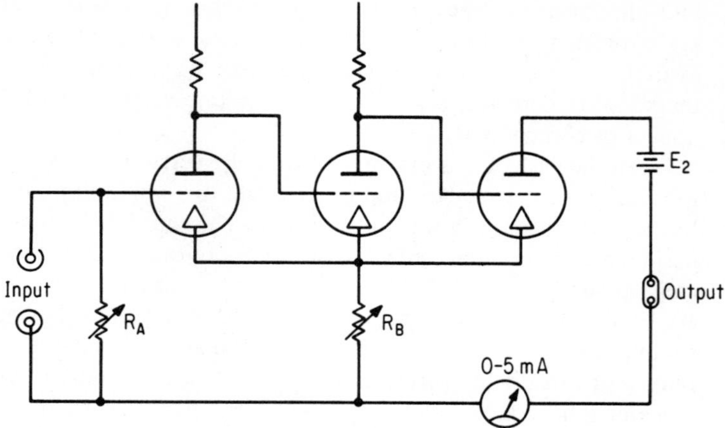

Figure 7-13. Functional circuit for high-sensitivity dc voltage measurements in electrometer-type VTVM. (*General Radio model 1230A.*)

output meter will operate either 5- or 1-mA graphic recorders, at which time the dynamic output resistance of the amplifier is less than 1 Ω, allowing great flexibility in recorder operation.

The system employed for minimizing drift uses subminiature tubes with 10-mA filaments for the first two stages of the amplifier. The filaments are in a resistor chain fed from a double-stabilized voltage-regulating system. Sensitive plate and screen voltages are also obtained from this same highly stabilized supply. As a combined result of the regulating system and the inverse feedback inherent in the cathode degeneration, line-voltage fluctuations have a negligible effect on performance, and drift is held down to around 0.2 mV/hr after warm-up.

For a comparison of this regulated system with a dc VTVM of the electrometer type employing a balanced system, the pertinent specifications of the *Keithley model 610B* electrometer read as follows.

Specifications

Direct-Current Voltage Ranges (*full-scale*)*:* 1 mV–100 V full-scale in nine ranges.

Input Impedance: 10^{14} Ω, shunted by approximately 30 pF.

Grid Current: Less than 2×10^{-14} A.

Zero Drift: Less than 200 μV/hr (after 1-hr warm-up).

Other Ranges: Direct-current down to 10^{-14} A (0.01 pA) full scale; ohms up to 10^{14} Ω, and 10^{-12} coulombs full-scale.

It is noteworthy that these specifications allow the measurement of the gate potential of a field-effect transistor without loading, even though the gate resistance is thousands of megohms.

Balanced-Type Electrometer Circuit

The circuit of this balanced-amplifier type of line-operated electrometer is basically a gain-augmented cathode follower driving an indicating meter, as shown in the block diagram of Fig. 7-14(b). The amplifier block consists of an input stage employing two dual subminiature electrometer tubes in a balanced-voltage amplifier circuit, feeding a single-ended output stage, whose output in turn feeds the cathode follower shown after the amplifier block (A). The amplifier in effect increases the gain so that a large negative-feedback factor (always greater than 100) can be used. This allows the order of accuracy to depend primarily on the precision of the measuring resistor R_a, which can be made very precise, contributing to the 1% of full-scale overall accuracy obtained, in spite of the extremely high input impedance of the measurement. When the measuring resistor is switched into the feedback loop, the electrometer also functions as a micromicroammeter (or picoammeter).

7-11. MODULATED TYPE OF DIRECT-CURRENT VTVM

The difficulty with drift that is inherent in a direct-coupled amplifier (e.g., of the electrometer type) is overcome in the dc modulated system by the use of a *chopper-vibrator* or *other modulating scheme*, to convert the dc input to an ac signal, which is then amplified by a stabilized multistage ac amplifier and reconverted to dc to activate the indicating meter.

Because of the inherent drift limitations, of the order of millivolts, the smallest full-scale range of the ordinary dc amplifier is limited to measuring an input signal of around 1–100 mV. In the chopper-modulator system, however, the drift can be cut down by a factor greater than 100, thus allowing an input signal range of around 0.01 mV (or 10 μV) full-scale (or less) to be handled. This advantage of sensitivity to smaller voltages, on the other hand, is obtained at the expense of a lower input impedance, necessitated by the dc-to-ac conversion. As a consequence, the overall ohms-per-volt sensitivity is not substantially altered, as may be seen from the comparison chart in Sec. 7-14 (Table 7-1), where both chopper and straight amplifier types wind up with similar ohms-per-volt ratings of the order of megamegohms (or tera-ohms) per volt (from 10^{11} to 10^{12} Ω/V). The important distinction that may be drawn from these figures is that the *electrometer type of instrument emphasizes the attainment of maximum input impedance* for dc measurement, while the *dc-modulated type provides the most sensitive full-scale range of measurement*, with both having similar ohms-per-volt capabilities.

(a)

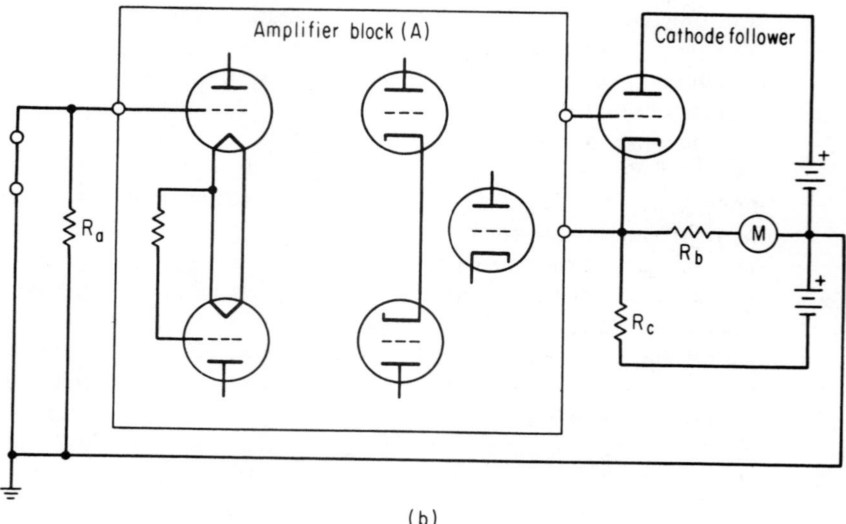

(b)

Figure 7-14. Balanced-tube-type electrometer: (a) view of *Keithley electrometer, model 610B*; (b) block diagram of balanced-circuit arrangement.

7-12. REPRESENTATIVE EXAMPLE OF CHOPPER-TYPE DIRECT-CURRENT VTVM

A commercial version of the chopper-amplifier type of dc-modulated instrument is illustrated in Fig. 7-15. This model (*Kintel 203*), with a zero-center scale, operates as a dc microvoltmeter (full-scale range 100 μV either side of zero). By measuring the voltage drop across standard internal resistors it also provides current measurements with the sensitive full-scale range as low as 100 $\mu\mu$A (100 pA or 1×10^{-10} A).

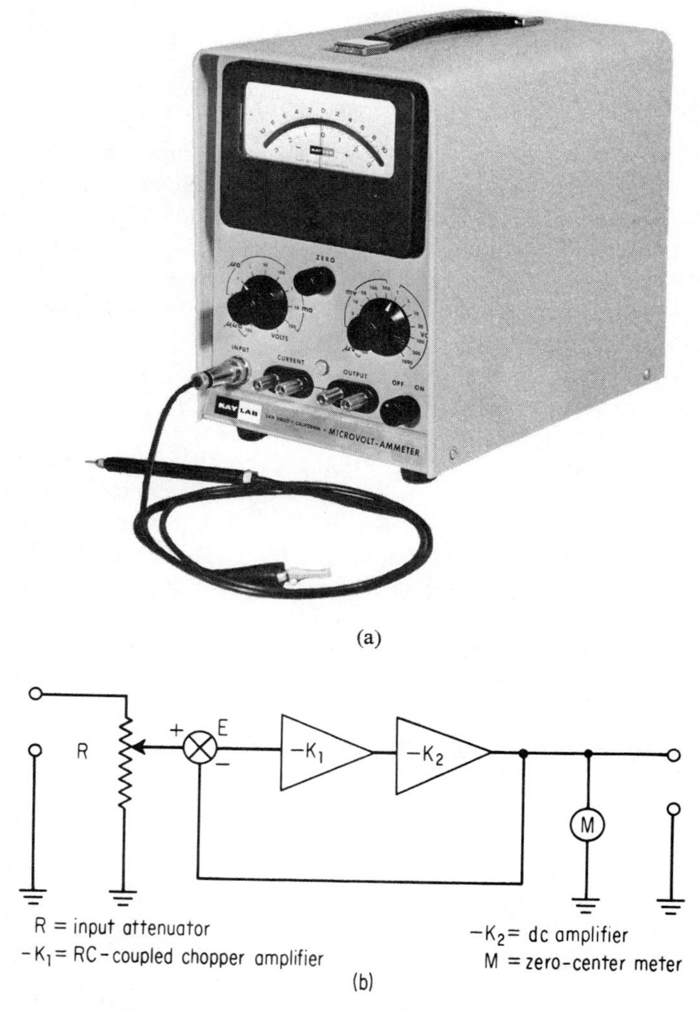

Figure 7-15. Chopper-modulated type of dc VTVM: (a) view of *Kintel K-lab model 203*; (b) functional circuit.

Typical specifications for the chopper-modulated type of dc instrument (in this case, for the *Kintel dc microvoltmeter, model 203*) are as follows (for zero-center indications):

Specifications

Voltage Range (end-scale): ±100 μV–100 V in 15 ranges.

Current Range (end-scale): ±100 μμA–100 mA in 10 ranges.

Accuracy: ±3% of full scale on all ranges.

Input Impedance: 10 MΩ below 10 mV, 30 MΩ at 30 mV, 100 MΩ above 30 mV.

Rating as Amplifier: 80-dB maximum gain; 1-V output across 1000 Ω; drift (after 15-min warm-up) = 10 μV equivalent input; output impedance is less than 2 Ω.

Chopper-Modulated Circuit

The modulation or "chopping" action interrupts the dc input signal and changes it into a pulse form suitable for an ac amplifier. Various forms of chopping are used; the main types are *mechanical vibrators, photoelectric* and *semiconductor switches*. Although the MOSFET type of field-effect transistor has proven to be quite popular, each type has its advantages and is in current use. The example discussed here employs the mechanical type of chopper (the other types are discussed in other sections).

The functional circuit of this chopper-type VTVM in measuring dc volts is shown in Fig. 7-16(b). It consists of an input divider R followed by a DPDT chopper converter and a four-stage RC-coupled ac amplifier K_1. One set of contacts on the chopper converts the dc input to a proportional ac signal, which is then amplified by the usual ac amplifier techniques previously described. The output of this ac amplifier is rectified synchronously by the other set of chopper contacts. This rectified signal, after filtering, is applied to the two-stage dc amplifier K_2, whose output actuates the indicating meter. By feeding a degenerative signal to the reed of the chopper, the input impedance of the chopper is kept relatively high with respect to the input divider R so that the input impedance looking into the instrument from the input terminal is kept constant. The large amount of overall feedback (up to 84 dB) helps to ensure high stability in the instrument, which is an example of a *chopper-stabilized* circuit.

7-13. VTVM CURRENT MEASUREMENTS

The vacuum-tube or solid-state voltmeter, its name notwithstanding, frequently has an important function as a current-measuring device. One can measure current, of course, with any kind of voltmeter, by measuring the voltage drop across a known resistor. This method assumes special importance with the high-sensitivity type of VTVM, because full-scale voltage sensitivities of the order of microvolts (as obtained in the previous dc example) make it possible to measure

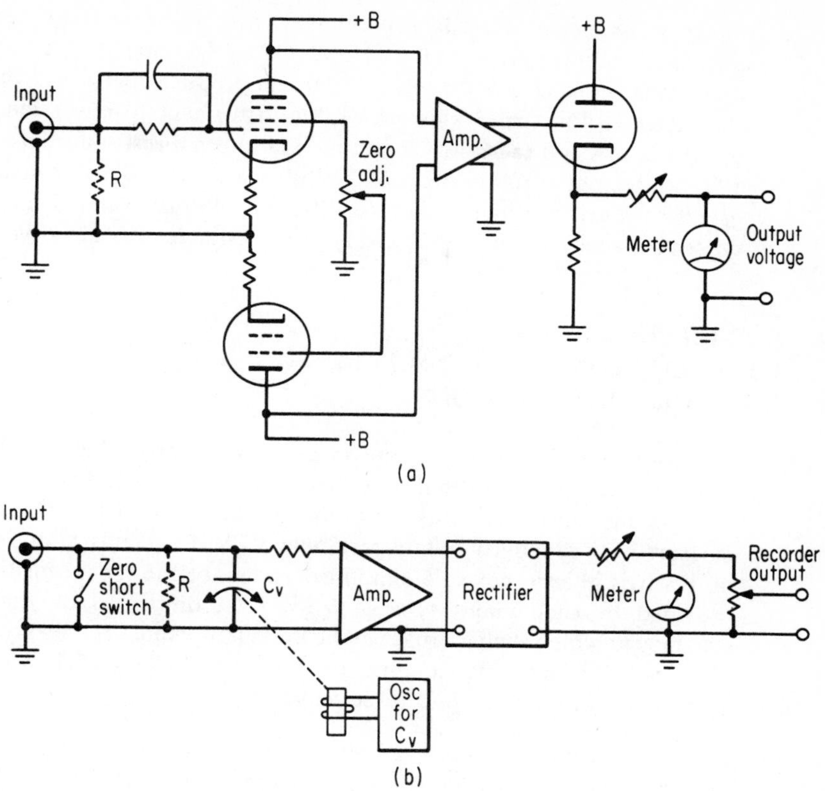

Figure 7-16. Comparing very-high-impedance dc electronic voltmeters: (a) basic electrometer-voltmeter circuit, using balanced input to dc amplifier and cathode-follower output; (b) vibrating-reed modulator with variable capacity (C_V) and ac amplifier preserves very-high-impedance feature. [Courtesy of *Instruments and Control Systems*, (J. F. Keithley, "Electrometer Measurements," Jan. 1962).]

very small currents of the order of micromicroamperes (or pA, 10^{-12} A) by this method with relative ease, compared with methods involving delicate current galvanometers. However, in every case where this method is employed, the amount of known resistance that may be tolerated in series with the current-measuring circuit becomes a special factor that must be considered.

In the many instances where the current is being measured in a circuit of inherently high resistance, such as would be the case for grid currents or ionization currents, the added series resistance presents no problems. Many electronic voltmeters are, accordingly, designated as electronic microvolt-micro-ammeters, and contain internal resistance standards used as shunts that are switched in for the particular current range desired. For example, in the *Kintel model 203 dc microvolt-ammeter*, described previously, when the current range selector is turned to the most sensitive 0–100-$\mu\mu$A (10^{-10} A) range, the current flows through an internal resistance of 10 MΩ (10^7 Ω) and produced a 1-mV drop, which deflects the meter to full scale. This allows the instrument to cover ten current ranges, up to 100 mA full-scale, at which time the internal resistance has been reduced to 1 Ω producing a 100-mV drop for full-scale deflection. Instruments having a *greater voltage sensitivity than the* 100 μV of this example allow even smaller currents to be measured and are usually designated as *electronic galvanometers*.

7-14. GENERAL SUMMARY OF ELECTRONIC-VOLTMETER MEASUREMENT RANGES

The foregoing discussion of the measurement capabilities of the various types of electronic voltmeters indicates that an extremely wide field can be covered by them in measuring dc volts, ac volts, dc current, and resistance. Since the total measurement capability of any particular VTVM depends primarily on the voltage ranges it is able to cover, a simplified review of the most sensitive voltage range provided by each major type of VTVM, accompanied by the impedance and accuracy at which such a range can be measured, would serve as a helpful summary of the VTVM field. Such a comparison is given in Table 7-1, where the electronic instruments are divided into three major types:

1. The general-purpose (or service-type) ac/dc VTVM
2. The high-sensitivity (laboratory-type) ac VTVM[4]
3. The high-sensitivity (laboratory-type) dc VTVM[4]

7-15. SUMMARIZED COMPARISON OF ADVANCED ELECTRONIC VOLTMETERS AND PICOAMMETERS

Although a degree of standardization has been attained in the conventional type of VTVM, such uniformity cannot be expected in the case of the more highly refined laboratory instruments because, by the nature of the delicate applications involved, there are many factors that might be considered negligible in one measurement situation but quite significant in another. However, the following brief comparisons[5] can prove helpful in summarizing the special capabilities of the main types of *highly refined dc electronic voltmeters* with their accompanying sensitive *picoammeter* (or *micromicroammeter*) *ranges*.

[4]Digital instruments are covered in Chapter 17.

[5]Adapted from an excellent discussion, "Electrometer Measurements," by J. F. Keithley, ***Instruments and Control Systems***, Jan. 1962.

TABLE 7-1. Comparison of Voltage Measurements by Electronic Voltmeter Instruments

Specification		*General-Purpose (Service-Type) VTVM**	*Laboratory-Type VTVM**	
			High-Sensitivity ac [Sec. 7-8]	*High-Sensitivity dc [Sec. 7-8ff.]*
Direct current	Most sensitive dc volts range	0–1.0 V	—	0–0.03 V (30 mV) [Sec. 7-10] 0–0.00001 V (10 μV)
	Input registance (for dc volts)	10 MΩ	—	100 MMΩ or TΩ (10^{14}) [Sec. 7-10] 10 MΩ (10^7 Ω) [Sec. 7-12]
	Accuracy (% of full scale)	Around 3%		±2% [Sec. 7-10] ±3% [Sec. 7-12]
Alternating current	Most sensitive ac volts range	0–1.0 V	0–0.001 V (1 mV)	
	Input impedance (for ac volts)	Around 1 MΩ†	10 MΩ shunted by 25 $\mu\mu$F†	
	Frequency range	Without probe to 1 MHz With probe to over 100 MHz	10 Hz–4 MHz	
	Accuracy	Around 5%	±2%	

*VTVM abbreviation includes both vacuum-tube and transistor voltmeters.
†At range given, and at mid/audio frequency.

1. *Very-High-Impedance Group*
   (a) *The dc electrometer VTVM* [Fig. 7-16(a)]
   Main advantage: *Input impedance* of 10^{14} Ω is 10 million times as great as that of the conventional dc VTVM (10^7 Ω).
   Main disadvantage: *Zero drift* of the order of 1 mV/hr.
   (b) *Vibrating-reed-modulator electrometer* [Fig. 7-17(b)]
   Main advantages: *Very high impedance* and *reduced zero drift* (around 10 μV/hr).
   Main disadvantage: *Equipment complexity* with correspondingly increased cost.
2. *Very-High-Sensivity Group*
   (a) *Chopper-modulator amplifiers* (Fig. 7-17)
   Main advantages: *Full-scale sensitivities* of 1 μV or less and *reduced zero drift.*
   Main disadvantages: *Sacrifice of very high impedance* and presence of *noise voltages,* whether the chopping is done mechanically, as in Fig. 7-17(a), or by photoconductive (or semiconductive) means, as in Fig. 7-17(b).

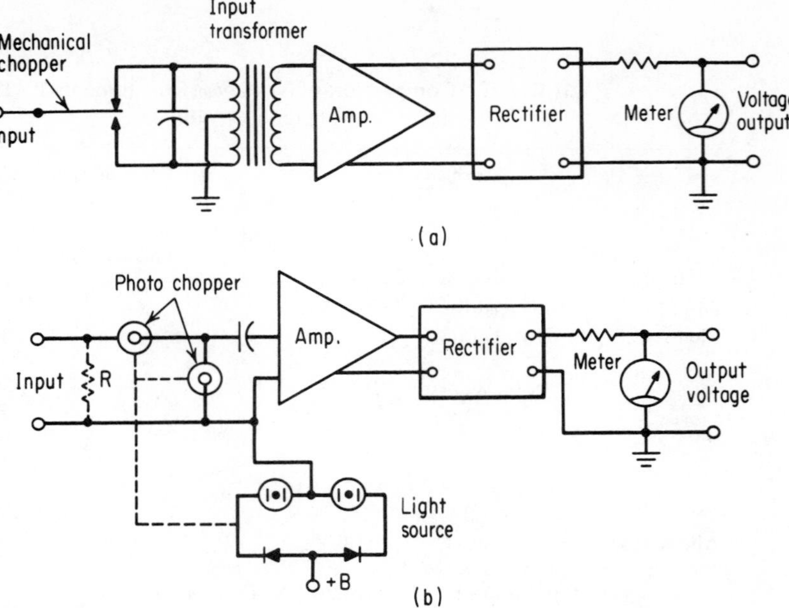

Figure 7-17. High-sensitivity chopper-modulated dc electronic voltmeter's basic circuits: (a) mechanical and (b) photo-chopper modulation systems achieve higher sensitivities with less zero drift than Fig. 7-18(a), at some sacrifice of very high imput impedance and noise voltages. [Courtesy of *Instruments and Control Systems*, (J. F. Keithley, "Electrometer Measurements," Jan. 1962).]

3. *Fast-Response Group* (Fig. 7-18)

(a) *Operational-type amplifiers*

Main advantage: The use of *negative feedback*, as shown, greatly *reduces the response time*, whether caused by large input *R*, or input *C*, or a combination of the two.

Main disadvantage: *Reduced sensitivity*, unless compensated by correspondingly larger open-loop gains than normally required.

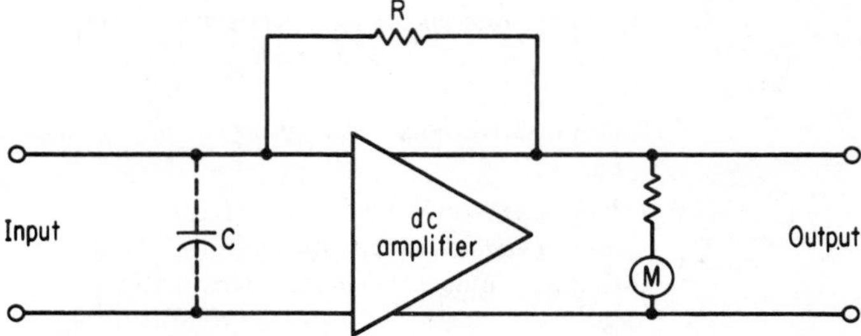

Figure 7-18. Operational amplifier type, employing heavy negative feedback to achieve fast response, where large input *C* would otherwise produce a large time constant.

QUESTIONS

Q7-1. (a) State two main advantages in using an electronic voltmeter rather than a nonelectronic moving-coil voltmeter (or part of VOM), for a given dc voltage measurement.
(b) State two disadvantages (one in terms of convenience of measurement).

Q7-2. State a clear-cut case (giving reasons) where it is preferable to make a dc voltage measured by:
(a) A nonelectronic voltmeter (part of VOM).
(b) An electronic voltmeter.

Q7-3. Explain why a random deflection of the indicator meter is obtained when a VTVM is used on its most sensitive dc range with no input signal, but the probe is not short-circuited.

Q7-4. (a) Explain the gradual change in the zero reading of a VTVM on its dc range.
(b) What precautions must be taken to avoid errors in reading resulting from this condition?

Q7-5. Explain why the resistance scale on a VTVM has increasing values of resistance to the right, as opposed to the scale of the nonelectronic ohmmeter (or VOM), where the higher resistance values increase to the left.

Q7-6. In using a VTVM, it is ordinarily not necessary to use an external blocking capacitor when measuring an ac voltage at the plate of an amplifier tube, while this precaution is almost always necessary when using the ac range of a volt–ohm–milliammeter (VOM). Explain.

Q7-7. What interpretation regarding the correctness of reading must be made when measuring the voltage of a sawtooth wave-form on:
(a) The ac (rms) scale of a VTVM?
(b) The ac (rms) scale of a VOM?
(c) The p–p scale of a VTVM?

Q7-8. Explain why separate ac scales are generally provided for low ranges (such as 0–1.5 and 0–3 V ac) in the VTVM.

Q7-9. State under what conditions it is preferable to use the p–p scale (rather than the rms scale) in using the ac volts range.

Q7-10. State under what conditions the decibel scale of a VTVM would be useful.

Q7-11. (a) State two important advantages (other than compactness) that favor the use of a transistor voltmeter (TVM) rather than a VTVM.
(b) State two disadvantages arising from the characteristic circuit action of the TVM.

Q7-12. Discuss the possibility of modifying a general-purpose VTVM to produce a substantially more sensitive range than 0–1.5 V dc usually provided on this service type of VTVM.

Q7-13. Distinguish electronic voltmeters of the rectifier/amplifier type from those of the amplifier/rectifier type, as to:
(a) Circuit action.
(b) Type of measurements made.

Q7-14. (a) What are two main advantages of the logarithmic-reading type of electronic voltmeter?
(b) What is a major disadvantage?

Q7-15. State why drift is a difficulty inherent in dc instrument amplifiers and not in ac instrument amplifiers.

Q7-16. Compare the electrometer type of electronic voltmeter with the general-purpose type, as to:
(a) Main advantage.
(b) Main disadvantage.

Q7-17. Distinguish the chopper type of dc VTVM from the chopper-stabilized type.

PROBLEMS

P7-1. An elementary from of VTVM is shown in Fig. P7-1(a).
(a) Draw the transfer curve I_b vs. E_c for the tube circuit shown in this figure, between the points $E_c = -5$ (Q point) and $E_c = 0$.
(b) What is the meter reading for the quiescent plate current I_{b0} at the Q point $E_c = -5$ when the signal voltage (E_s) is zero?

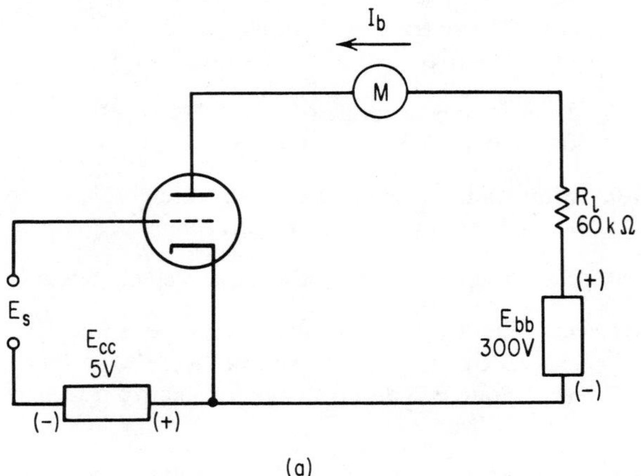

Figure P7-1. (a) Circuit; (b) tube curves.

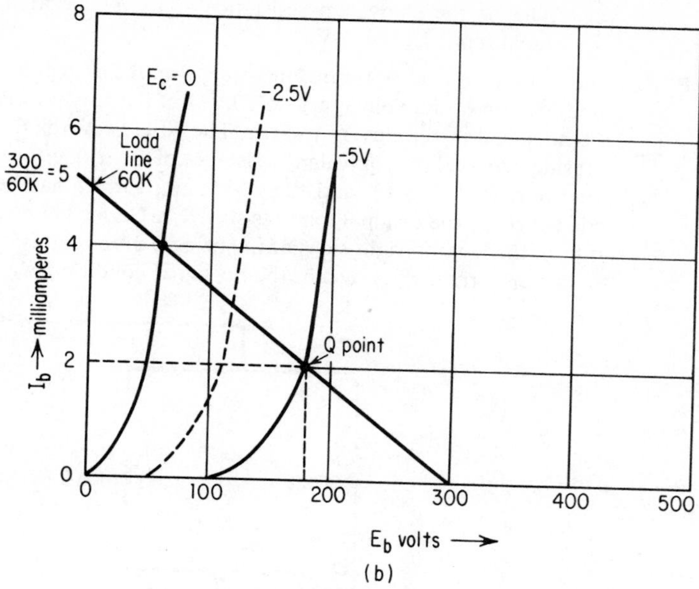

Figure P7-1. *Continued.*

P7-2. The circuit and curves of Problem P7-1 result in the transfer curve shown in Fig. P7-2.

(a) Find the equivalent resistance of the triode at the quiescent point Q.

(b) Find the plate-current change as the signal voltage E_s is increased from 0 up to +5 V (corresponding to change in E_c from −5 V to $E_c = 0$ V, respectively).

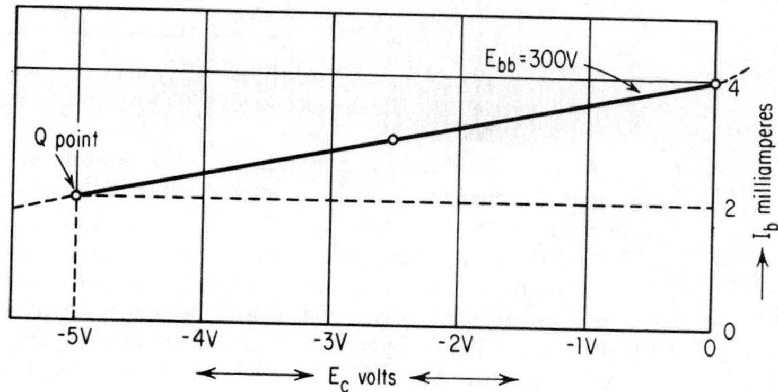

Figure P7-2. Transfer curves.

(c) What is the equivalent resistance of the triode at the condition when the signal input E_s is $+5$ V?

P7-3. The resistance-bridge arrangement in Fig. P7-3 represents a current-bucking modification of the elementary dc VTVM of the circuit of Problem P7-1, to reduce the no-signal meter current to zero. The 90-kΩ resistor R_1 has been added to the left side (to equal the equivalent resistance of the triode R_{tr} at the original quiescent condition $E_c = -5$ V), and the 60-kΩ resistor R_2 has been added to the right side (to equal the original load resistor R_l). Under these conditions of 1 : 1 ratios in both the left and right sides of the bridge, the circuit is obviously balanced, and the meter current I_m is zero at the no-signal condition $E_s = 0$.

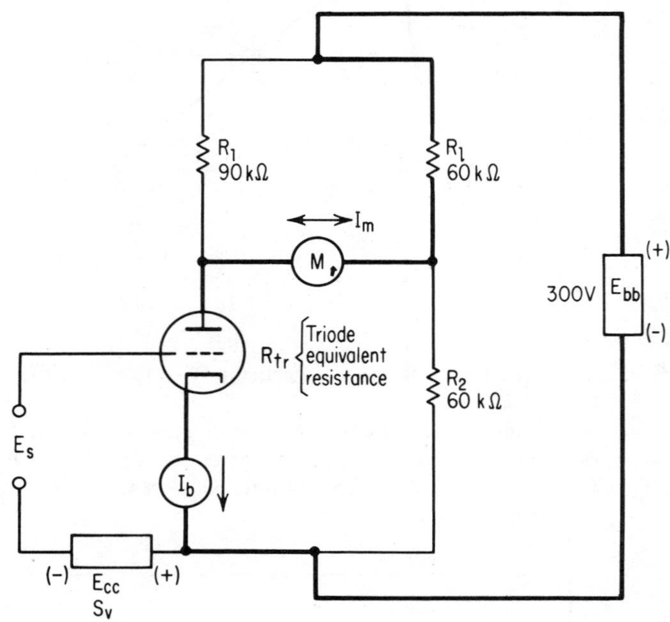

Figure P7-3. Resistance-bridge VTVM arrangement.

Assume that when a signal $E_s = +5$ V is applied, the equivalent triode resistance R_{tr} changes from its quiescent value (90 kΩ, when $E_c = -5$) to the lower value of 15 kΩ (when $E_c = 0$), as in Problem P7-2.

(a) Find the meter current I_m corresponding to $E_s = +5$ V and $R_{tr} = 15$ kΩ, calling the resistance of meter M negligible. (*Suggestion:* Find the Thévenin equivalent of the circuit dashed at the meter terminals.)

(b) Based on linear operation, draw a calibration for the resistance-bridge arrangement of this VTVM, plotting meter current I_m vs. signal input E_s from 0 to $+5$ V signal input.

P7-4. Check the results of Problem P7-3 by means of the general relation

$$I_m = A - BI_b$$

Suggestion: The coefficients are determined, according to the text, as follows. *Coefficient A* is determined at cutoff $I_b = 0$ as equal to the meter current I_m flowing at that time and should be found to be $\frac{5}{4}$ mA. *Coefficient B* is determined at balance condition $I_m = 0$ and should be found to be $\frac{3}{4}$ mA. Therefore, the resulting equation will be

$$I_m = \tfrac{1}{4}(5 - 3I_b)$$

Using this equation, check the meter current I_m by first finding the meter current I_b for the following conditions known from Problem P7-3.

(a) When $E_s = 0$ (making quiescent $E_c = -5$, at which time R_{tr} is 90 kΩ); hence, I_b should be found to be $\frac{5}{3}$ mA for this input.

(b) When $E_s = +5$ (making $E_c = 0$, at which time R_{tr} is 15 kΩ); hence, I_b should be found to be 5 mA for this input.

P7-5. In the circuit in Fig. P7-5, a volt–ohm–milliammeter (VOM) having an input resistance of 20 kΩ/V and a VTVM having an input resistance of 11 MΩ (on all ranges) are each used separately on a range of 0–3 V dc, to measure the output voltage E_{out}. Compare the percentage error in voltage reading caused by the loading of each instrument.

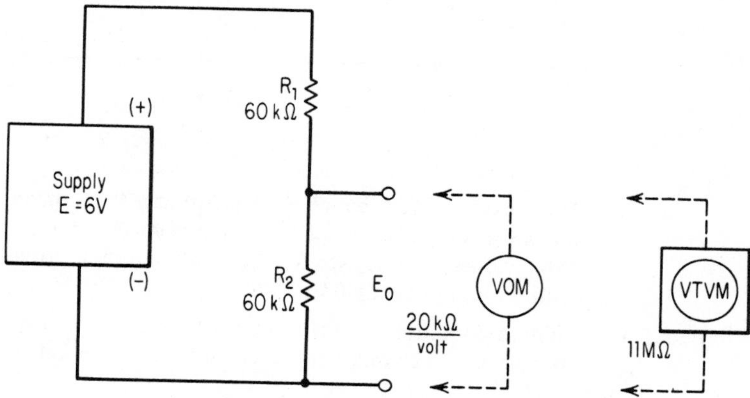

Figure P7-5

P7-6. If Problem P7-5 is changed from $E = 6$ V to 600 V, and each meter is used separately on its 0–300-V dc range, compare the percentage error in the voltage reading caused by the loading of each instrument in this case.

P7-7. In the dual-triode bridge circuit of Fig. P7-7 for the values shown, a dc input of +1.5 V at the active grid causes full-scale deflection of indicating meter *M*. (Assume the current taken by the combination of meter *M* and the 36-kΩ calibrating resistor to be negligible.) Find the dc input required for full-scale reading:

(a) If a combination of a 200-μA meter in series with a calibrating resistor of 18-kΩ is used instead of the combination shown.

(b) If the 500-μA meter and 36-kΩ calibrating resistor remain the same, but resistors R_{l_1} and R_{l_2} are each changed to 25 kΩ. (See the text for an expression for the voltage gain of the cathode-coupled amplifier.)

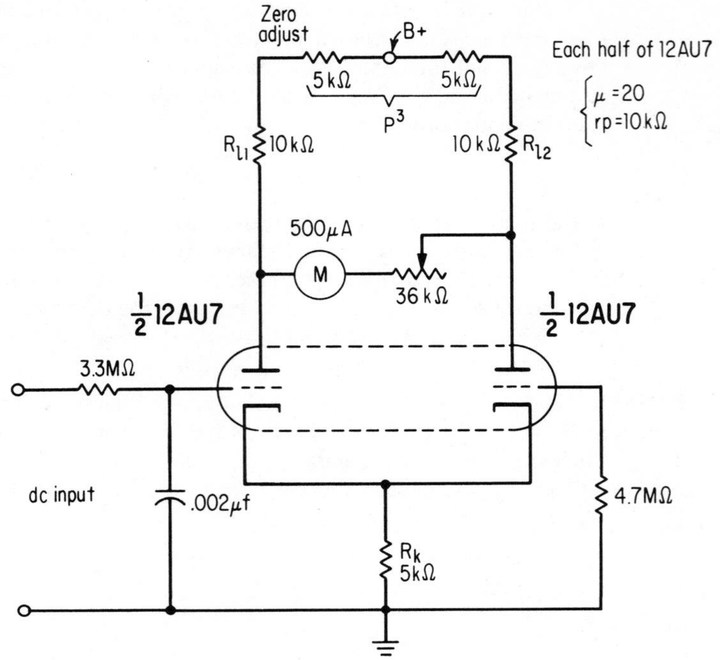

Figure P7-7

(c) If the load resistor change to 25 kΩ in part (b) is made at the same time that a change is made in the combination of meter M and its calibrating resistor. What values must be used for the combination to provide full-scale deflection with an input of only 0.06 V (60 mV)?

P7-8. If the circuit conditions of Problem P7-7 remain as shown in the figure, but the 5-kΩ resistor R_k is changed to 25 kΩ, what is the effect on the original 0–1.5-V dc range?

8

Recording Systems

8-1. FEATURES OF ELECTRONIC GRAPHIC RECORDING SYSTEMS

Recorders generally provide a *graphic record* of the variations in the quantity being measured, as well as a large, easily visible scale on which the *indication* is displayed. Many recording instruments include an additional provision for some sort of controlling action, where this extra function is desired. If the control function is the primary one, the measuring instrument is called a "controller." From the standpoint of electronic measurement, however, the control action can be considered optional. Hence, the term *indicating/recording instrument* in this text will be used to cover the general class of recording (or "direct-writing") instruments, regardless of whether or not a control function is included.

Electronic recording instruments are of two main types: the *null type*, operating on a comparison basis (in both the potentiometer and bridge forms) and the *galvanometer type*, operating on the deflection principle. The null type will be discussed first, including both the elementary form, where the balance is achieved manually, and the more common commercial form, which features self-balancing action.

Manual and Self-Balancing Null Systems

Although the null type of measurement method provides greater accuracy and superior features in general, it was mainly confined, in the past, to the precision laboratory. The prime reason for this was the careful manipulation required by the operator to obtain a balance as each measurement was made. When the means became available for obtaining this balance condition automatically—at first by mechanical methods—the null instrument emerged from the laboratory into limited industrial use. The subsequent replacement of the inherently slow *mechanical* method of balancing by an *electronic* method, using electronic amplifiers, provided a fast and smooth response in the self-balancing action and resulted in wide commercial acceptance of the instrument.

Self-balancing potentiometer types of recorders (shown in Fig. 8-1) are now in almost universal use in industry, as well as in laboratory and research work, and provide practical measurement of a host of physical variables. The servomotor, which is actuated by the electronic amplifiers of modern recorders, now provides a straightforward means for adding the functions of *recording* and *control* to the original *indicating* action. All these features combine to make the electronic self-nulling recorder a highly acceptable instrument, well suited for both sensitive laboratory use and rugged industrial instrumentation.

The other main group of graphic recording instruments is the so-called "direct-writing oscillograph" or galvanometer system, which does not operate on a balance system but depends instead on the actuation of a comparatively high-torque galvanometer or "pen motor" to produce a written record. This type is discussed separately in a later section.

Self-Balancing Action by Servomotor

Returning to the null type of recorder, the self-balancing feature is obtained by servomotor action. Figure 8-1(a) (*Honeywell Electronik Recorder*) shows the external appearance of the circular-chart form of this self-balancing recorder. The arrangement for obtaining the required torque through servomotor action is illustrated in the internal views in Fig. 8-1(b) and (c). The interior view in Fig. 8-1(b) with the door open and chart plate removed shows the position of the *balancing motor* (mounted with its axis horizontal), which drives the helical *slide-wire*, shown in the upper right-hand corner, with its axis vertical. The interior view with the door swung open [Fig. 8-1(c)] shows the balancing-motor gear drives the taut stainless-steel cable on the rear of the door. The rear of the cabinet contains the *Electronik Continuous Balance* amplifier unit and terminal block connections.

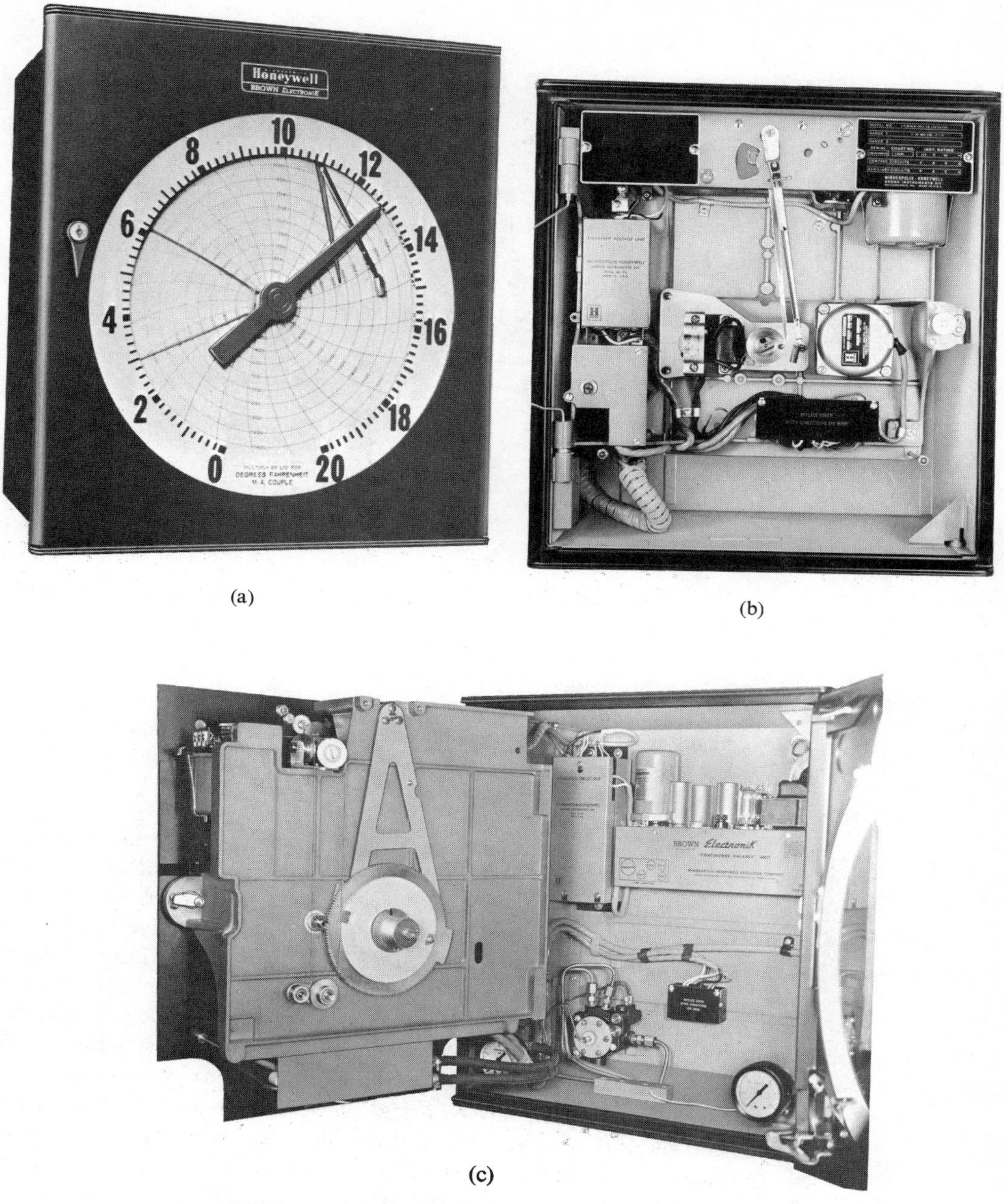

(a)

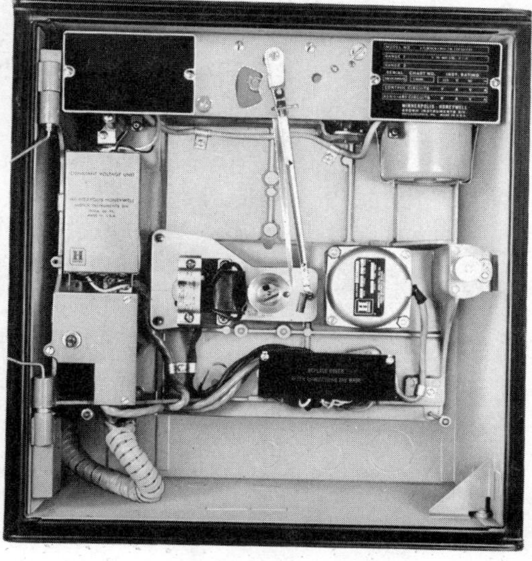

(b)

(c)

Figure 8-1. Circular-chart self-balance recorder: (a) external view; (b) view with door open and chart plate removed, showing enclosed slide-wire at upper right and balancing motor below it; (c) view of inside of case and back of chassis, showing amplifier unit at upper right. (*Honeywell, Brown Electronik, class 15.*)

8-2. TYPES OF TRANSDUCER INPUTS

As was explained in the discussion of null-type instruments in Chapter 4, both the manual and the self-balancing instruments start with the same basic comparison balance circuit, repeated for convenience in Fig. 8-2, where the unknown quantity, in the form of an electrical input, is balanced by a known voltage. In this illustration a heated *thermocouple* is the *transducer* used to derive an electrical signal from the thermal quantity being measured; in this case, the electrical signal is a dc voltage proportional to *temperature*.

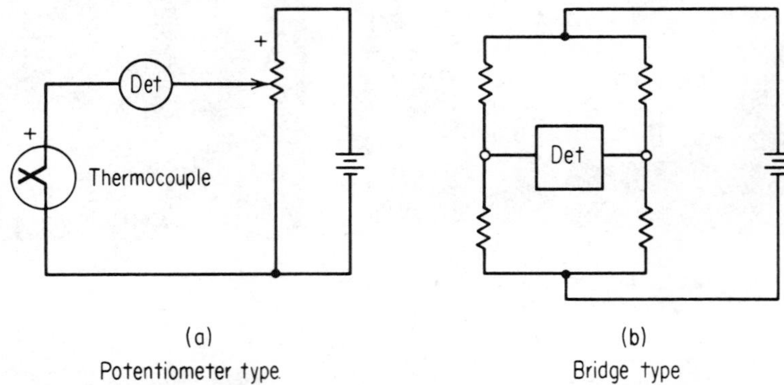

Figure 8-2. Similarity between null-type circuits: (a) potentiometer type; (b) bridge type. In both types of null circuits, zero current is provided in detector D at balance.

This, of course, is but one instance of the many physical variables capable of being monitored in a similar fashion, and the resulting electrical signal can be ac as well as dc. We shall take note of a few types of transducers at this point (before the detailed discussion in Chapter 9). Temperature-sensitive resistors (or thermistors) can be arranged in the form of a bridge, as shown in Fig. 8-3. A rise in temperature in the thermistor arm of the bridge is accompanied by a corresponding change in thermistor resistance, which causes the bridge to be unbalanced. The bridge unbalance, in turn, produces an electrical signal corresponding to the temperature being measured. In this case, the electrical signal serving as the recorder input derives not from any *electrical generating* (*or thermovoltaic*) *action* as in the thermocouple, but rather from a *change in resistance* (*or thermoconductive action*). The conductivity change allows the bridge supply voltage to become active at unbalance and thus produce the resulting voltage signal. Hence, the output of the thermistor bridge type of transducer can be chosen to be dc or ac, depending upon what kind of supply voltage is used for the bridge.

In similar fashion, either dc or ac output can be obtained from *strain gages* arranged for *force or pressure* measurements. Other transducers are inherently ac output devices, such as *variable-inductance transducers for detecting small dis-*

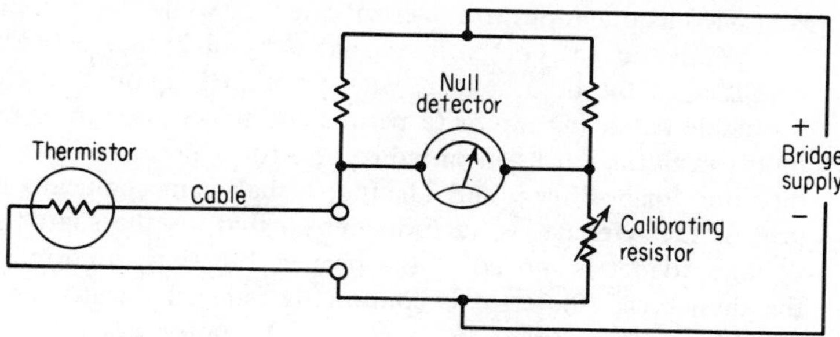

Figure 8-3. Thermistor transducer, used in bridge circuits for temperature measurement, produces a change in electrical resistance, corresponding to a change in temperature.

placements. This type is characterized by the production of an ac output when an initial electrical balance is disturbed by the slight displacement of an iron core. The initial balance in the case of a linear-variable differential transformer (LVDT) is obtained from two equal and opposite induced voltages and results in an original balance condition, even though the familiar type of bridge circuit is not present. In any of the above transducer examples that provide an ac output, this ac voltage can be rectified, whenever desired, to serve as a dc input signal to the recorder. Accordingly, a *dc input signal* will be used for a general example.

8-3. PRINCIPLE OF SELF-BALANCE OPERATION (ALL-DIRECT CURRENT SYSTEM)

A basic method for achieving an automatic, rather than a manual, balance in the simplified dc comparison circuit of Fig. 8-2 can be shown in block-diagram form in Fig. 8-4. The dc signal voltage obtained from the quantity being measured

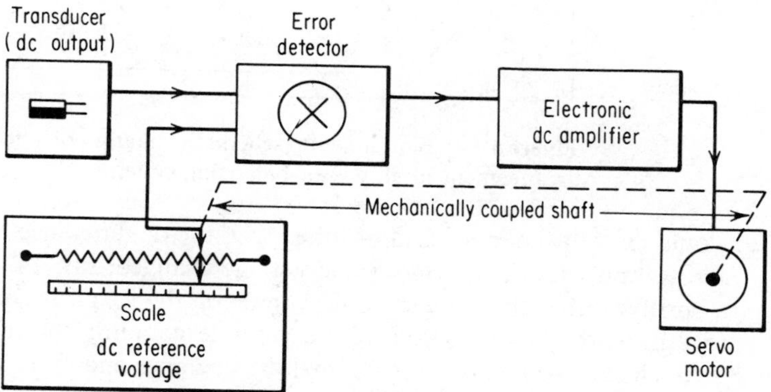

Figure 8-4. Block diagram of all-dc self-balancing system.

is applied as one input to the error detector, while the reference voltage opposing it is applied as the comparing signal. Any difference between the two signals is amplified in the dc amplifier, whose output is applied to the armature of a dc reversible balancing motor (a permanent-magnet dc motor in this example). The motor armature turns in one direction for a negative input and in the opposite direction for positive input. The motor shaft is mechanically linked to the moving arm of the reference-voltage divider so that, as the shaft turns, the arm of the voltage divider is moved in the proper direction to correct the error. Thus, as the shaft turns, the error is continually reduced toward zero, at which point the input to the armature is also zero, and the motor stops.

8-4. ANALYSIS OF ELECTRONIC DIRECT-CURRENT SELF-BALANCE CIRCUIT

In the schematic diagram of Fig. 8-5, the operation of each of the blocks is shown for this electronic dc self-balance circuit. The electrical signal from the thermocouple transducer is opposed by the reference signal from the potentiometer voltage-divider circuit. In the original balance condition, a no-signal condition

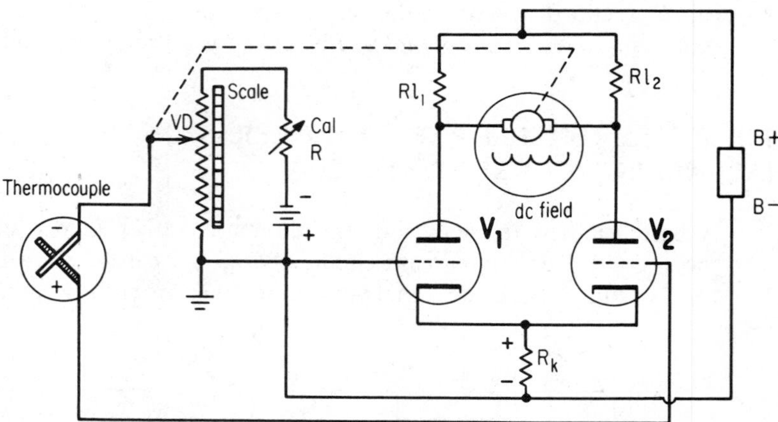

Figure 8-5. Fundamental schematic diagram of a dc amplifier arrangement in all-dc self-balancing system.

would exist at the input grid of tube V_2 of the dc difference amplifier, resulting in zero output. As the thermocouple voltage changed, say, to an increased voltage, a positive difference voltage would appear on the V_2 grid, causing its plate voltage to fall. Through the action of the cathode coupling, more current would flow through the cathode resistor R_k, making its upper end more positive and its lower or grid end more negative. This negative-going voltage appearing at the V_1 grid would result in less plate current through the tube V_1, causing its plate voltage to rise. Since the V_1 plate voltage has gone up while the V_2 plate voltage has gone down, a corresponding value of current is caused to flow through the armature of the reversible dc balancing motor. The motor then turns in such a direction as to position the moving arm of the voltage divider (which is mechanically linked to it) to a higher voltage, thus counteracitng the error that had originally caused the motor action. When the arm has moved to the position that makes the reference voltage equal to the new input voltage, the error signal becomes zero, and the no-signal condition produced as a result of the zero difference causes the motor to stop.

The new value of the input voltage from the thermocouple can be read directly on the calibrated scale of voltage divider *VD* each time a balance point is reached for any new input. Of course, if the change in thermocouple input voltage should be in a negative direction instead of positive, the motor would turn in the opposite direction and achieve a new balance in a similar manner.

8-5. COMPARING DIRECT-CURRENT AND ALTERNATING-CURRENT INPUT SIGNALS

The scheme illustrated in the elementary dc system of Sec. 8-4 has a serious weakness: it is the difficulty in the dc amplification of small signals, such as are obtained from many otherwise satisfactory transducers. The inherent tendency of any dc amplifier to change its output as the dc operating conditions change—even though the input signal remains constant—constitutes the well-known *drift* problem in the best of dc amplifiers. This drift tendency causes difficulty even in the balanced type of dc amplifier shown in Fig. 8-5; it becomes an increasingly serious problem as one attempts to measure smaller and smaller dc inputs, since the dc amplifier produces a changed output because of its drift, exactly as if the dc input signal of comparable magnitude had changed. This situation imposes a severe practical limit on the sensitivity of a dc amplifier—a limit that is not present in a corresponding ac amplifier.

On the other hand, when an unrectified ac signal is used as the input voltage, even though the annoying drift problem is avoided, a new difficulty arises in the measurement of the signal from the transducer, because of the vulnerability of a high-gain ac amplifier to 60-Hz hum and other stray ac pickup voltages. Sometimes the choice between ac or dc for the input system is dictated by the type of output produced from the most suitable transducer available, and in that case, if the output is large, the matter of overcoming the circuit difficulty becomes secondary. However, when either type of transducer is suitable, the preference generally is for the use of the dc system in high-gain circuits to avoid the pickup difficulty; the drift problem is then tackled by means of chopping techniques, as described below.

8-6. CHOPPER-TYPE INPUT SYSTEMS

The method widely used for overcoming the drift problem in a dc amplifier is to interrupt the dc signal from the transducer by various forms of chopping action. The interruption produces a rectangular-wave ac signal, which can be fed into an ac amplifier. The chopping action may be performed in a variety of ways, such as by mechanical action, by transistor switching, or by optical means, all of which are generally referred to simply as *choppers* (or less frequently as modulators or converters). Whichever chopping method is used, some means must be provided to preserve the original dc polarity sense of the transducer signal so that, as in the previous thermocouple example, for instance, the motor will turn in one direction as the difference or error signal becomes positive and in the other direction as it becomes negative. In spite of these added requirements for modulation of the input and some method of phase-sensitive demodulation of the output, the chopper-type system produces overall results that are much superior to all-dc or straight ac systems, with respect to a better combination of sensitivity and stability. It is therefore employed extensively in the many cases where the transducer output is a low-level signal (whether dc or ac).

8-7. REPRESENTATIVE EXAMPLE OF CHOPPER-TYPE SELF-BALANCING SYSTEM

A chopper-type potentiometer system that is representative of the large class of *dc millivolt instruments* is illustrated in the block diagram of Fig. 8-6 for the *Honeywell class 15 Electronik Continuous Balance* potentiometer system. Here, as in the previous elementary dc self-balance system, the dc output voltage from the thermocouple-transducer T/C (at A++) is compared with a portion of a dc reference voltage picked off by the side-wire (at A+), to produce a difference signal between these two points. (The double plus indicates a higher positive polarity than the single plus.) Instead of being used directly as a dc input, this difference signal is applied instead to the mechanical chopper-converter, which generates a square-wave ac signal, having a phase-polarity that corresponds to the dc polarity of the error (in a manner later described). The ac signal is then amplified by a multistage ac amplifier consisting of voltage and power amplification stages. (In these stages, there is no difficulty with the drift problem encountered in the elementary dc amplifier.) The ac output of the power amplifier is applied to the balancing motor, which is a two-phase servomotor that responds by producing a different direction of rotation for each of the two possible phase senses of the chopped ac signal (analyzed below). This motor rotation, as before, drives the slide-wire arm—through the mechanical linkage—in the proper direction to correct the error and so produces a new null balance, which is indicated on the calibrated scale of the instrument.

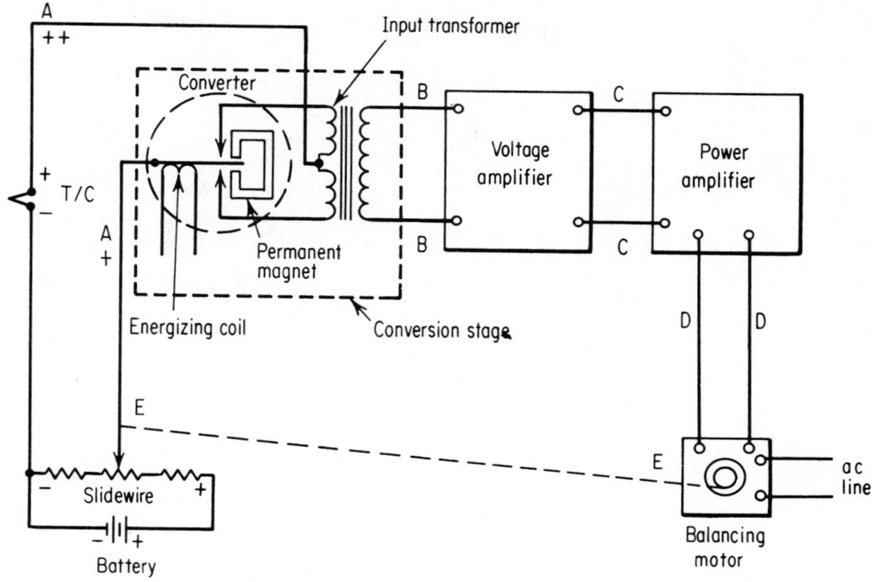

Figure 8-6. Functional block diagram of a chopper-type dc millivolt recording instrument, potentiometer type. (*Honeywell class 15 Electronik Recorder.*)

Chopper-Converter Action

The action of the chopper in converting the dc difference signal into an ac signal having two possible phase senses (in-phase and 180° out-of-phase) is shown in Fig. 8-7. In Fig. 8-7(a), the thermocouple voltage is assumed to be *higher* than the reference (owing to a temperature *increase*) and is accordingly labeled double positive at the center tap of the input transformer, with respect to the slide-wire voltage, which is labeled single positive and is applied to the vibrating reed. This reed is caused to vibrate between two fixed contacts by the 115-V, 60-Hz excitation voltage, whose relative phase is indicated at the top of the diagram. The dc current flow in the top half of the input transformer is indicated by the arrow pointing upward for the first half cycle, during which the reed rests on the upper contact. During the second half cycle the reed rests on the lower contact, producing dc current flow in the lower half of the input transformer, as shown by the arrow pointing downward. The resulting voltage induced in the secondary of the transformer over a full cycle is plotted in the wave-form shown below the diagram, and is seen to consist of an essentially square-wave ac signal, having an in-phase polarity with respect to the exciting voltage used as a timing or reference wave.

In Fig. 8-7(b) are depicted the conditions corresponding to a thermocouple

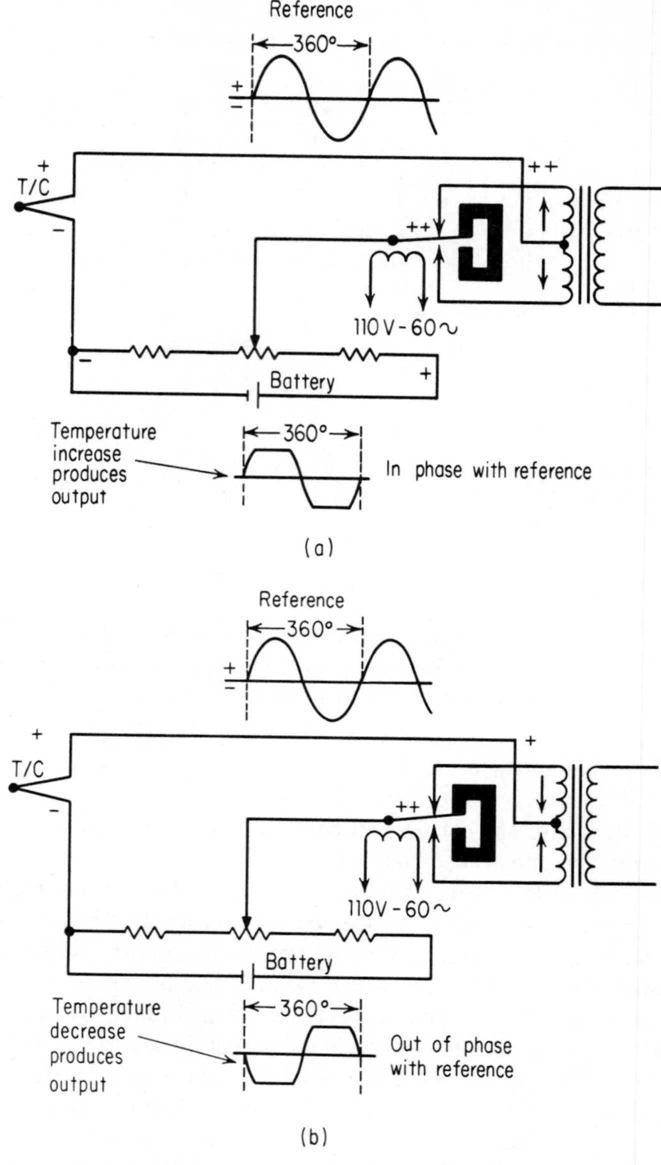

Figure 8-7. Chopper action produces a phase-sensitive ac output, corresponding to the polarity of the dc input: (a) when thermocouple (T/C) output is greater than the reference voltage, producing an ac signal, in phase with the reference voltage; (b) when T/C output is smaller than the reference, producing an out-of-phase output.

output *less* than the reference voltage (owing to a temperature *decrease*). Analyzed in a similar manner, the arrows showing the direction of dc current in the input transformer for each half cycle will be seen to be opposite to the current-direction arrows of Fig. 8-7(a). Accordingly, the wave-form diagram for the secondary output voltage is shown at the bottom of the diagram as 180° out-of-phase with the timing wave that is shown at the top of the diagram.

By the action of the chopper-converter, the dc output developed by the transducer results in an ac output at the transformer secondary, having a relative phase (or a phase polarity) either in or out of phase with reference to the 60-Hz timing wave. The ac output wave has an *in-phase* polarity when the dc voltage *rises* and a 180° out-of-phase polarity when the transducer voltage falls. (Note that there are no significant phase conditions between 0° and 180°; for this reason the expression "phase polarity" is used in preference to the relative phase angle, and this phase polarity changes abruptly from the in-phase to the out-of-phase condition, as the difference voltage reverses in its dc polarity.) The value of the dc difference voltage, whether in one direction or another, determines the height or magnitude of the corresponding ac wave-form. Thus, if we trace a condition where the thermocouple is at a temperature higher than the reference at the start and then progressively cools off to a temperature lower than the reference, we can follow the corresponding ac output as starting with a large value of ac in-phase voltage, gradually becoming smaller in value—passing through zero output—and then gradually building up to a large value of out-of-phase voltage, corresponding to the cooler-than-normal temperature. In this manner the original dc difference or error signal from the transducer is converted by the chopper to an ac signal having a phase polarity that corresponds to the dc polarity of the original difference signal.

Chopper Modulation Methods

With the trend toward replacing mechanical devices by electronic ones, wherever feasible, the mechanical chopper has also been partially superseded by semiconductor devices. It is still widely used in those cases where it is necessary to retain its property of being either completely on (practically zero resistance) or completely off (infinite resistance), in spite of its relatively high contribution of noise. On the other hand, in cases where freedom from noise is the governing factor, semiconductor chopping (with the MOSFET, for example) has taken its place.

In the instance of high-performance amplifiers, an optical-photocell combination is used, as explained in the chapter on high-sensitivity voltmeters (for example, *Hewlett-Packard model 425*) where it effectively isolates input from output circuits. In other cases, the insulated-gate field-effect transistor (IGFET or MOSFET) is widely used, to take advantage of its extremely high-off resistance (over 100 MΩ). In any case, regardless of which type of chopper modulation is used, the amplifier into which it feeds is practically always of the solid-state type.

Adjustable Ranges and Spans

In order to increase the flexibility of general-purpose recorders, provision is made for a control to *change to various ranges*, as well as a control to *vary the zero position* (or zero offset). Two such recorders are illustrated in Fig. 8-8. The *Leeds and Northrup* instrument shown is described as AZAR, meaning adjustable zero, adjustable range. Where the recorder is of the two-pen type, the controls for varying range and zero are repeated for each channel.

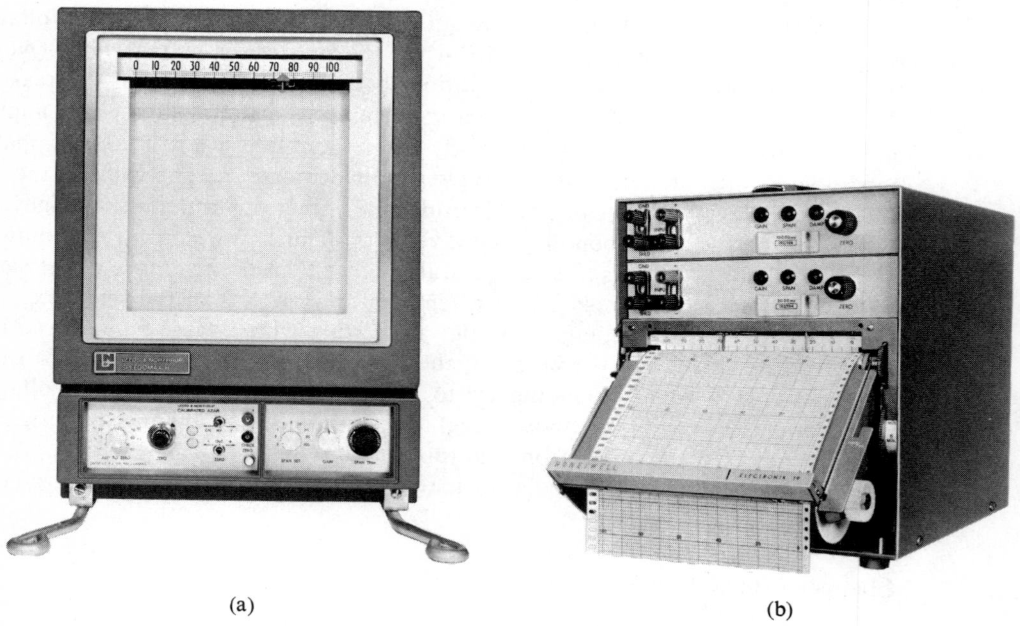

(a) (b)

Figure 8-8. Recorders providing means for varying range and zero suppression (or off set): (a) *Leeds and Northrup AZAR recorder*; (b) *Honeywell dual-pen model Electronik 19.*

Amplifier and Servomotor Action

The square-wave signal produced in the secondary of the input transformer as a result of the chopping action, is applied to an ac voltage amplifier. This type of amplifier is capable of high-gain amplification, practically free from drift, and very stable, with proper shielding precautions taken against excessive hum pickup. The ac signal, at a greatly amplified voltage level, is then applied to a power-amplifier stage, which is arranged to produce a high-current output, suitable for turning the balancing servomotor in either direction, depending on the phase polarity of the incoming signal.

The *servomotor* is of the two-phase type. The ac reference voltage is applied to one of the windings (the *reference phase*), while the amplifier output feeds the other, which becomes the *control winding*. When the signal applied to the control winding is in phase with the reference, the motor turns in one direction, and the motor reverses its direction when the control signal is out of phase with the reference. In this manner, the *phase sensitivity* of the chopped input signal is preserved.

For any given condition, whether in phase or out of phase, the force causing rotation will depend upon the magnitude of the error signal causing it. Since the direction of rotation is such that the arm of the slide-wire moves to a position that continually reduces the error, the magnitude of the error signal continually decreases and becomes zero at balance—at which time, as we have seen, there is ideally no force on the motor and it stops.

In summary, the action of the phase-sensitive power amplifier combined with that of the reversible balancing servomotor results in a *self-balancing action*. As each null is automatically produced, the pointer on the moving arm of the slide-wire indicates the value of the original millivolt dc signal that was generated by the transducer. Thus, although various instruments of this type may have different calibrations on the scale, depending on what kind of transducer is employed in place of the thermocouple used in this instance, the instrument in all cases functions as a *dc millivolt self-balancing potentiometer system*.

8-8. COMPARISON OF ELECTRONIC INDICATOR/RECORDER INSTRUMENTS

The picture that comes to mind most frequently when instrument recorders are being considered is that of a large panel in some central control room—a panel large enough to occupy perhaps one complete wall of the room and covered with panel meters and recording instruments that are continually bringing monitored information into the central control point. This picture, which is realized more and more as instrumentation facilities grow, brings together many varied types of measurement and test systems. Even if we leave out of the picture the simple indicating meters, such as voltmeters and ammeters, and if we also ignore the various manual switches and automatic controllers, we still find ourselves confronted with a host of different recorders. Some write with a *single pen*, many others have *dual pens* to record the simultaneous comparison of, let us say, temperature and pressure, while still others have the capability of recording as many as 36 simultaneous signals. Variations also occur in the form of writing; some write with ink, others may write without ink on papers that are sensitive to pressure or heat. The moving-pen recorder, in general, can be expected to handle variables of the slowly changing type up to a *frequency* in the neighborhood of 100 Hz. (Optical and cathode-ray types of oscillographs can be used to much higher frequencies.)

Of course, all sorts of transducers are involved in feeding the assortment of recorders. A typical setup might, for example, be making simultaneous monitoring records on one chart of wind velocity from a dc tachometer input and wind direction from a position-conscious ac selsyn arrangement. This system could be adapted to handle these different signals as a *basic ac instrument*, where the dc signal is converted to an ac signal is converted to an ac input by chopping or other modulating means. On the other hand, a *basic dc instrument* might be used by rectifying the ac signal to supply an appropriate dc input. Either method can be made to work, with the preference depending to a great extent on the preponderance of a given transducer type. As the system grows, however, and additional instruments are added for monitoring new variables, the panel frequently ends up bearing both types of basic instruments. If carried far enough, it can be seen that a complicated operation might well be the result, requiring many *different calibration and operating procedures for the different instruments.*

8-9. STANDARDIZING RECORDER SYSTEMS

As a means of reducing unnecessary complication, it is often advantageous to concentrate on one basic type of recording instrument and to accept the added requirement of adapting transducer outputs to produce only one type of input to the basic recorder, using preamplifiers as necessary to accommodate various *sensitivity* and *impedance* levels. This often proves to be quite practical and is particularly worth the effort whenever a new system is being planned.

Which type of basic recorder to employ—that is, whether to concentrate on the ac or dc type—is a question that does not lend itself to an arbitrary answer, since so much depends on individual circumstances. (As a special example, the need for a radio link instead of a customary cable connection, involving some form of carrier modulation such as AM/FM transmission, would alter the situation.) It will be instructive, however, to examine the choice of the dc millivolt potentiometer instrument as a standard, bearing in mind that such selection need not apply in all cases, some of which might well favor an ac signal system.

Features of General Direct-Current Millivolt System

The features of an instrument circuit that measures dc millivolts in providing a common basis for a general-purpose recording instrument have been critically summarized as follows.

"Almost all users of control equipment are familiar with the use of millivolts. They have used thermocouples and resistance thermometers and strain gages (for pressure measurement) for years; they understand the equipment; it does not often fail. Even most chemical analyzers generate a millivolt-type dc signal.

"There are many advantages in using millivolt systems in interchangeability of units, purchasing, warehousing, and engineering. For instance, people using potentiometers are familiar with the fact that by changing range cards they can move through any desirable ranges of temperature span; for example, they can choose any 300° span, from 0 to 2000°F. Standard dc potentiometers can read as low as 1.0-mV or 2.0-mV spans. Strain-gage type pickups can be purchased so that their total range is 0 to 50 mV. Hence, by using a standard potentiometer circuit, any part of this 0–50 mV (such as 0–5 mV) can be read. Also, this 0–5-mV signal can be read with a high degree of accuracy. This means that strain-gage transmitters need not have many adjustments; for instance, a 0–300 in. range can be scaled down from 0–50 mV to 0–5 mV to read 0–30 in. of water, with a high degree of accuracy. This flexibility of scaling is pointed out to illustrate that all required ranges can be met with a transmitter of only one span in a given point."[1]

Summarizing the advantages of a *dc millivolt system:*

1. It requires minimum maintenance and a minimum of understanding of multiple electronic systems.
2. It can be battery operated with transistorized systems and thus can easily be made to ride through power failures.
3. It requires a minimum amount of engineering by having one design apply to many situations.
4. By providing flexibility in the use of standard components, it provides a system compatible with the requirements of other data-processing and computer equipment.

Features of Basic Alternating-Current Recorder System

Standardizing on a basic ac recorder system might well offer some benefits similar to the above. Lacking the advantage of easy incorporation of a battery supply for excitation voltages, it could, on the other hand, be designed to take advantage of the flexibility of ac operation and its freedom from drift problems. Obviously, if all the transducer inputs are relatively high-level ac signals, ac amplification is the easiest and least costly method.

In any case, there are undeniable benefits to be gained by *standardizing on a single basic recording system*, whichever one is chosen, as far as is practical in a given instrumentation situation. For general-purpose use, there is a tendency to specify the dc millivolt recorder of the potentiometer type, having a span in the order of 0–10 mV.

8-10. X–Y RECORDERS

The X–Y recorder is a specialized version for recording variables on two axes, in the Cartesian X and Y coordinates. The model illustrated in Fig. 8-9, *Moseley model 135 X–Y Recorder*, can plot the required curve on standard $8\frac{1}{2} \times 11$-in. notebook-size graph paper. It forms an extremely flexible general-purpose instru-

[1]W. F. O'Conner, "A Critical Review of Electronic Control Systems," *Instruments and Control Systems*, **33**, 1960.

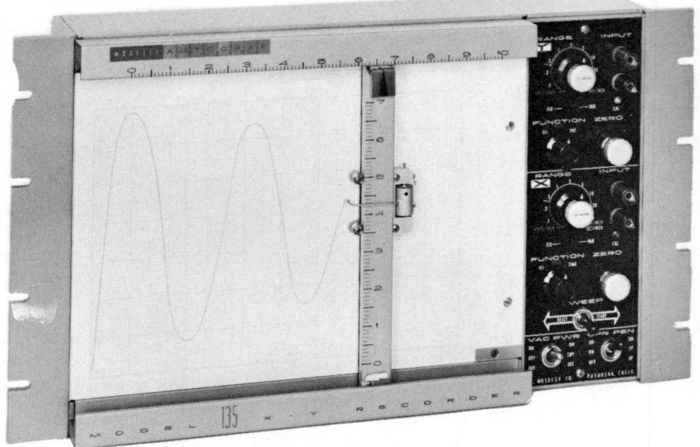

Figure 8-9. X–Y recorder (*Moseley model 135*). The variable on each axis records independently, or if a graph as a function of time is required, the X axis may be used as a time sweep. (Courtesy of Moseley Autograf Instruments.)

ment by providing 16 calibrated voltage ranges, from a minimum of 0.5 mV/in. up to 50 V/in. on either axis. For use of the X axis as a time base, seven calibrated sweeps are provided, from 0.5 to 50 sec/in. Independent servomotors drive each axis, and the two servo systems do not react with each other.

8-11. GALVANOMETER (DIRECT-WRITING OSCILLOGRAPH) RECORDERS

In a galvanometer recording system, the recording pen is actuated directly by a galvanometer unit; accordingly the system is sometimes called a "direct-writing oscillograph) or "graphic galvanometer." The null-balance recorders previously described have been based primarily on a dc millivolt input, with provision in some cases for rectifying the ac output from transducers in order to provide this dc input. The galvanometer-type recorders are well suited for low-frequency ac inputs, obtained from quantities varying slowly at frequencies up to around 100 Hz (and in special cases up to around 1000 Hz). Because of the compact nature of the galvanometer units (or pen motors) this type of recorder is particularly suitable for *multiple-channel operation*. It thus finds extensive use in the *simultaneous recording of a large number of slowly varying transducer outputs*.

A six-channel oscillograph recorder is illustrated in Fig. 8-10 by the *Sanborn 350 series recorder*. This basic assembly has six recording channels that are 5 cm wide (graduated in millimeters). The writing is done by an inkless method on

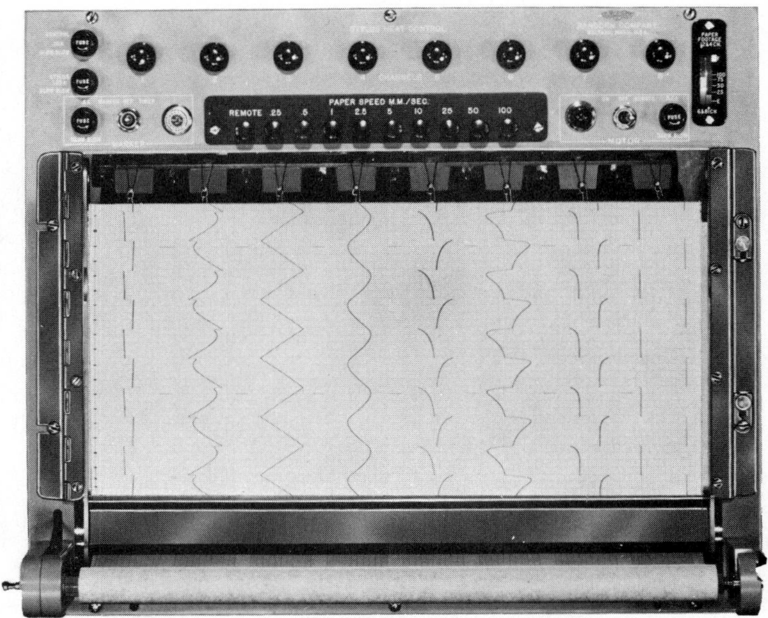

Figure 8-10. Multichannel galvanometer-type recorders: (a) *Brush recorder Mark II* provides two analog channels, plus event or time marker for each channel; (b) *Sanborn six-channel 350 series oscillograph recorder* provides six analog channels and timer-marker records for each side of the chart; the eight stylus heat controls for each pen are seen across the top.

thermosensitive plastic-coated paper, and each pen has its individual stylus heat control. In addition, time-marker records are provided by a pen at each side of the chart, making eight pens in all. To make a complete recording system, individual power amplifiers for each of the six channels and power supplies must be added, usually in a cabinet having a rack-and-panel arrangement. The recorder-power-amplifier assembly (without any preamplification) has 0.1-V division sensitivity. For each 50-division channel, a 5-V signal is required for full deflection. A separate winding on each galvanometer provides negative voltage feedback for velocity damping.

The essential principle of the direct-writing records is quite straightforward; the input to each channel is separately amplified and applied to its separate galvanometer unit. The pen attached to each galvanometer then traces its own curve on a chart, generally made wide enough to accommodate many channels.

A good example of the usefulness of such an arrangement is the direct-writing galvanometer recorder system that tells at a glance the various conditions telemetered from space experiments. Here, the slowly varying inputs from the telemetry receivers are applied to as many galvanometers as are needed to give a simultaneous tracing of temperature, pressure, acceleration, biological, radiological, and other conditions encountered.

8-12. INK AND INKLESS RECORDING METHODS

There are many writing methods used for producing the recorder tracing, of which the most common is the ink-pen combination, where the ink is fed to the pen by capillary action. Inkless methods are also widely used, especially for unattended recording over prolonged periods. In such cases, the trace may be variously produced by heat-sensitive, pressure-sensitive, light-sensitive, or electrical methods. The heat-sensitive method used in the *Sanborn* illustration is shown in Fig. 8-11. Each stylus tip is heated by an electrical current flowing through the

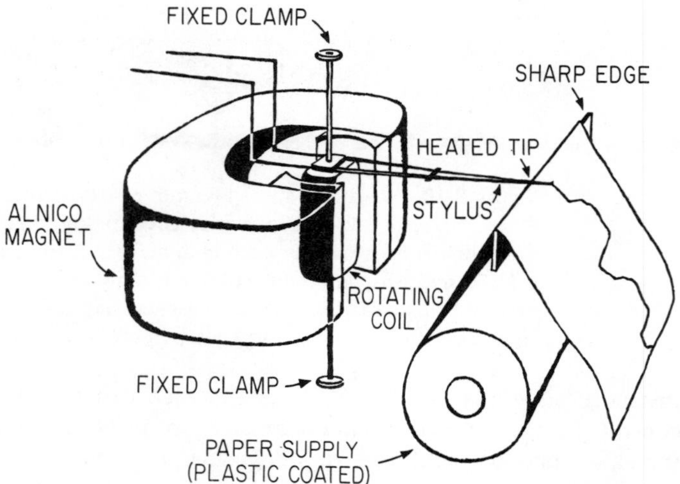

Figure 8-11. Inkless recording, obtained by the action of the electrically heated stylus tip on the plastic-coated chart paper. (*Sanborn 350 series recording system.*)

stylus. The chart paper is coated with wax (or its plastic equivalent) and is fed from the chart roll over a sharp edge, providing a good point contact with the heated stylus tip. As the chart moves forward, the action of the heated tip on the plastic coating produces a clear, thin line for the recorder tracing.

Another heat method uses the Peltier effect to heat only the tip of the stylus "*Thermotip*."

A method that dispenses entirely with a moving pen is employed by the Varian "*Statos*" recorders. This system uses digitized electronic pulses that are progressively switched across 100 motionless styli in each channel (40 mm wide).

The analog input thus produces a digital output of closely dotted clean electrographic traces of any signal from dc up to 1500 Hz.

8-13. OTHER TYPES OF PERMANENT RECORDINGS

Supplementing the permanent records provided by the widely used graphic recorders that have been mentioned in this chapter, there are some other specialized forms of recordings worthy of mention, as listed below.

For the centralized control panels of industrial automated systems, where it may be required to monitor perhaps many dozens of variables at one time, there are *miniaturized graphic recorders* for central control. In the example illustrated in Fig. 8-12, a great many of these monitoring devices are grouped together in view of the operator at the central station. Each of these miniature monitors displays a desired set point, with colored lights showing green when the variable is within limits and turning to amber and red as the value becomes excessive. Permanent records can thus be obtained of the status of the many variables in the industrial process at any given time.

Magnetic-tape recordings are a major form of permanent storage of digital

Figure 8-12. Center monitor station as used for industrial-process instrumentation. (Motorola Veritrak, Controls Systems Division.)

information in computer systems. This form of record is essentially similar to the magnetic-tape recording of audio speech and musical renditions [as well as the video-tape recording (VTR)], but it differs substantially in important details. Not only does the wider instrumentation tape provide room for many more tracks, but it also involves extensive refinements in fast and precise tape transport. Moreover, the requirements for the recording of multiple instrumentation data in telemetry (Chapter 19) call for a much broader frequency response to handle the great variety of signals involved. A typical example of a portable instrumentation tape recorder (*Honeywell model 5600*) has a 14-channel capacity on interchangeable reels sizes 5–$10\frac{1}{2}$ in.) and tape widths ($\frac{1}{4}$–1 in.) at seven selectable speeds ($\frac{15}{16}$–60 in./sec). The electronics accepts any of three types of input data: direct, FM, and digital inputs.

The recording of *transient phenomena* is another special situation, where a permanent record is made possible by the process of *photographing the display of a fast-rise oscilloscope* (usually with storage facility), as discussed in Chapter 10.

QUESTIONS

Q8-1. How does the self-balancing potentiometer type of recorder differ in principle from the direct-writing galvanometer type?

Q8-2. In a dc self-balancing recorder system, explain the function of each of the following blocks: transducer, reference voltage, error detector, amplifier, and servomotor.

Q8-3. In a recording system where only low-level (millivolt-range) signals are available, the dc signal system might be preferred because of its freedom from ac pickup of undesired signals, in spite of the difficulty of amplifying dc signals.
- (a) Discuss *drift* as a major difficulty encountered in the use of such a dc signal system.
- (b) State *two methods* that may be used to overcome this difficulty.

Q8-4. State three applications in which electronic forms of indicator/recorder systems are commonly found in industrial work.

Q8-5. What principle is used to obtain reversing action in a two-phase servomotor?

Q8-6. A dc self-balancing potentiometer system is used to record the output of a temperature-monitoring thermocouple.
- (a) Describe the action of the system when the thermocouple output increases, with a rising temperature, by tracing the dc signal through the chopper amplifier, servomotor, and indicator, paying particular attention to the relative phase between the amplifier output voltage and the reference voltages.
- (b) Repeat for a reduced thermocouple output as the temperature drops.

Q8-7. In recorders of the direct-writing galvanometer type, distinguish between *rectilinear* and *curvilinear* chart records.

Q8-8. State how damping is accomplished in:
(a) Servomotor recorders.
(b) Galvanometer recorders.

Q8-9. In selecting the speed of a chart drive, what factors favor the use of each of the following speeds:
(a) 24 hr per revolution?
(b) 3 min per revolution?

Q8-10. State two examples where magnetic tape is commonly used as the recording element.

Q8-11. State the kind of applications where an X–Y recorder is more suitable than the common form of single-pen recorder.

Q8-12. Explain two methods for obtaining inkless recordings.

9

Transducers

9-1. SCOPE OF TRANSDUCERS IN MEASURING SYSTEMS

In an electronic measuring instrument, the transducer is the *sensing or detecting element that converts the quantity being measured into a corresponding electrical quantity*. It is thus the first element in a general measuring system, as shown in Fig. 9-1, considered as comprising three functional blocks: (1) *detection* (or sensing) by a tranducer to produce an electrical signal; (2) *signal modification*, as necessary to amplify the signal to a proper level (or perhaps to shape it to a suitable wave-form); (3) *indication* to provide a readout of the quantity being measured, often (but not necessarily) in the form of a meter reading.

These elements take on many different forms, depending on the nature of the measurement. In the audiometer for measuring sound levels, shown in Fig. 9-2(a), the familiar microphone is used as a transducer but can display the result on an oscilloscope rather than on an ac output meter (or on both, as desired). Similarly, a remote indication of engine speed (electronic tachometer) might employ one of the two methods shown in Fig. 9-2(b) and (c): in one case, a dc tachometer-generator would produce a dc signal voltage having an amplitude proportional to engine speed, and this value would be displayed on a dc voltmeter calibrated in revolutions per minute, without the need of an amplifier; in the alter-

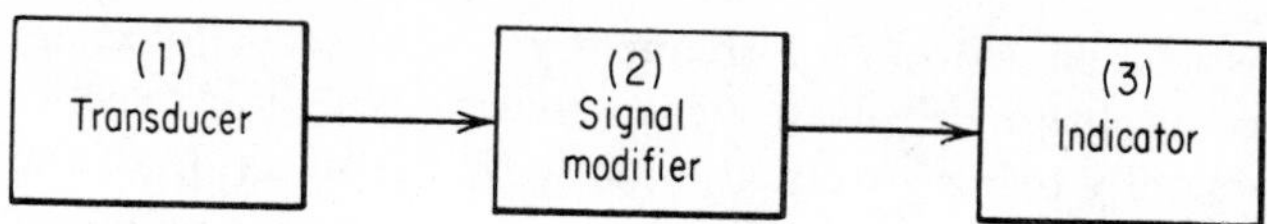

Figure 9-1. Block diagram of major units in an electronic measuring system.

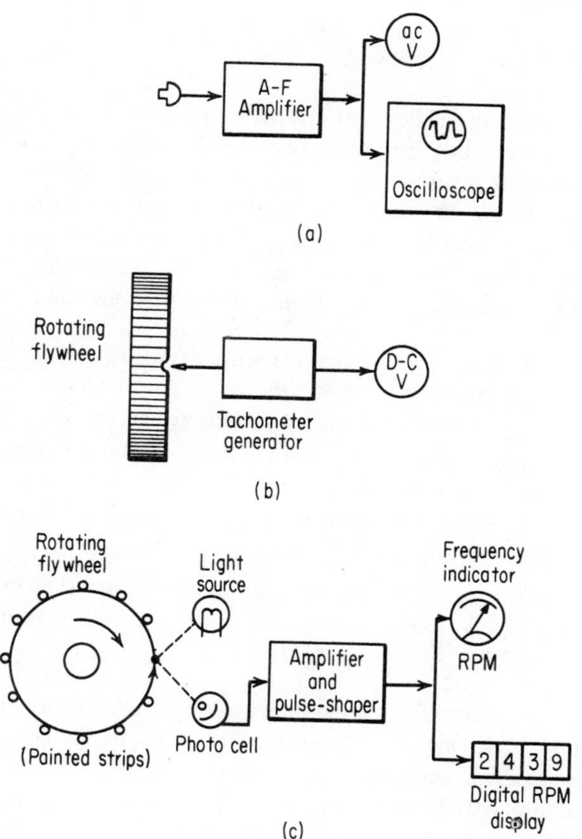

Figure 9-2. Examples of transducer action: (a) microphone transducer in measuring sound intensity; (b) and (c) alternative ways of measuring engine speed.

nate case, a photocell detecting reflections from dots painted on a flywheel would produce a signal frequency proportional to the engine speed, and this low-level ac signal would then be amplified and shaped into pulses to actuate a frequency counter, where again a choice of either a meter or a digital display of rev/min or rpm might be made.

It is well to reemphasize a point shown in these examples: namely, that instrument work very frequently presents a choice among several methods for accomplishing a given measurement—with the selection of a particular method generally being governed initially by the selection of the most suitable transducer for the job. A control room of a power plant might employ various transducers to enable the operator to *remotely monitor many variables at one central point. Finding the suitable transducer* is the starting point for effective instrumentation.

9-2. CLASSES OF TRANSDUCERS

The list of quantities that can be converted into an electrical output by a transducer is so long that it seems to include almost anything measurable: starting from the standard measurements of temperature and pressure in physics through the determination of pH of chemical solutions, and on to measurements, of heart sounds and the detection of nuclear radiation in medical electronics (or bionics).

A mention of some areas in which transducers find important use would serve to provide a useful listing of transducers, as follows:

Mechanical: Strain-gage type for force, weight, or torque, pressure and flow gages, accelerometer, humidity indicator, and the like.

Thermal: Resistance thermometer (or thermistor), thermocouple.

Optical: Photovoltaic (sun battery or solar cells) and photoconductive cells.

Acoustical: Microphone.

Magnetic: Permeameters and Hall-effect semiconductor.

Chemical: pH and conductivity cells.

Biological: Electrocardiograph and electroencephalograph.

Nuclear: Geiger tube, ionization chamber, scintillation and semiconductor detectors.

A better way of classifying the various transducers for the purposes of this study is to arrange them according to the basic *electrical measuring principle* involved in converting the varying quantity into a corresponding closely proportional electrical variation. A classification on the basis of the electrical principles used in transducers follows (selected types of these classes are discussed in separate sections later).

Classification of Main Electrical Principles Used in Transducers

1. *Variable-Resistance Types*
   (a) Strain and pressure gages
   (b) Thermoconductive thermometers (resistance bulbs and thermistors)
   (c) Photoconductive sensors (photocells)
   (d) Chemical conductivity meters

2. *Variable-Inductance Types*
   (a) Linear-variable differential transformer (LVDT) for displacement
   (b) Variable-reluctance pickup
   (c) Selsyn generators and receivers
3. *Variable-Capacitance Types*
   (a) High-frequency *LC* and *RC* sensing
   (b) Reactance-tube type producing frequency modulation
4. *Voltage-Divider Types*
   (a) Simple form of potentiometer position sensor
   (b) Simple pressure-actuated voltage divider
5. *Voltage-Generating Types*
   (a) Piezoelectric pressure sensor
   (b) Rotational-motion (rpm) tachometer
   (c) Thermocouple sensor
   (d) Photovoltaic (solar) cell

Transducer Classes in Force and Pressure Measurement

The principles involved in measurements of force may properly be expected to be employed also in the measurements of quantities *proportional to force*, such as displacement, pressure, torque, and acceleration. Since *pressure* is one of the quantities most frequently monitored, it is profitable to examine the operating principles of some transducers used for pressure measurement.

Resistive Pressure Transducers

The measurement in the resistive type of transducer is based on the theory that a change in pressure results in a resistance change in the sensing element. This principle is used in the *strain gage*, very commonly employed for stress and displacement instrumentation. In the general case of pressure measurements, the sensitive resistance element may take many forms, depending on the mechanical arrangement on which the pressure is caused to act. In Fig. 9-3 two ways are shown by which the pressure acts to influence the sensitive resistance element: part (a) shows the *bellows type*, and part (b) shows the *diaphragm type* of action. (The familiar Bourdon-tube type of pressure gage would furnish still another modification.) In each of these cases, the element moved by the pressure change is made to cause a change in resistance. This resistance change, when made part of a bridge circuit, can then be taken as either an ac or dc output signal to determine the pressure indication.

Inductive Pressure Transducers

A simple arrangement where a change in the inductance of a sensing element is produced by a pressure change is shown in Fig. 9-4. Here, the pressure acting on a movable magnetic core will cause an increase in the coil inductance corresponding

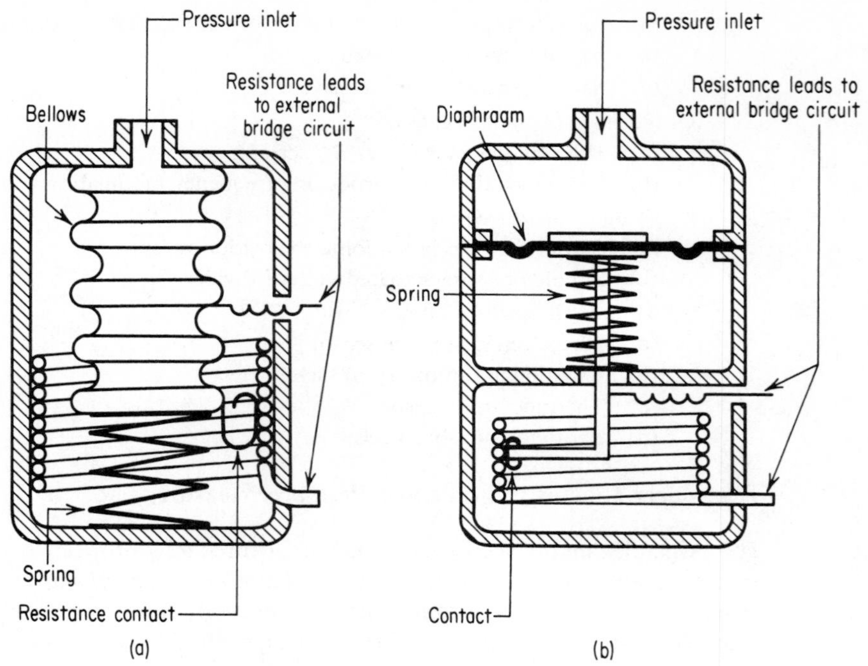

Figure 9-3. Resistive pressure transducers: (a) bellows movement coupled to movable resistance contact; (b) sensitive diaphragm moves the resistance contact.

to the acting pressure. The inductance change can again be made the basis of an electrical signal, this time in an ac inductance-bridge arrangement. An advantage of the inductive (and also of the capacitive) type over some of the previous resistive arrangements is that no moving contacts are present, thus providing continuous resolution of the change, with no extra friction load imposed on the measuring system.

In a slightly modified form, this principle is used to obtain a change in mutual inductance between magnetically coupled coils, rather than in the self-inductance of a single coil. Where a change in an induced voltage is involved, the transducer is sometimes called a variable-reluctance sensor (or magnetic pickup).

A very important example of the mutual-inductance type is the *linear-variable differential transformer* (LVDT), in which the induced voltages in two secondaries originally oppose each other to produce an initial null; a linearly proportional ac output is produced when the pressure changes from its original condition.

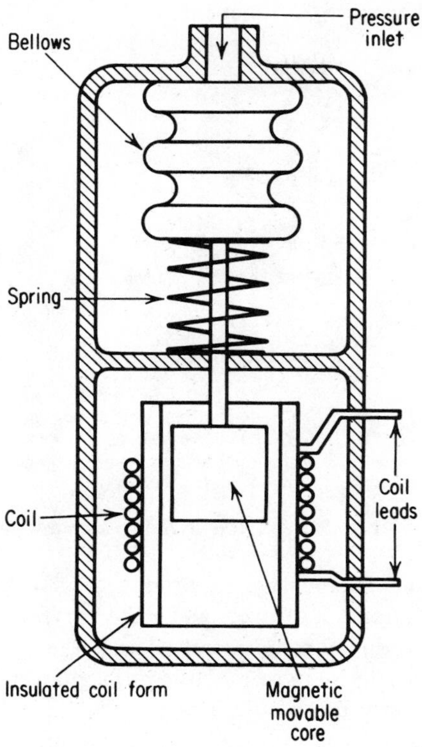

Figure 9-4. Inductive type of pressure transducer; pressure changes actuate the magnetic movable core.

Capacitive Pressure Transducers

Changes in pressure may be easily detected by the variation of capacitance between a fixed plate and another plate free to move as the pressure changes, as illustrated in Fig. 9-5. The resulting capacitance variation follows the basic formula for capacitance:

$$C = 0.0885K\frac{(n-1)A}{t} \qquad \text{pF}$$

where A = area of one side of one plate, cm^2
n = number of plates
t = thickness of dielectric, cm
K = dielectric constant

The capacitive transducer, as in the capacitive microphone, is simple to

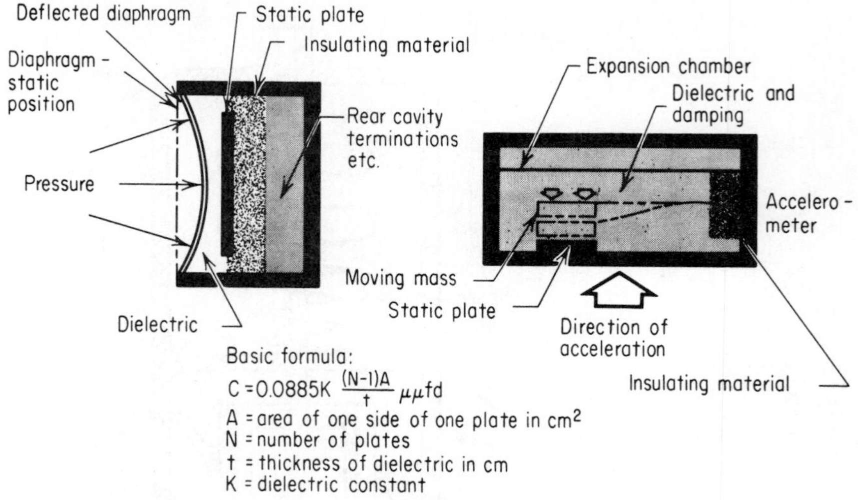

Figure 9-5. Capacitive pressure and acceleration transducers.

construct and inexpensive to produce and is particularly effective for high-frequency variations. However, when the varying capacitance is made part of an ac bridge to produce an ac output signal, the conditions for resistive and reactive balance generally require much care in guarding against unwanted signal pickup in the high-impedance circuit and equally careful compensation for temperature changes. As a result, the receiving instrument for the capacitive sensor usually calls for a more advanced and complex design than is needed for other transducers.

Voltage-Divider Pressure Transducer

One of the simplest transduction methods uses the motion of a pressure-sensitive bellows to actuate the arm of a potentiometer voltage divider. The voltage-divider action is similar to that shown in Fig. 9-3, but in this case the moving arm of the voltage divider produces a substantial voltage output proportional to the acting pressure, rather than a simple resistance variation. Though it may be energized by either an ac or dc supply, the voltage-divider potentiometer is particularly suited to simple dc indication and recording systems, often requiring no amplification. As in the case of the resistive pressure transducer, it is subject to moving-contact irregularities. These, however, can usually be made small enough to be neglected for simple applications.

Voltage-Generating Pressure Transducer

The action of generating a voltage by the application of pressure is illustrated in the *piezoelectric crystal* transducer, shown in Fig. 9-6. Crystalline materials widely used for the production of electric potential by means of stress are quartz, tourma-

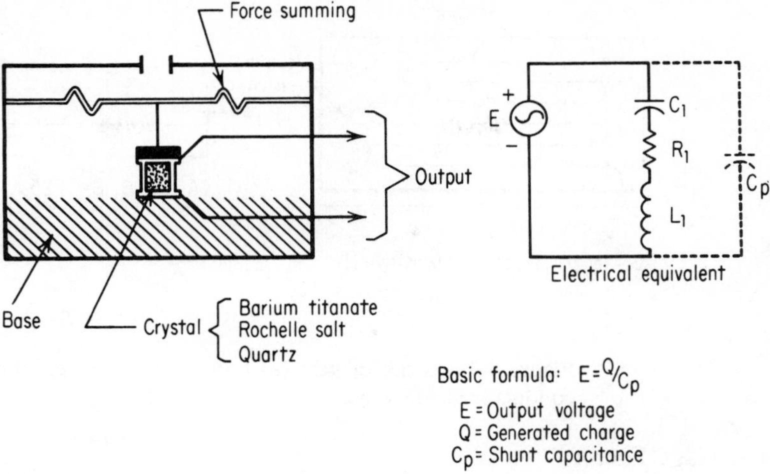

Figure 9-6. Voltage-generating (piezoelectric) pressure transducer.

line, Rochelle salt, and ceramics, such as barium titanate. The familiar crystal and ceramic microphones and phonograph pickup cartridges are common examples of devices using such piezoelectric crystals. The basic expression for the output voltage E is

$$E = \frac{Q}{C_p}$$

where Q is the generated charge and C_p the shunt capacitance.

9-3. THE STRAIN GAGE

The strain gage is a commonly used transducer utilizing *resistance variation* to sense the strain or other results of a *force*. It can be employed as a very versatile detector for measurements of torque, weight, pressure, displacement, and similar quantities.

The construction of a *bonded* strain gage [Fig. 9-7(a)] shows a fine-wire element looped back and forth on a mounting plate, which is usually cemented to the member undergoing stress. The extra length obtained by the hairpin looping increases the effect of a stress applied in the direction of length l. Thus, a tensile stress would stretch the wire, increasing its length and therefore its resistance. In addition, the combined act of the wire's getting thinner and also undergoing a so-called "Bridgman effect" (an inherent increase of resistance with strain) are mutually aiding and thus reinforce the increase of resistance caused by the strain.

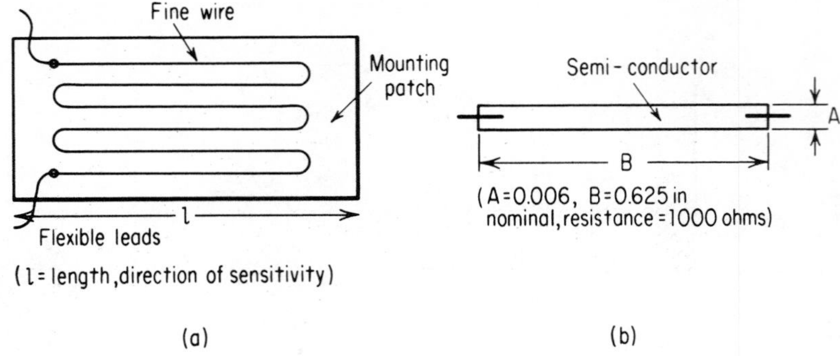

Figure 9-7. Strain gages: (a) bonded strain gage; (b) semiconductor strain gage.

Gage Factor

The measure of the sensitivity of a gage to the strain is called the *gage factor* and is given by the ratio of the fractional change in resistance ($\Delta R/R$) to the fractional change in length ($\Delta l/l$), or

$$GF = \frac{\Delta R/R}{\Delta l/l}$$

Initial resistance R is typically around 120 Ω, and the magnitude of the gage factor may have values from -12 (for nickel) to $+6$. A *typical gage factor of 2* is quite reasonable for most wire strain gages. (The more recent semiconductor strain gages with greater sensitivity will be discussed later.)

The strain-gage output is derived from a bridge arrangement in which the gage forms one of the four arms of a bridge, which may be either ac- or dc-actuated. A simple dc arrangement is shown in Fig. 9-8, using only one active element to produce an output proportional to strain. Since the resistance of the fine-wire element is quite sensitive to temperature as well as stress variations, some compensation must be provided. The dummy gage in the figure accomplishes this in a simple manner, when it is placed in the same temperature environment as the active gage, but without being subject to the strain. Then, with the other two arms R_1 and R_2 selected as resistors with negligible temperature coefficients, the bridge retains its balance under initial conditions of no strain, at any temperature within its operating range.

Strain-Gage Features

The commercial wire strain gage is a comparatively rugged sensor and very straightforward in its installation and use. In ac applications, it is amply capable of measuring vibrations at high frequencies, up to 100 kHz or more when necessary. Excellent linearity and hysteresis characteristics are obtained. While its com-

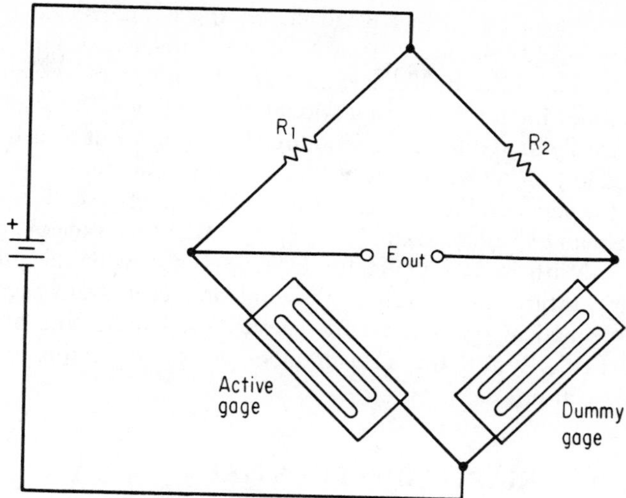

Figure 9-8. Temperature-compensated dc bridge for strain-gage measurement.

paratively low sensitivity, compared to silicon, limits its span (full-scale magnitude) and temperature range of operation, it is, by the same token, comparatively less sensitive to shock and undesired vibration.

Unbonded Strain Gage

In measuring situations where a relatively freely moving member is involved, the unbonded strain-gage construction is employed. Here the fine strain-sensitive filaments are wired between a fixed and a movable frame, usually in such a way that the gage forms all four bridge arms. By nature, it has a small volumetric displacement but is still capable of continuous resolution and possesses substantially all the other important features of the bonded type, although it is generally less sensitive.

The Semiconductor Strain Gage

A very large improvement in the gage factor (and hence, the sensitivity) of the strain gage has been made possible by the recent development of semiconductor strain gages [Fig. 9-7(b)]. Many previous efforts had been made to improve the low gage factor of metallic strain gages, but the development of the semiconductor type, which has long been known to have a higher gage factor, was impeded by their unacceptable thermal properties and also their physical brittleness.

The development of a *flexible silicon strain gage* has overcome these difficulties and has resulted in a practical semiconductor strain gage (bonded or unbonded), which can be used in exactly the same manner as the prior metallic strain gage, but with a gage factor approximately 65 times larger (130 for the silicon gage compared with 2 for the metallic gage). This large value of gage factor is accompanied by a thermal rate of change of resistance also approximately 65 times higher than conventional gages; hence, the semiconductor strain gage is relatively as stable as the metallic strain gage, but at a much higher output level. In addition, straightforward temperature compensation methods are applicable to the semiconductor strain gage, enhancing its usefulness and allowing the measurement of "microstrain," formerly beyond the range of convenient measurement.

With its low figure for hysteresis, and with the use of regular methods of temperature compensation, the semiconductor strain gage has proven to be stable for practical operation with conventional indicating and recording equipment, particularly for the measurement of small strains in the range of 0.1–500 microstrain.

9-4. LINEAR-VARIABLE DIFFERENTIAL TRANSFORMER

A transducer in general use for providing an electrical output that is linearly proportional to a mechanical displacement is the linear-variable differential transformer (Fig. 9-9), commonly abbreviated LVDT (also designated by some

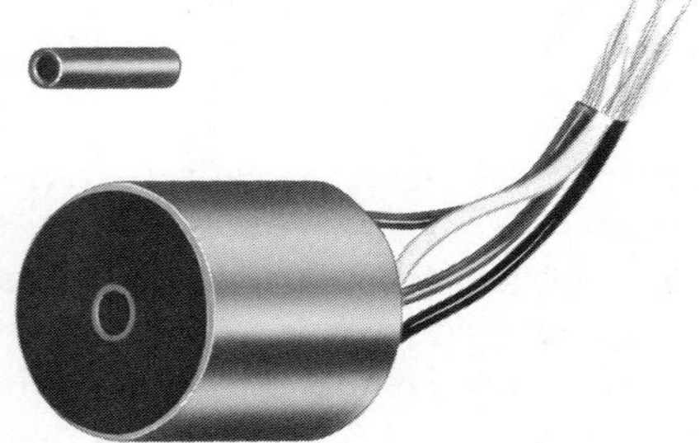

Figure 9-9. Direct-current differential transformer, having self-contained conversion of 6 V dc to ac excitation and demodulator for dc output.

manufacturers as LDVT linear differential voltage transformer). It is an electromechanical transducer of the inductance-variation type, in which the mechanical displacement acts on a movable core. As shown in Fig. 9-10(a), three coils are wound on a cylindrical coil form, with the center coil acting as a primary, inducing a voltage in each secondary coil on either side of it. The magnetic core is free to move axially inside the coil assembly, and the motion being measured is mechanically coupled to this movable core.

When the primary coil is energized with alternating current, as shown in the schematic diagram of Fig. 9-10(b), voltages are induced in the two outer coils.

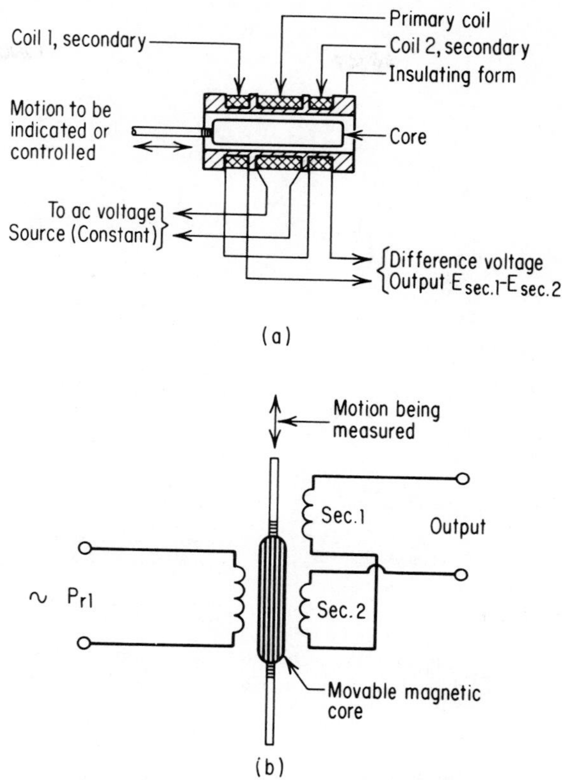

Figure 9-10. The linear-variable differential transformer (LVDT): (a) construction; (b) schematic diagram.

The secondary coils are connected in series opposition, so that the two voltages in the secondary circuit are opposite in phase, and the net output of the transformer is the difference of these voltages. For the central position of this core, the output voltage will be zero. This is called the balance point or null position.

When the core is moved from this balance point, the voltage induced in the secondary coil toward which the core is moved increases, while the voltage induced in the other secondary coil decreases. This produces a differential voltage output from the transformer which, with proper design, varies linearly with change in core position. Motion of the core in the opposite direction beyond the null position produces a similar linear voltage characteristic, but with the phase shifted 180°. A continuous plot of voltage output vs. core position (within the linear range limits) appears as a straight line through the origin, if opposite algebraic signs are used to indicate opposite phases, as shown in Fig. 9-11. The relative positions of the core inside the transformer are shown for the three output conditions just described.

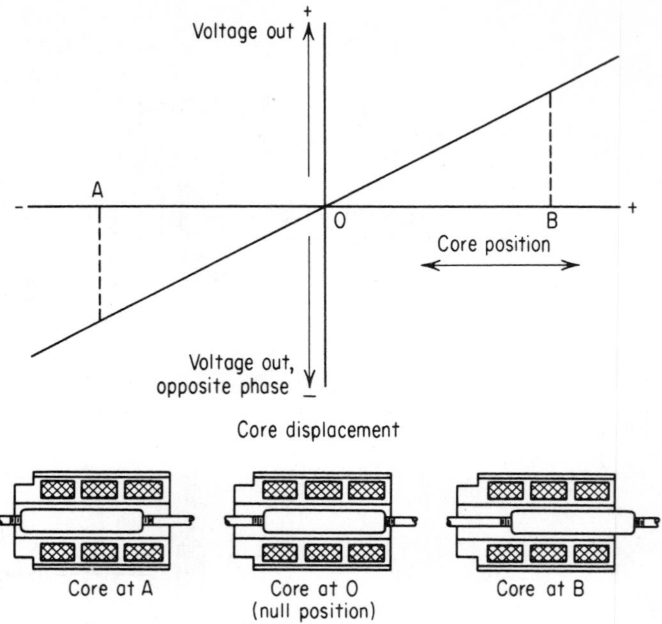

Figure 9-11. Output voltage as a linear function of position of core in a linear variable differential transformer (LVDT).

9-5. FORCE MEASUREMENT APPLICATIONS (THE ACCELEROMETER)

Since the measurement of the amount of displacement or deflection of an elastic member can also serve as a basic method for measuring *force*, both the strain gage and the linear-variable differential transformer (LVDT) can be used as transducers for other measurements depending on force, such as *weight*, *pressure*, and *acceleration*. There is a great variety of different arrangements employed by manufacturers of commercial transducers, each accomplishing the basic measurement in special forms suited to various industrial processes. The *accelerometer* will be selected as one example which, in this case, employs the LVDT for the basic measurement of elastic deformation.

The elastic deformation, instead of being produced directly by an externally applied force, may be caused by acceleration. In this case, the total of the mass of the resilient member plus any attached mass will determine the position of a point A with respect to a reference B, and this new position can be taken as a measure of the accelerating force. A form of accelerometer is shown in Fig. 9-12. This is a common application of the LVDT. As the accelerating force may be a

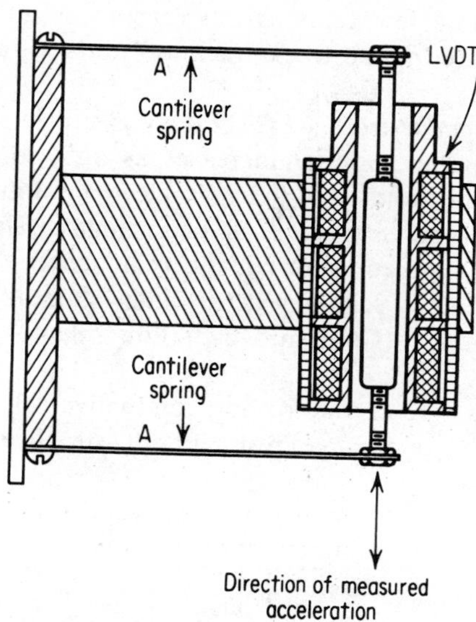

Figure 9-12. Linear-variable differential transformer (LVDT) used as an accelerometer.

variable quantity, the rate of acceleration may be determined from this same structure.

9-6. SEMICONDUCTOR PHOTOELECTRIC AND TEMPERATURE DEVICES

With the development of low-priced semiconductor photoelectric and temperature devices, many new applications have been created. Semiconductor photoelectric and temperature devices have advantages such as small size, light weight, low cost, low power requirements, and high dependability.

Types of Sensors

The four basic types of photosensors in use today are photovoltaic, photoemissive, photoconductive bulk, and the junction types.

Photovoltaic cells are devices that generate a voltage. They require no external power source. Photovoltaic devices are used in some photographic light-meters and as a source of voltage to operate some types of low-powered electronic equipment. These cells are sometimes refered to as *solar* cells.

Photoemissive types detect the presence and the intensity of light by emitting electrons from a surface into space (usually inside a vacuum tube). Photo tubes were popular a few years ago as sensors in automatic door openers, alarms, for motion-picture sound-track pickup, and in a variety of industrial control applications.

Photoconductive bulk effect devices utilize the principle that conductance of a semiconductor material increases as light intensity increases. The light gives energy to the atoms and releases electrons into the conduction band, creating electron–hole pairs which serve as current carriers. The greater the total surface area of the exposed material, the more effect the light changes will have upon conductance. Any semiconductor material could be used to make photocells; but cadmium sulfide (CdS) and cadmium selenide (CdCe) are probably the most widely used.

Figure 9-13 shows a photoconductive cell (bulk-effect type) connected in a circuit consisting of a load resistor and a supply voltage. As light intensity increases

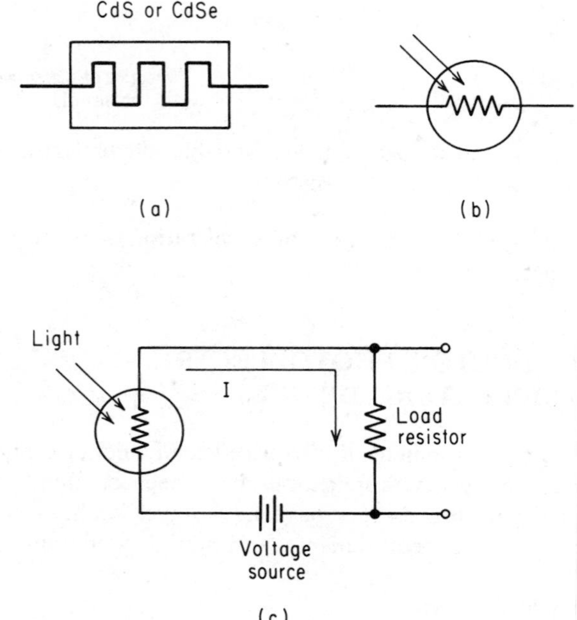

Figure 9-13. Photoconductive cell: (a) construction; (b) symbol; (c) simple circuit.

on the photocell, the current increases and thus the output voltage will increase. Changes in this output voltage may be used directly or the voltage may be amplified where greater voltage changes are needed. It should be noted that the cell is a resistance and thus the supply may be connected with either polarity.

Bulk-effect cells have wide usage because of their low cost, small size, high reliability, and sensitivity. Figure 9-14 shows a few commercially available cells

Type	Sensitive material	Peak spectral response (Angstroms)	Resistance @ 2 ft-c (ohms)	Min. dark resistance
CL5P4	CdSe	6900	2.7 K	720 K
CL5P4L	CdSe	6900	0.45 K	120 K
CL5P5	CdS	5500	9.0 K	450 K
CL5P5L	CdS	5500	1.5 K	75 K
CL7P4	CdSe	6900	54 K	14.4 Meg
CL7P4L	CdSe	6900	1 K	268 K
CL7P5	CdS	5500	166 K	8.3 Meg

Figure 9-14. The *Clairex* 5P and 7P Photocell series. (Courtesy of Clairex Electronics.)

and also the resistance of the cell with no light (dark resistance) and the resistance at 2 foot-candles (or 2 lumens per square foot). The light/dark ratio can be calculated by

$$\text{ratio} = \frac{R_L}{R_D}$$

where R_L is the light resistance and R_D the dark resistance.

Photoconductive junction types of cells need a p-n junction. These are usually photodiodes and phototransistors. As with the bulk-effect devices, the junction cells require a voltage source. The photodiode is a p-n junction physically arranged so that the light is focused on the junction area. Figure 9-15 shows photodiode

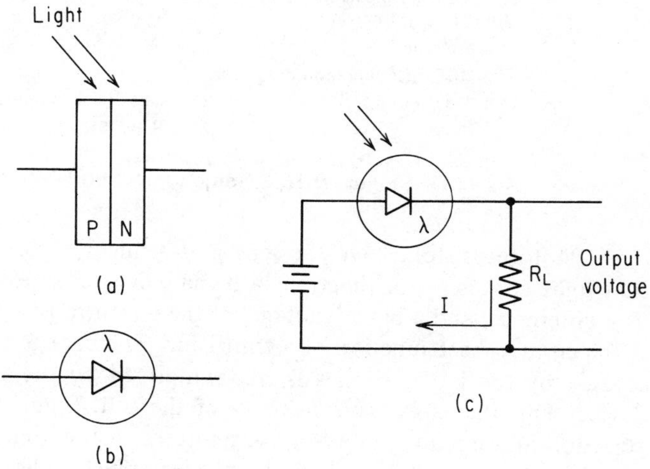

Figure 9-15. The photodiode: (a) construction; (b) symbol; (c) simple circuit.

construction and circuitry. When a single-section photodiode such as the one shown in Fig. 9-15 is used, the diode must be reverse-biased. This gives the diode very low dark resistance and dark current. As the light intensity increases, circuit current increases, thereby increasing the output voltage. As before, this signal may be used directly or it may need to be amplified. Figure 9-16 shows a few typical photodiodes. The devices are small in size, an advantage in the manufacture of smaller and lighter electronic devices. Another advantage of the junction-type photoconductor is its fast switching time. This is of extreme importance in today's digital and microcomputer circuits.

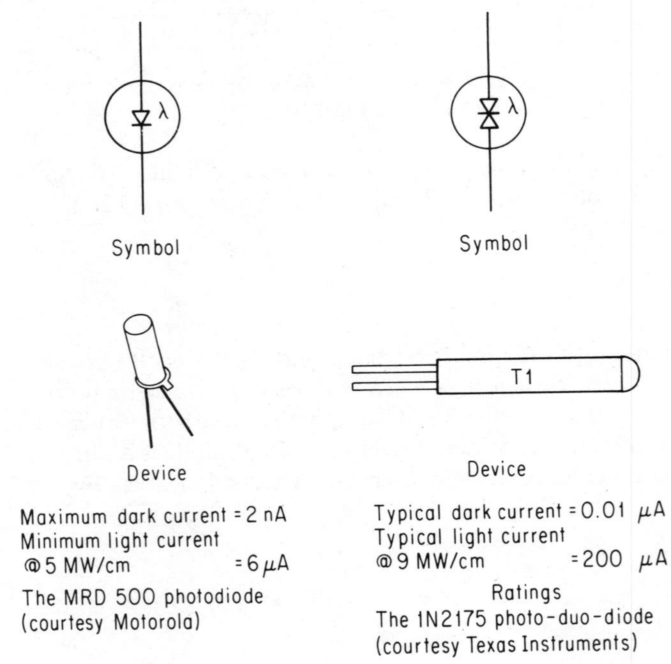

Figure 9-16. Some typical photodiodes.

Phototransistors have the lens and semiconductor material arranged such that light strikes a p-n junction in a manner similar to that of the photodiode. The difference, and a big advantage, of the phototransistor is that the p-n junction is the emitter–base junction of a transistor. The change in the current in the base created by the light changes is now amplified by the transistor portion of the device. This increases the sensitivity of the cell. Figure 9-17 illustrates the construction and circuit symbol of the phototransistor, Fig. 9-18 shows some phototransistor types, and Fig. 9-19 shows clear plastic phototransistor construction. The device contains transistors connected in the Darlington-pair configuration.

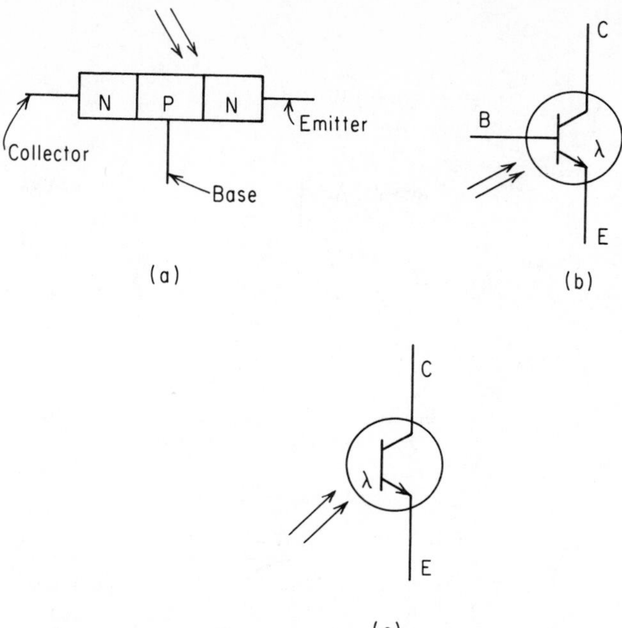

Figure 9-17. Phototransistor: (a) construction; (b) symbol with base lead; (c) symbol without base lead.

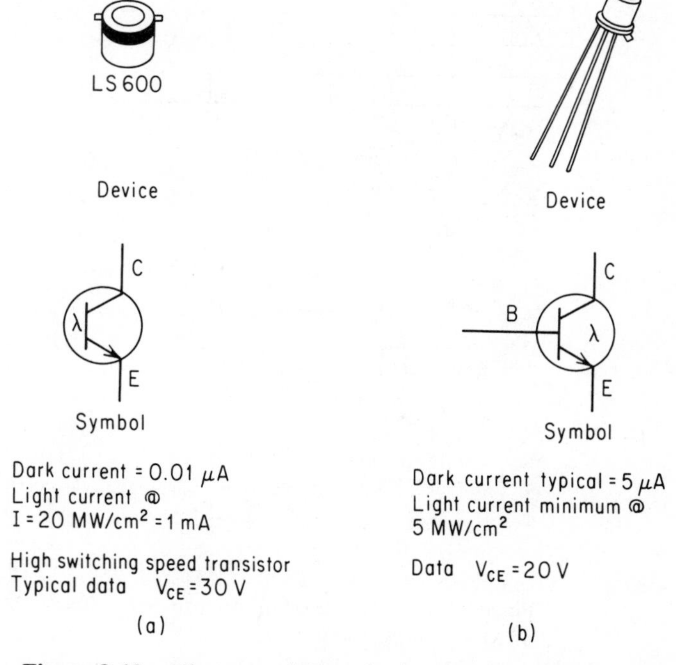

Figure 9-18. Phototransistors: both are sensitive to visible light and the near-infrared range. [(a) Courtesy of Texas Instruments, Inc.; (b) courtesy of Motorola Semiconductors.]

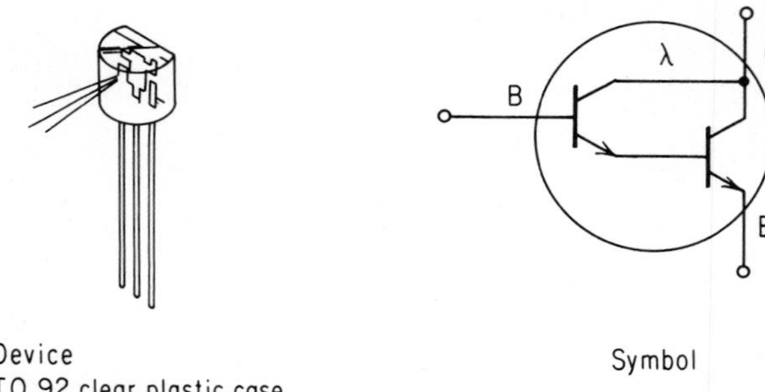

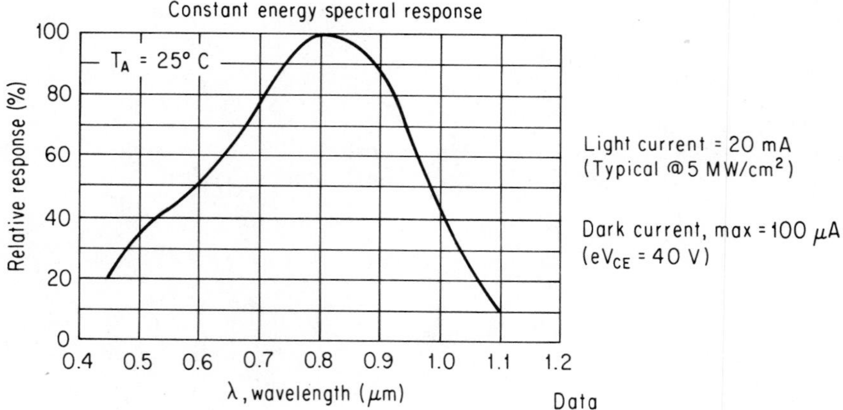

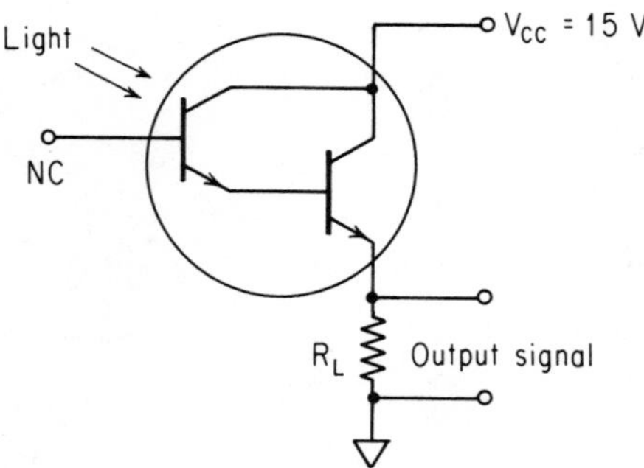

Figure 9-19. The Darlington transistor 2N5780. (Courtesy of Motorola Semiconductors.)

The Light-Emitting Diode

When a p-n junction is illuminated, it conducts current. The opposite is also true; when current flows through a p-n junction, light is generated. The light-emitting diode (LED) has a lens and a junction arrangement such that light can be emitted from the junction and utilized. The wavelength of this emitted light is determined by the width of the forbidden band of the junction. Figure 9-20 gives the data for some LEDs.

The most commonly used materials for LEDs are gallium arsenide (near infrared), gallium phosphide (green), and gallium arsenide phosphide (red). The LED is a diode; thus, to get current to flow, the device must be forward-biased. LEDs are used as photocell exiters or drivers, lamps, optical communication devices, and in seven-segment devices and digital displays. Since the LED has no filament, it has almost no heat generation and a life expectancy of over 100,000 hr.

Temperature-Sensing Devices

Temperature-sensing devices are made in such a way as to cause a characteristic of the device to change with temperature. When these exact variations are known, the device can be used as a predictable temperature sensor. We will concern ourselves in this text only with devices whose resistance changes because of temperature changes.

As temperature increases, the motion of electrons increases. Metals have many free electrons; therefore, increased motion causes increased resistance. This is called a positive temperature coefficient. Resistance increases with temperature. Semiconductors, on the other hand, have a negative temperature coefficient. This is because as temperature increases, there are more electrons released into the conduction band.

Figure 9-21 illustrates some types of thermistors (temperature-sensing devices). Most thermistors are made from carbon, carbon compounds, silicon or silicon compounds, or metal film deposited on a nonconducting surface. Metal-oxide-semiconductor materials are also used to make thermistors.

Figure 9-22 lists some commercially available Keystone thermistors that have a negative temperature coefficient of resistance, and Fig. 9-23 illustrates some Texas Instruments Sensistors that have a positive temperature coefficient. All values given in the tables are "free air" resistance values. When these devices are used in a circuit the I^2R heat must be considered. This heat will have an effect on the operation of the device.

Thermistor Applications

The circuit of Fig. 9-24 illustrates how a bulk-effect device can be used to control a relay. The circuit might be used to turn a lamp on or off at dusk or dawn. If we assume the lamp to be on at night, we see that at dawn the light will increase in

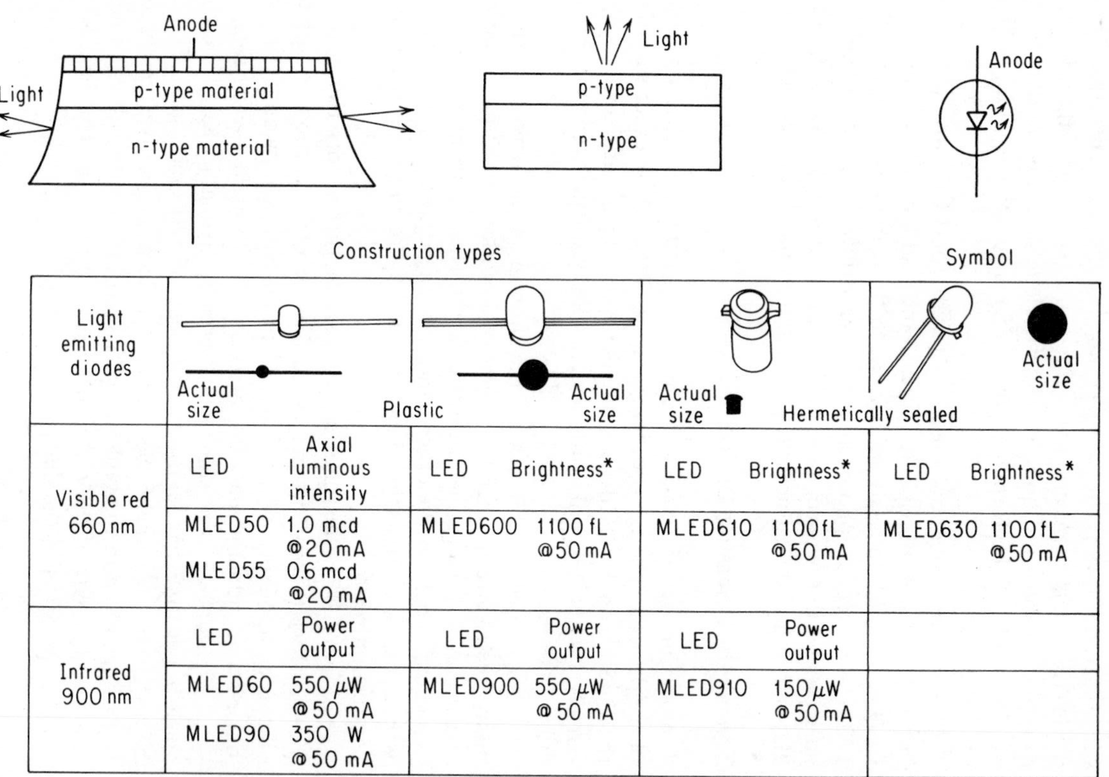

Light emitting diodes	Actual size	Plastic	Actual size	Actual size	Hermetically sealed	Actual size		
Visible red 660 nm	LED	Axial luminous intensity	LED	Brightness*	LED	Brightness*	LED	Brightness*
	MLED50 MLED55	1.0 mcd @20 mA 0.6 mcd @20 mA	MLED600	1100 fL @50 mA	MLED610	1100 fL @50 mA	MLED630	1100 fL @50 mA
Infrared 900 nm	LED	Power output	LED	Power output	LED	Power output		
	MLED60 MLED90	550 μW @50 mA 350 W @50 mA	MLED900	550 μW @50 mA	MLED910	150 μW @50 mA		

Figure 9-20. Light-emitting diode (LED): construction, symbol, and some typical commercially available units.

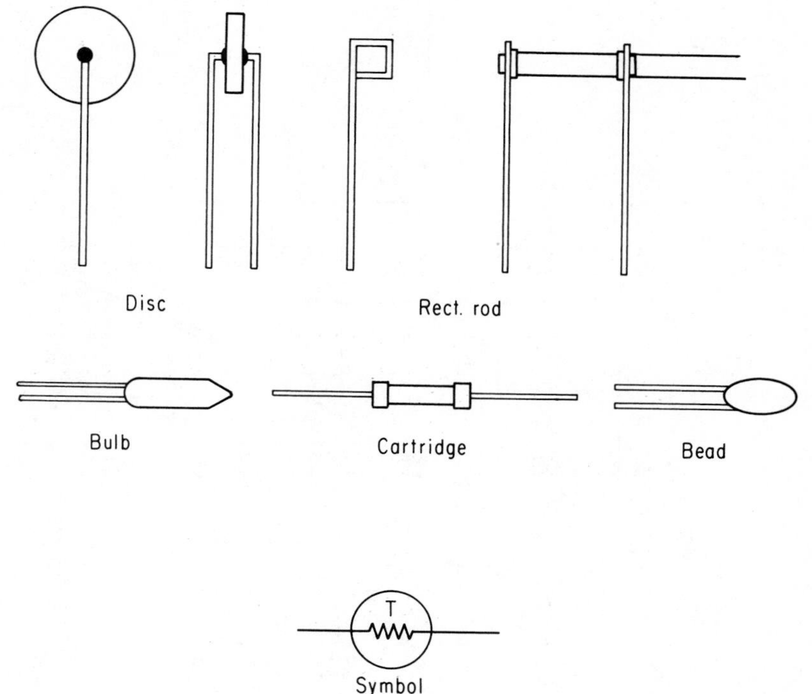

Figure 9-21. Some thermistor types.

Type	Resistance at °C:						
	0	25	37.8	75	104.4	150	260
L-060637-GK-85-M2	31.8 K	10.1 K	6000	1640	706	242	40.7
L-060637-22K-95-M2	122 K	37.5 K	22 K	5700	2315	739	104
L-060637-60K-103-M2	355 K	104 K	60 K	14.8 K	5830	1750	222
L-060637-200K-120-M2	1.3 Meg	363 K	200 K	45 K	16.7 K	4520	520
L-060637-350K-125-M2	2.35 Meg	637 K	350 K	76.5 K	28 K	7630	793

Figure 9-22. Typical thermistors (negative temperature coefficient). (Courtesy of Keystone Carbon Company.)

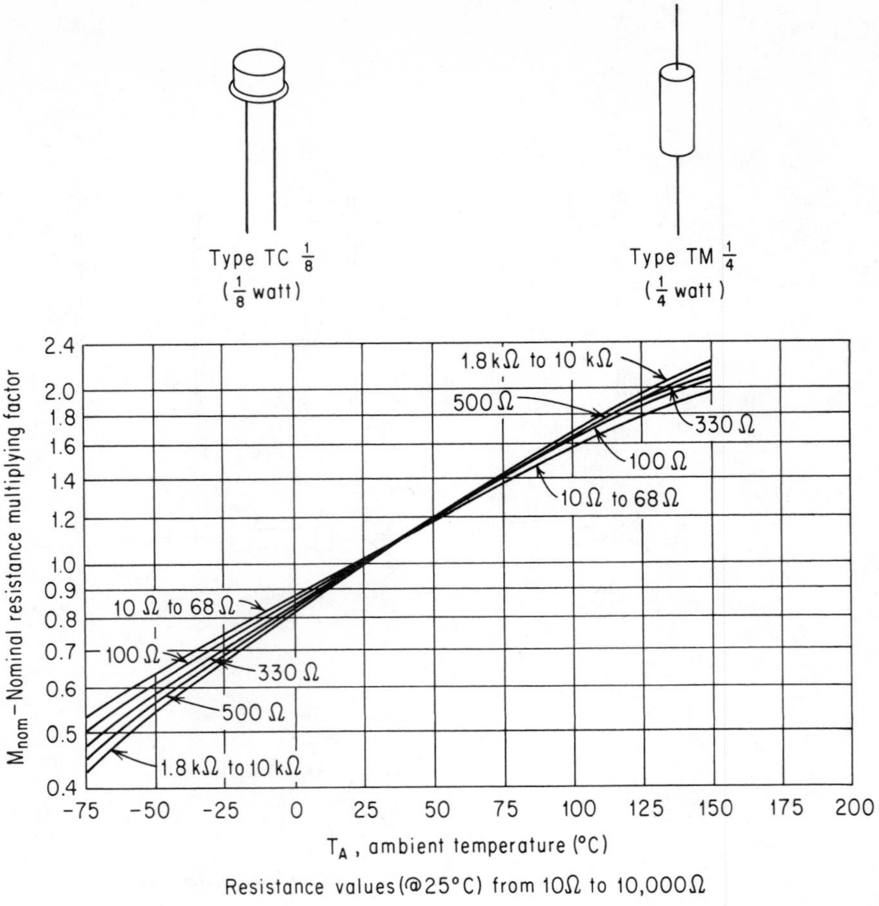

Figure 9-23. Typical sensistors (positive coefficient) from

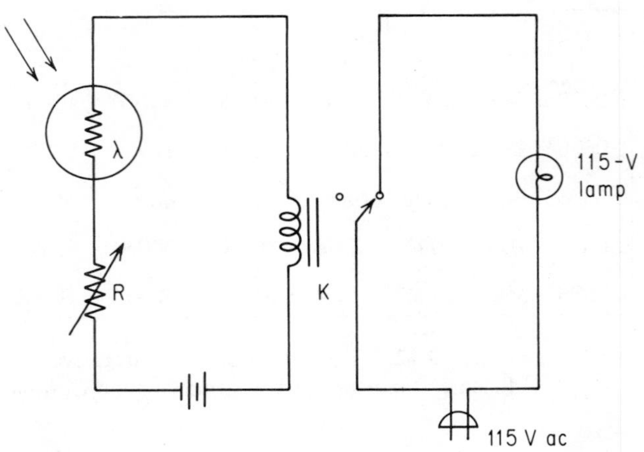

Figure 9-24. Relay(photocell) controlled lamp circuit.

intensity, which will decrease the resistance of the photocell. As this happens, current will increase and the relay will become energized. This will turn the lamp off. Resistor R is used to control the turn-off intensity required.

Figure 9-25 shows some devices where the lamp and detector are in the same package, usually a light-tight case. When the lamp is connected to a voltage source, the cell will be illuminated.

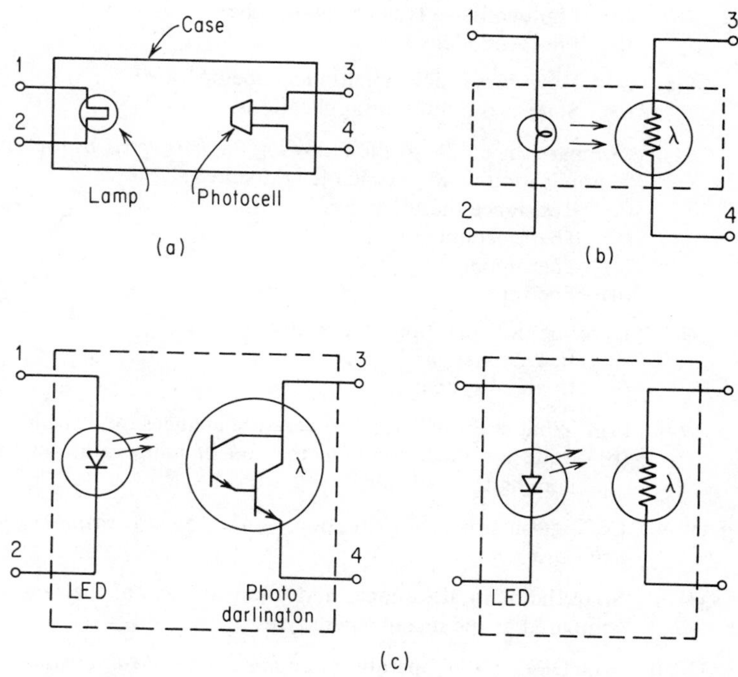

Figure 9-25. Photocell lamp assemblies: (a) construction; (b) circuit (incandescent lamp); (c) some other circuits.

QUESTIONS

Q9-1. How do each of the following devices conform to the definition of a transducer:
(a) Microphone.
(b) Thermistor.
(c) Thermocouple.
(d) Tachometer.

Q9-2. Describe the principle of operation of a *pressure transducer* employing each of the following principles:
(a) Resistive variation.

(b) Capacitive variation.
(c) Inductive variation.

Q9-3. Explain the difference in operating principle between the following two *pressure transducers:* the potentiometer type and the piezoelectric type.

Q9-4. Give examples to distinguish the following kinds of *photoelectric transducers* (*or photocells*):
(a) Photovoltaic type.
(b) Photoemissive type (or phototube).
(c) Photoconductive type.

Q9-5. (a) How does a photomultiplier tube operate?
(b) State two important applications.

Q9-6. Compare the merits of the following *thermoelectric transducers*, stating the kind of application to which each is best suited:
(a) Resistance thermometer.
(b) Thermocouple.
(c) Thermopile.
(d) Thermistor.

Q9-7. Describe the operation of a strain gage:
(a) In a dc system.
(b) In an ac system.

Q9-8. (a) What is the effect of temperature changes on a *strain gage*?
(b) State the most common method of temperature compensation for overcoming this difficulty.

Q9-9. Distinguish between a situation using a *bonded strain gage* from one *using an unbonded type.*

Q9-10. State the main advantages and disadvantages of the *semiconductor strain gage* compared to the metallic-wire type.

Q9-11. (a) Describe the operation of a *linear-variable differential transformer* (LVDT).
(b) State two important applications of the LVDT.

Q9-12. Describe the essential difference between a *variable-reluctance type* of transducer and the LVDT.

Q9-13. Give two examples situation where a change in *oscillator frequency* may be used as the basis for transducer operation.

Q9-14. Describe the difference in the *action of light* on the following devices:
(a) Photodiode.
(b) Phototransistor.
(c) Photosensitive gas discharge tube.

Q9-15. (a) Describe the principle of operation of an *accelerometer.*
(b) Give two important applications.

Q9-16. State one important application of a:
(a) Selenium cell (sun battery).
(b) Solar (silicon) cell.

PROBLEMS

P9-1. A strain gage having a nominal (unstressed) resistance of 240 Ω is connected to act as one arm of an equal-arm bridge (shown in Fig. P9-1), powered by $E =$ 12 V dc. If its gage factor $[GF = (\Delta R/R)/(\Delta l/l)]$ is 2.5 and the voltage detector has a full-scale range of 15 mV, find the approximate value of the microstrain (μin./in.) required for full-scale deflection.

(*Note:* Assume negligible current taken by the millivoltmeter detector, and use the approximate relation for voltage output at small unbalance in the bridge.)

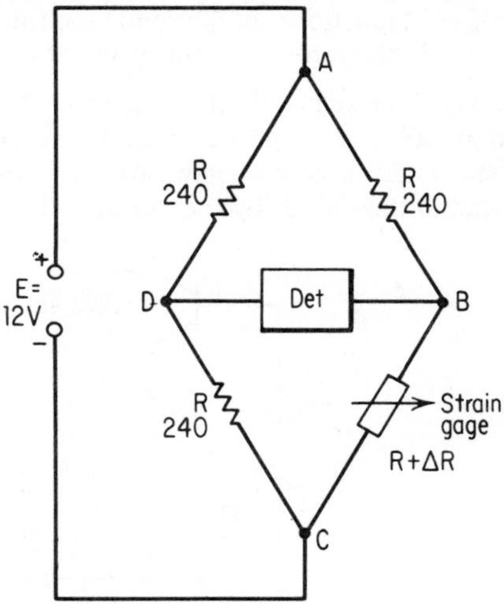

Figure P9-1

P9-2. With all other conditions remaining the same in Problem P9–1, find the microstrain required for full-scale deflection when:

(a) A second unstressed strain gage identical to the first is used for temperature compensation.

(b) Two active and two inactive gages, all having the same properties, are used in the bridge.

P9-3. The following table shows the EMF obtained from three types of thermocouples (differing in composition to accommodate their operating temperature ranges) when the cold junction is kept at 0°C (in ice water) and the hot junction is 100°C.

Thermocouple Type	*Operating Temperature Range (°C)*	*Millvolts at 0–100°C*
Copper to constantan	To 300	4.28
Chromel to alumel	To 1000	4.10
Platinum to platinum–rhodium	To 1600	0.64

Source: M.B. Stout, *Basic Electrical Measurements,* 2nd ed. (Englewood Cliffs, N.J.: Prentice-Hall, 1960).

A temperature measurement made with a *copper-to-constantan* thermocouple gives a reading of 250°C on a null-balance potentiometer (full-scale 300°C), calibrated directly in degrees and millivolts. Ignoring the correction factor given for reference-junction changes, what would be the expected millivolt reading when substituting:

(a) The chromel-to-alumel couple?
(b) The platinum-to-platinum–rhodium couple?

P9-4. If a thermopile is constructed from 12 thermocouple elements connected in series, what millivolt output voltage may be expected at a temperature difference (between hot and cold junctions) of 50°C, for each thermopile constructed from the materials given in the table of Problem P9-3?

P9-5. The sensitive relay in Fig. P9-5[1] operates when as little as 5 foot-candles of illumination fall on the photodiode. If the effective current amplification of the transistor is 50, and the relay closes at 1 mA, find the effective output (in μA) produced in this circuit by the photodiode at this minimum level of illumination.

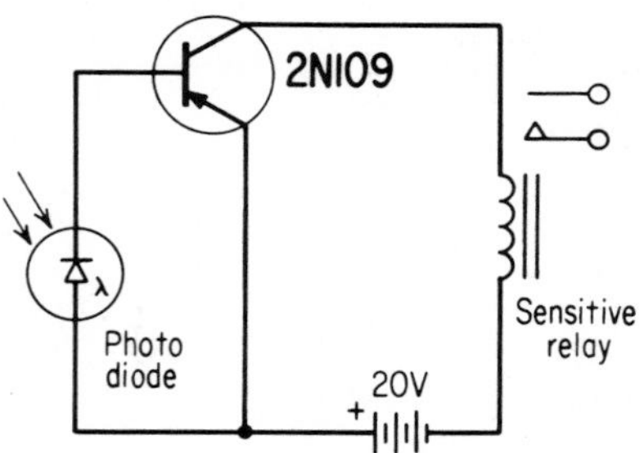

Figure P9-5

P9-6. Draw a suggested electrical arrangement for an automobile fuel-level indicator, where a float immersed in the tank mechanically operates the arm of a voltage-divider potentiometer. The ends of the potentiometer are connected to the 12-V supply of the automobile. Specify the type of voltmeter to be used as the fuel indicator.

[1]H. Allen, "Photoelectrics in Computers," *Industrial Electronic Engineering and Maintenance*, Aug. 1962.

10

The Cathode-Ray Oscilloscope

10-1. BASIC CATHODE-RAY-TUBE PRINCIPLES

The cathode-ray oscilloscope is a highly versatile indicating instrument. Its principles of operation are of such importance and its range of applications so extensive that it will be best, at the outset, to subdivide the topic. The discussion will start with the basic principles of *electron deflection in the cathode-ray tube*, will go on to investigate the *fundamental types of wave-form patterns* obtainable with the oscilloscope, concluding with the outstanding *applications* suitable for laboratory work.

The patterns formed on the scope screen result basically from the effect produced when a narrow beam of electrons is shot from the "electron-gun" assembly and impinges on the phosphorescent coating of the tube face that forms the screen of the cathode-ray tube. The glow resulting from the energetic impact of the electrons on the phosphor produces a *single spot of light on the prepared face* (or screen) of the tube. Although we rarely see this spot of light in its stationary form (indeed we try to avoid leaving the spot stationary, to avoid "burning up" the screen material), it is the fact that this spot is formed, and that the beam forming it can be deflected horizontally and vertically by applied voltages, that makes possible the presentation of wave-form patterns on the screen.

Electron-Gun Arrangement

The arrangement of the "electron-gun" assembly, as used in the electrostatic-deflection type of cathode-ray tube of the oscilloscope, is shown physically in Fig. 10-1, and its electrical connection circuit is shown in Fig. 10-2. The electron

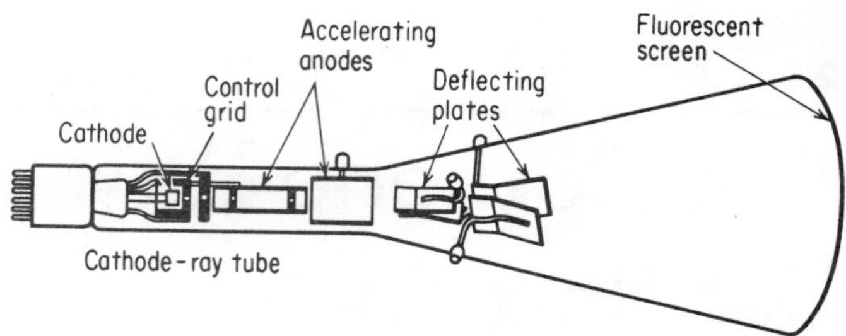

Figure 10-1. Cathode-ray-tube construction.

gun in these diagrams is seen to consist of the heater with its indirectly heated cathode, the control grid, and the first and second anodes. The control grid in the CRT, unlike the ordinary electron-tube grid, is cylindrical in shape with a small aperture in line with the cathode. The electrons emitted from the cathode emerge from this aperture as a slightly divergent beam. The quantity of electrons making up this beam (and therefore the beam current) is controlled by the amount of negative bias applied to the grid, so that the cylindrical grid exerts its control in essentially the same manner as the conventional grid structure, with the added provision of beam formation. Since the intensity (or brightness) of the phosphorescent spot depends upon the value of the beam current, the knob controlling the grid bias is called the *intensity control.*

The diverging beam of electrons emerging from the control-grid aperture is converged and brought to a focus on the screen by the two anodes, which produce a lens action electronically. In line with the opening in the grid cylinder is a corresponding opening in the narrow cylinder that constitutes the *first anode*, and which is highly positive with respect to the cathode. A wider cylinder following the first anode also has its aperture in line with the electron beam and is operated at a still higher positive potential. This *second anode* is known as an accelerating anode. The combination of the first anode cylinder and the wider and more positive second anode cylinder produces a configuration of the electric field that exerts its force to bend the electron beam and produce a thin beam of electrons that gradually converge at the screen. This action is strictly comparable to that of a converging lens when it bends a diverging beam of light, so that the light is converged to a single point-focus. In fact, this arrangement of first and second anodes is known as an "*electron-lens*" arrangement.

The control by which the electron lens is caused to focus the electron beam exactly on the screen surface is the potentiometer providing a variable voltage to

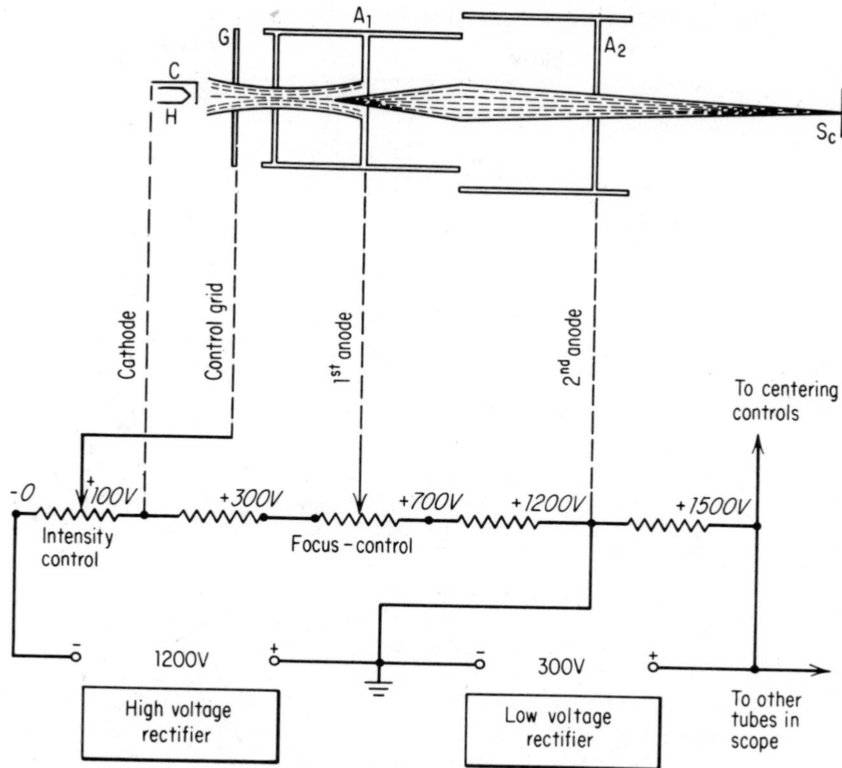

Figure 10-2. Electron-gun arrangement for electrostatic-deflection cathode-ray tube. The variable voltage applied to the grid controls intensity, and the variable voltage applied to the first anode controls focus. [From J. F. Rider, *The Cathode-Ray Tube at Work* (New York: Hayden–Rider, 1935).]

the first anode, which is therefore termed the *focus control.* When this control is turned to either side of its correct focusing position, it will be found that the phosphorescent spot becomes larger and fuzzier. Returning the control to its correct position, for each particular electron–lens combination, will give a concentrated, bright spot. When the spot is properly focused, it will later produce the sharp narrow lines that trace the pattern on the cathode-ray-tube screen, as a result of the action of the deflecting voltages on the beam of electrons.

The Deflecting Plates

The two sets of deflecting plates (for electrostatic deflection) are shown in Fig. 10-3. The first set (from the gun side) is for vertical deflection and the next set for horizontal deflection. With no potential applied to the deflecting plates, the

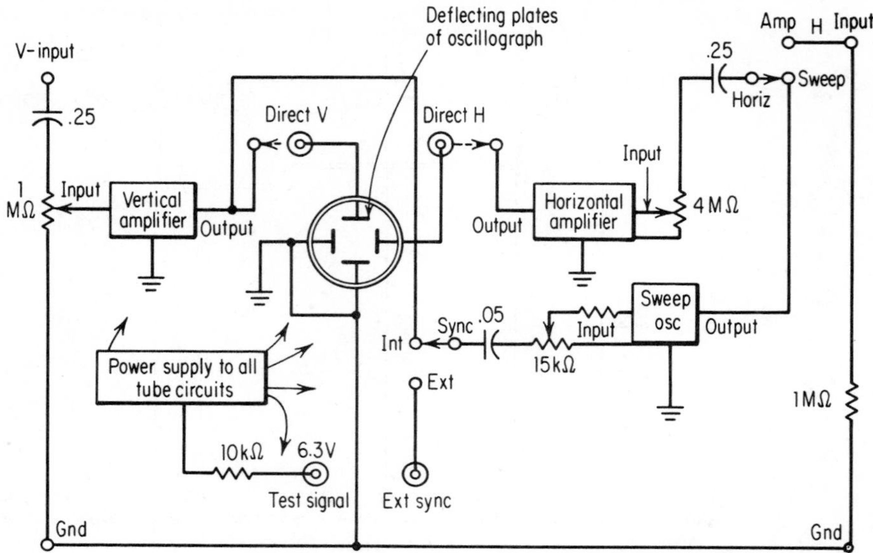

Figure 10-3. Block diagram of typical cathode-ray oscilloscope. (From S. D. Prensky, "Know Your Oscilloscope," *Radio-Electronics*, 1940.)

cathode-ray beam passes between both sets of plates and appears as a bright spot on the center of the tube screen, as previously discussed. If a positive potential is placed on the top one of the vertical-deflecting (or *V*) plates (the other plate of the pair being grounded in single-ended operation), the beam—and thus the spot—will move upward by an amount depending on the potential applied. That this will be so, is obvious when we consider that the beam is a stream of electrons (negatively charged particles) and is electrostatically attracted toward an oppositely charged plate. By the same token, if a negative charge is applied to the same top plate, the spot will travel downward owing to electrostatic repulsion between like charges.

The general procedure, in the common, general-purpose type of oscilloscope, is to ground one of each of the pairs of deflection plates, so that deflection voltages are supplied to only one plate in each pair. It will thus be seen that an ac voltage applied to the vertical-deflection plate will cause the spot to move alternately up and down from its central position, with this motion being continually repeated for each ac cycle. Since the application of 60-Hz ac will cause 60 such up-and-down movements in a second, the visual indication will be a straight vertical line, owing to the effect of the persistence of vision. In fact, an intermittent image that registers even as little as 16 times/sec will appear continuous because of this *persistence-of-vision characteristic* of the eye. This effect is also aided by the medium-persistence characteristic of the phosphorescent screen, which causes the glow produced by the electron impact on the screen to remain for a very small fraction of a second after the impact. As a result of the combination of both of these persistence effects, the line of motion produced by all but the very lowest frequencies will appear to be a continuous line, with flicker becoming noticeable only at frequencies less than about 20 Hz.

Horizontal deflection is produced in precisely the same way, when a deflection voltage is applied between one of the horizontal-deflecting (or *H*) plates and ground. As a general rule, the voltage applied to the horizontal plate produces a *horizontal base* (*designated as the x axis*), and is made to represent time by sweep methods that will be discussed later.

The *vertical, or y-axis, deflection* generally represents the applied signal voltage, and its peak-to-peak amplitude will thus represent the maximum positive and negative swings of the applied signal.

10-2. HORIZONTAL TIME BASE (SWEEP VOLTAGE)

The production of a horizontal-deflection voltage, which will sweep the trace uniformly from left to right in a controllable time and thus act as a *horizontal time base*, is accomplished by the use of a *periodic voltage in the form of sawtooth wave*. To see the reason for this more clearly, consider a condition where no voltage is applied on the horizontal plates while an ac voltage, having the familiar 60 Hz sine-wave form, is supplied to the vertical plates. The result is a vertical trace, where the vertical displacements up or down from the center represent the maximum positive and negative peaks of the applied ac voltage. Thus, for the duration of one cycle ($\frac{1}{60}$ sec), the trace will move up to its positive peak, back to zero, down to its negative peak, and back to zero again; it will do so for each cycle of the ac voltage. The result of this beam movement appears to the eye simply as a vertical line.

The problem now is to devise a means for spreading out (or sweeping) this vertical movement to produce a screen pattern in the form of a graph showing vertical deflection vs. time and so depict the wave-form of the signal under observation.

Electronic Sweep Action

Satisfying the requirement for a sweep motion having a uniform horizontal speed in one direction, followed by an abrupt flyback to the starting point—all done in synchronism with the vertical signal—would seem to be a quite difficult task, and it is, in fact, very hard to accomplish by mechanical means. However, by electronic means it is comparatively easy to satisfy the required conditions by using a *sawtooth wave-form*.

In the oscilloscope, a voltage having this sawtooth form is applied to the horizontal-deflecting plates. As this sawtooth voltage increases in a substantially linear manner, it will move the electron-beam spot uniformly to the right by an amount depending on its ultimate amplitude (this has the same effect as moving our imaginary paper to the left). When the sawtooth voltage falls abruptly to zero, the spot experiences no horizontal deflection and flies back almost instantly to its original position. The generator of this sweep voltage can be a unijunction transistor relaxation oscillator, producing the periodic sawtooth voltage at some initial repetition rate (or frequency); moreover, it can easily be made to fall into syn-

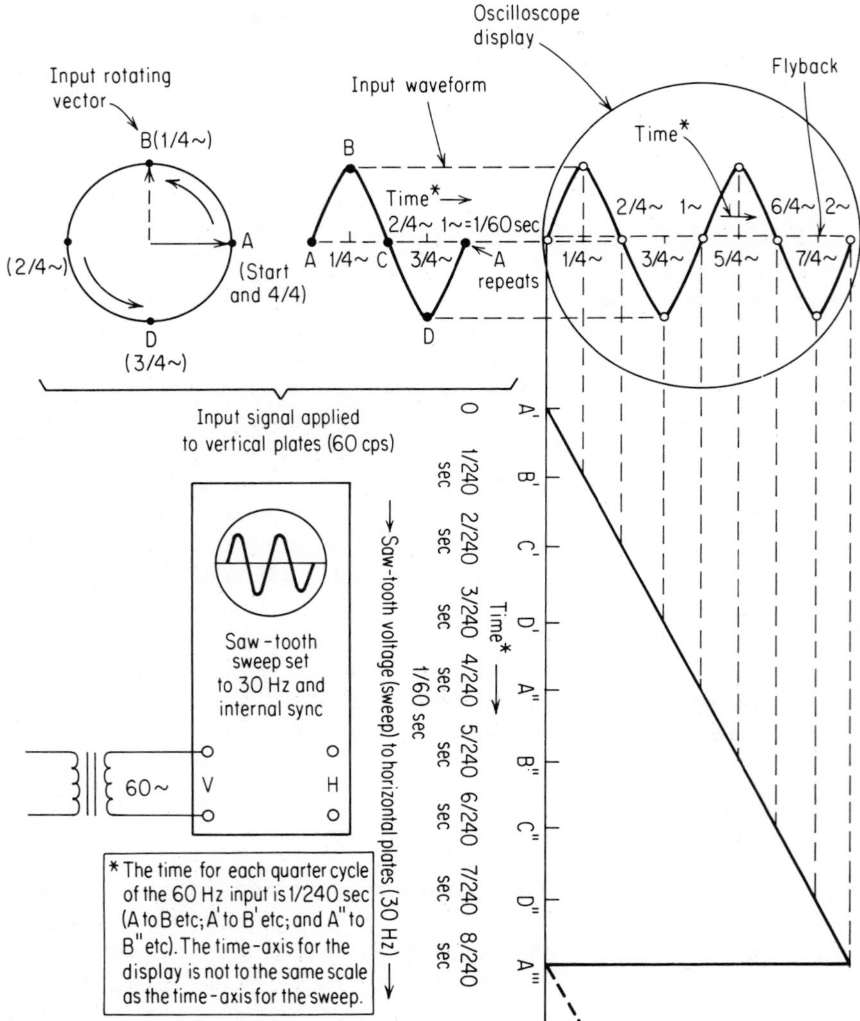

Figure 10-4. Oscilloscope screen pattern of 60-Hz signal with a sawtooth voltage (repetition rate 30 Hz) applied to horizontal plates.

chronization with the applied vertical signal by the simple expedient of injecting a portion of the applied signal into the base 2 circuit of the sawtooth (or sweep) generator. Then, as long as the free-running frequency of the sweep generator is adjusted to be approximately within the vicinity of a submultiple of the frequency of the applied signal, the sawtooth voltage will be forced into exact synchronization with the applied signal. Since the synchronization is obtained by the internal wiring connections, this action is called "*internal synchronization.*" As a result, the successive images being observed will superimpose in exact step, and will appear as a stable stationary pattern, owing to the persistence of vision.

As an example, with a 60-Hz signal connected to the vertical input post of the oscilloscope, assume that the sweep-generator frequency has been set to produce a free-running repetition rate for the sawtooth sweep voltage in the neighborhood of, say, 25 to 30 Hz. The internal synchronization arrangement would then force the sawtooth generator into a whole-number submultiple of 60 Hz or, in this case, into a stable 30-Hz *repetition rate*. This would result in 30 superimposed traces per second, and the resulting screen pattern would display *two complete cycles* of the applied 60-Hz signal. This condition is analyzed in Fig. 10-4, which shows the action resulting in a stationary pattern that displays two cycles of the applied signal.

The different appearance of the display when the sawtooth is swept at the same frequency as the applied 60-Hz signal is shown in Fig. 10-5(a), contrasted with the display for the same signal when swept at the slower 30-Hz rate shown in Fig. 10-5(b). By the same token, if the approximate frequency of the sawtooth sweep were made much lower, so that the sweep would lock in at *one-fourth of the applied 60-Hz frequency*, or at 15 Hz, we would observe a pattern of *four*

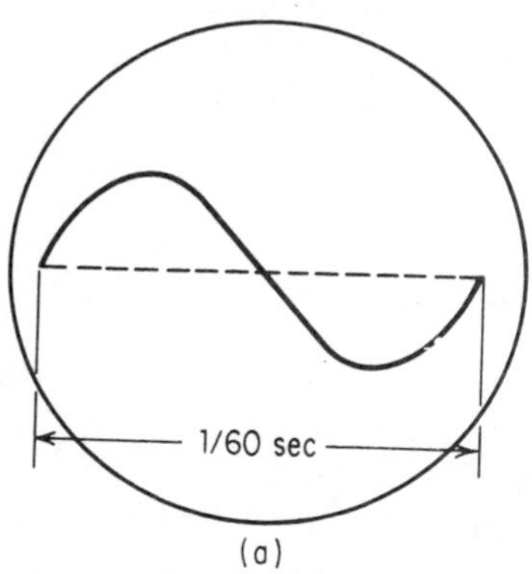

(a)
Single cycle of 60∼ voltage with 60∼ sawtooth sweep

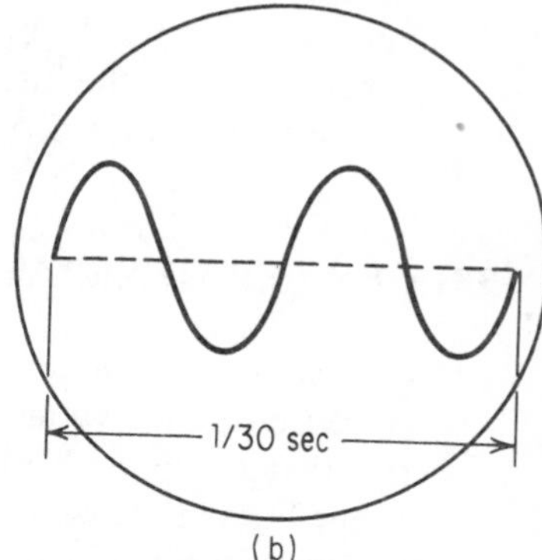

(b)
Two cycles of same 60∼ voltage using 30∼ sawtooth sweep

Figure 10-5. Wave-form display with different sweep repetition rates; (a) single cycle of 60-Hz voltage with 60-Hz sawtooth sweep; (b) two cycles of same 60-Hz voltage using 30-Hz sawtooth sweep.

complete cycles. Since this pattern would be the result of 15 complete traces superimposed on each other every second, we would then be able to observe a *flickering of the stationary pattern due to the relatively slow repetition rate of 15 patterns presented per second.*

10-3. HORIZONTAL DEFLECTION BY SINE-WAVE VOLTAGE

It would be instructive, at this point, to investigate the result of applying a *sine-wave voltage to the horizontal plates* (rather than a sawtooth sweep voltage), while a similar sine-wave signal is being applied to the vertical input terminals. The action resulting from a 60-Hz sine-wave signal applied to *both vertical and horizontal plates* is shown in Fig. 10-6. The display produced is seen to be a straight slanting line. A conventional graph of voltage vs. time is not obtained in this case, because our time base on the screen is no longer a linear one (as we would have when a sawtooth voltage is used for the sweep), but is instead a sinusoidally varying sweep that keeps time with the sine-wave variations of the applied signal on the vertical plates. This pattern, produced by sine-wave forms on both the vertical and horizontal plates, is termed a *Lissajous figure*. The figure, resulting from the sine-wave sweeping of a sine-wave input signal, will obviously depend on the frequency and phase relations between the vertical and horizontal signals. In this case, the straight slanting line is an indication that the two signals have the same frequency and are in phase with each other. As will be seen in later discussions of frequency and phase measurements, the line would open up into an *ellipse* if the *frequencies of the two signals were the same but their phase was different.* Other patterns would be obtained if the two frequencies were different, and in later discussions, the interpretation of the various Lissajous figures will be shown to be very useful for certain frequency and phase measurements.

10-4. THE SWEEP-GENERATING CIRCUIT

Returning to the production of the more usual type of wave-form presentation, where the applied voltage is displayed on the screen as a function of time, we have seen that a sawtooth sweep voltage, which rises linearly with time, is required for the linear time base on the screen. A simplified diagram of such a sweep circuit is shown, in functional form, in Fig. 10-7, where a unijunction transistor is used as a relaxation oscillator to generate the sawtooth voltage output. The generation of this output by the UJT depends for its action on charge and discharge of a capacitor, in the relaxation oscillator circuit shown. Here a capacitor is selected by switch S_1 and is charged by the supply voltage, while the UJT is nonconducting. The comparatively slow rise of this charge occupies the major part of the generated cycle and is confined to a practically linear portion of the charging curve by the

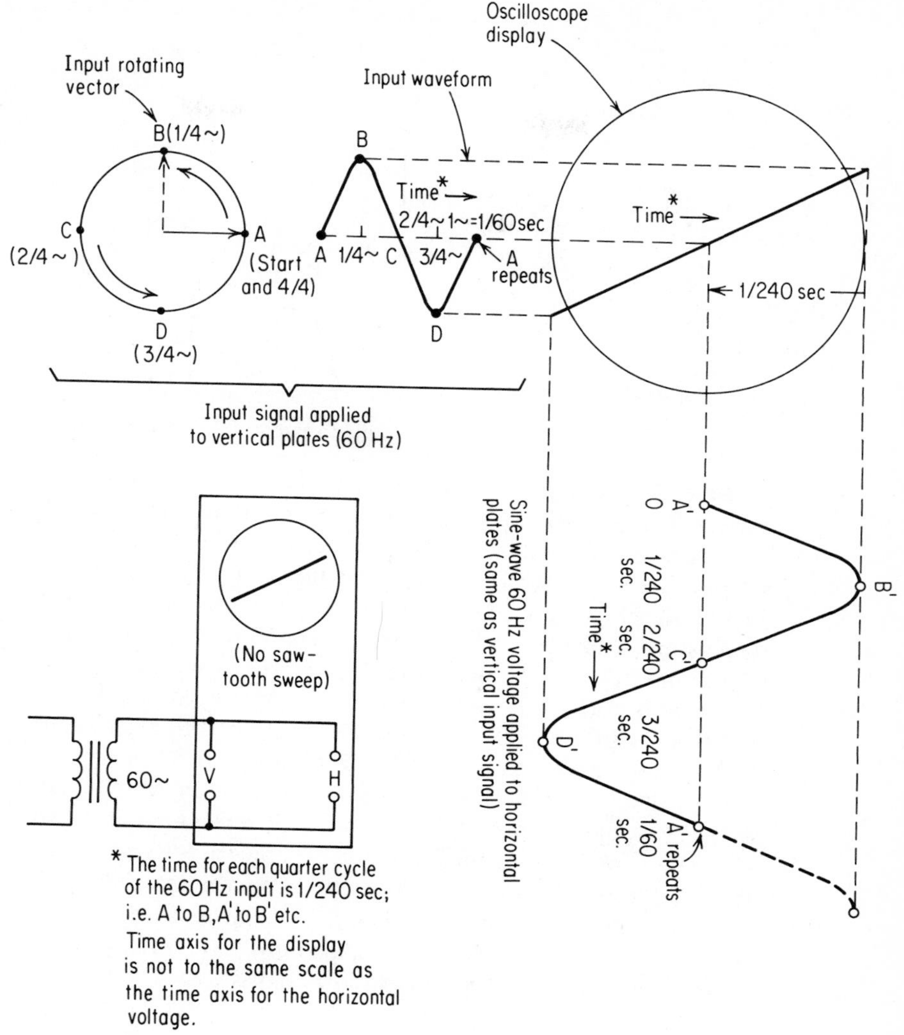

Figure 10-6. Oscilloscope screen pattern of 60-Hz signal with a sine-wave voltage applied to horizontal plates.

circuit constants. Then, when the voltage across the capacitor has risen high enough to cause the emitter of the UJT to conduct, the capacitor discharges rapidly through the UJT, causing an abrupt fall of output voltage toward zero, and also dropping the voltage from emitter to base 1 so that the UJT turns off. This starts the new cycle with the capacitor voltage slowly rising again to its peak voltage before the UJT fires once more. The time (A to B) for the voltage rise of the sawtooth wave-form should occupy as large a portion of the whole cycle as

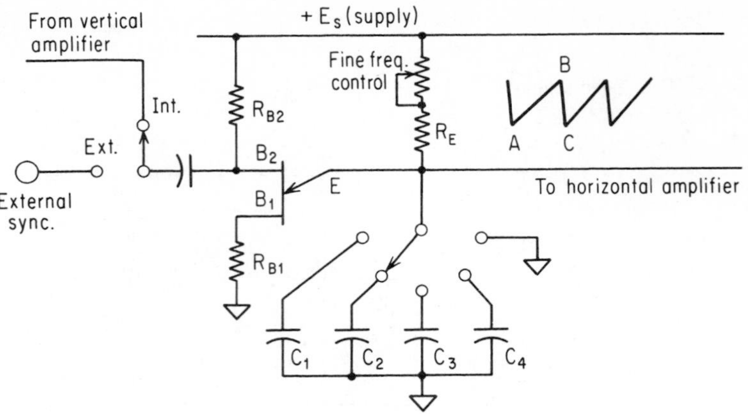

Figure 10-7. Simplified schematic of sweep circuit.

possible; it can easily be made to be well over 90% of the time for the complete cycle and to be fairly linear by the proper choice of circuit constants.

A necessary requirement for proper operation of the oscilloscope circuit, as we have seen, is that the frequency of the sawtooth generator should be in exact step (or synchronization) with the voltage being observed, in order for the pattern to remain stationary on the scope screen. This exact synchronization is accomplished in the base 2 circuit of the UJT. It will be seen, in the sweep circuit of Fig. 10-7, that a portion of the vertical signal can be fed into the base 2 circuit when switch S_1 is in the INT position. The effect of this injected signal is to "lock" the frequency of the oscillator in with the injected frequency, whenever the oscillator frequency gets close to a submultiple of the injected signal. A closer control of the generated frequency is provided by the variable resistor labeled FINE FREQ control. This locking-in action becomes very pronounced as a submultiple frequency is approached,[1] so that, at the proper point, the rapidly moving pattern will quite suddenly become stationary. (For this reason it is also called a "HOLD" control, particularly in television applications.) The EXT position of the S_1 switch allows synchronization from an external signal.

10-5. CLASSES OF CATHODE-RAY OSCILLOSCOPES

The many types of cathode-ray oscilloscopes that are commercially available can be roughly divided into two main groups: the basic *general-purpose type*, widely used for the display of fairly simple wave-forms and in service work, and the more advanced types, which will be grouped together in a *laboratory class* of scopes.

[1]The synchronization is effected when the sync signal at base 2 causes the UJT to fire a short time before it would have fired without synchronization, thus forcing the relaxation oscillations to take place in step with the sync signal. Further details of this process are given in Jacob Millman and Herbert Taub, *Pulse, Digital, and Switching Waveforms* (New York: McGraw-Hill, 1965).

Before attempting a detailed description of either class of oscilloscope, it will be helpful to consider the reason for the existence of such a wide variety of oscilloscope models (and the correspondingly wide price range that they encompass). For general-purpose viewing of ordinary wave-forms, such as the audio-frequency variations produced in either a microphone or a loudspeaker, the requirements imposed on the scope specifications as to sensitivity and frequency response, for example, are relatively easy to meet. Moreover, even when the desired frequency range extends to 4.5 MHz, (as it does in television servicing), an inexpensive general-purpose scope of the maintenance-service type can still meet the requirements (with some sacrifice in sensitivity, but still adequate for general use). Consequently, there are a large number of these service-type scopes commercially available at low cost, making them quite popular for general use (see Fig. 10-8).

The popularity of these *service-type scopes* tends to obscure the fact that the inexpensive, free-running sweep arrangement requires careful adjustment for synchronizing signals at discrete frequencies. For educational use (as in the study of topics in sound, for example), a *triggered-type sweep* arrangement is much more preferable and is obtainable, in some instances, at only a slight increase in cost. Examples of such triggered-sweep types (see Fig. 10-12) and of other still more advanced features (Figs. 10-13 and 10-14) are discussed in Sec. 10-9 as *laboratory-type scopes*.

In the laboratory class of scopes, the great versatility of the combination of the cathode-ray tube and electronic amplifiers has permitted very extensive developments and refinements, making it possible to display highly complex and rapidly varying wave-forms and pulses. The laboratory class of scopes, as a result, possesses advanced capabilities in many specialized areas, such as precise time measurements for radar use and stable amplification of very-low-level signals in biomedical applications. The distinguishing features of each major class are discussed next.

Advanced Features in Laboratory Scopes

Among the factors that characterize the more advanced class of scopes, as compared with the simple type, we can distinguish the *increased sensitivity* and *wide-band frequency response* of the vertical amplifiers. In the horizontal-amplifier portion, we can observe the kind of synchronizing arrangement for the time-base sweep, whether it is a manually adjusted *recurrent* (*or free-running*) *sweep* or an automatic type of *trigger* (*or driven sweep*), and whether a *delayed sweep* is available. Finally, we can note whether provisions have been made for obtaining *calibrated amplitude* for the vertical deflection and *calibrated sweep time* for the time-base arrangement. As is to be expected, the price class of the advanced types of laboratory scope will naturally differ widely, depending upon how many of the major advanced features are desired or needed. As a result of the rapid technological developments in the electronic areas of space telemetry, digital data processing, and pulse techniques in general, the oscilloscope has been called upon for increasingly greater capabilities in sensitivity, accuracy, and fast response. In keeping

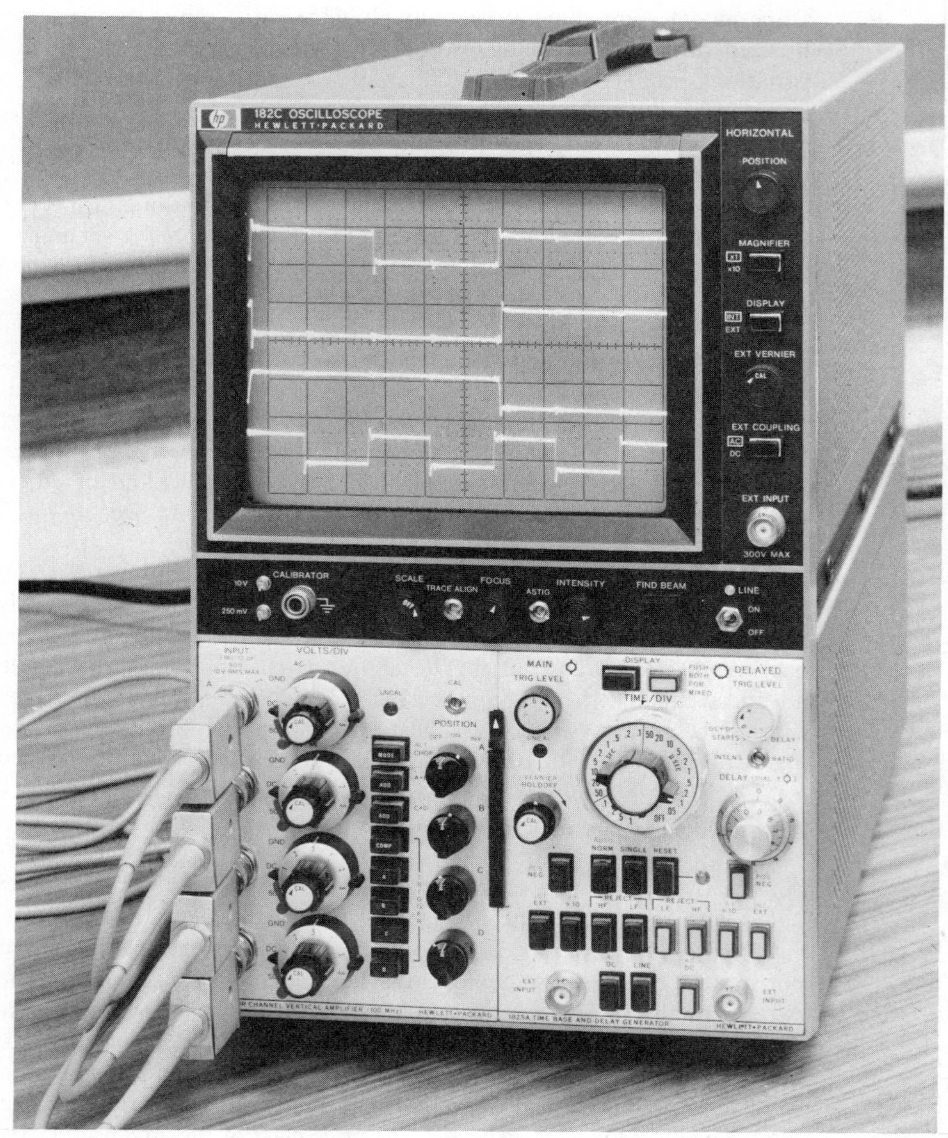

Figure 10-8. A modern dual-trace oscilloscope. (Courtesy Hewlett-Packard.)

pace with these advanced developments, many new specialized models have been produced, some providing fast *rise times measured in millimicroseconds* (*nanoseconds*), others having built around the basic cathode-ray-tube block a variety of *plug-in units* for specialized purposes such as, for example, dual-trace displays and various other features.

Extended Applications of General-Purpose Scopes

Concurrently with the development of specialized oscilloscope applications the usefulness of the simple, general-purpose scope has been extended by *sensitive transducers*, which change the nonelectrical observed quantity into an electrical signal that can be displayed on the oscilloscope. Thus, the general-purpose scope, even without any of the advanced features discussed later in greater detail, is still a highly versatile instrument not only in displaying electrical phenomena, but also in monitoring such nonelectrical quantities as the pressure variations in an automobile cylinder and a host of other nonelectrical industrial variations. More particularly, in the wide field of *science teaching, the combination of suitable transducers and the general-purpose scope* forms a highly effective tool for class demonstrations of scientific principles in such various areas as physics, chemistry, biology, and general science.

In the discussion of typical oscilloscope models that follows, considerable attention will first be given to the basics of the general-purpose scope of the service type and a clarification of its simple operation, to emphasize its availability as an effective electronic instrument to science workers both in and out of the strictly electronic field.

10-6. REPRESENTATIVE GENERAL-PURPOSE OSCILLOSCOPE

In the block diagram of a general-purpose scope in Fig. 10-3, it will be observed that there are *four main functional blocks* that enable the cathode-ray tube to display the input signal wave-form:

1. *The vertical amplifier*, to provide ample vertical deflection of the input signal;
2. *The sweep oscillator*, combined with
3. *The horizontal amplifier* where, for the ordinary wave-form presentation, the combination of the sweep oscillator with the H amplifier provides a suitable time-base deflection, in synchronism with the input signal;
4. *The power supply*, made up of a low-voltage section and a high-voltage section (usually a negative supply of around a few thousand volts).

Vertical Amplifier

The vertical amplifier is the main factor in determining the sensitivity and frequency response characteristics of the oscilloscope. In general, *greater sensitivity* must be obtained at the expense of a *smaller bandwidth* (and vice versa), since the *product of gain times bandwidth is a constant* for a given amplifier. Therefore, in describing the sensitivity of the scope, it should be specified as so many volts required per centimeter (or per inch) of vertical deflection at midband frequency, and with the response down (say, −3-dB or 70.7% of the original) at some specified band-edge frequency, as 500 kHz. In the case of the representative example (*H.P. type 182C*, Fig. 10-8), the sensitivity is given as 18 mV rms (or 50 mV p-p) per inch of vertical deflection when the (−1-dB) bandwidth is 500 kHz. It should be realized that while the −3-dB point is generally known as the *high-frequency cutoff point* (f_c), there is actually no sharp cutoff, only a gradual decrease in amplitude at frequencies higher that f_c. Thus, in this particular scope, the high-sensitivity position (18 mV/in.) is called the narrow-band position and is −6-dB down at 1.5 MHz; or, in other words, its amplitude is down only 50% at 1.5 MHz; or stated in still another way, it is still eminently usable at 1.5 MHz with 50% of its maximum sensitivity, even though its −3-dB point (or high-frequency cutoff) would be given as around only 1 MHz.

The general interrelation between sensitivity and bandwidth is well illustrated by the *182C scope model*, since it allows a choice to be made by switching between a wide-band position and a high-sensitivity position. In the wide-band position, with a high-frequency response out to 4.5 MHz (at the −1-dB point), the sensitivity is 53 mV rms/in., while in the high-sensitivity position, it is 18 mV rms/in., but the −1-dB point is then only at 0.5 MHz or 500 kHz (i.e., narrow-band). Quantitatively, this does not show an exactly constant gain-bandwidth product because of circuit complexities, but the general idea is illustrated: the sacrifice of gain for greater bandwidth, or vice versa.

The input impedance is 1 MΩ shunted by around 75 pF for shielded cable input. Methods of reducing the equivalent input capacity will be discussed in later sections on high-frequency scopes.

Sweep Oscillator

The sweep oscillator for generating the time base in a representative general-purpose scope has a frequency range of from about 10 Hz up to around 50 kHz (100 kHz in the example), controlled by a coarse sweep range control, and by a fine-control sweep vernier knob.

Synchronization of the time base is provided by a switch allowing the choice between internal and external synchronization. In the INT SYNC position, the fine-frequency control is adjusted to set the sweep frequency as close as possible to the desired submultiple of the incoming signal, and then a small amount of synchronizing voltage from the input signal is sufficient to lock the pattern into a stationary display. Although it is easier to get the pattern to lock in by applying a larger amount of sync signal, it is not good practice to do so, since excessive sync signal will distort the resulting display.

[For television-service work, SWEEP PRE-SET positions are provided in this scope for 30 Hz (TV "V") and 7875 Hz (TV "H"), as is a line-voltage sweep with a phasing control.]

Methods providing automatic lock-in by a trigger-actuated (or -driven) sweep are discussed in the more advanced type of scopes, as are also delayed sweeps for observing transients.

Horizontal Amplifier

The horizontal amplifier serves two purposes: (1) in ordinary wave-form presentation, it amplifies the time-base voltage from the sweep oscillator, providing a control for the width of the resulting pattern; (2) when switched out of the sweep position, it accepts a signal applied to the H terminals and amplifies it to a desired amount for application to the horizontal-deflection plates. This would be done when it is desired to compare two separate signals (one on the *V* and the other on the *H* terminals), as to relative frequency or phase by the resulting Lissajous figures.

Power Supply

The positive low-voltage supply is the familiar conventional one, to supply between +300 and +400 V to the plates of the receiving-type tubes in the various blocks (much lower in solid-state models). The high-voltage supply for the cathode-ray tube is a separate supply, providing a negative voltage of from −1000 to −1500 V at a few milliamperes through a high bleeder resistance to ground. The intermediate voltages for the various electrodes of the CR tube are obtained from this bleeder resistance, allowing for the functions of INTENSITY, FOCUS, and VERTICAL or HORIZONTAL POSITIONING.

It is important to observe the **warning notice** given in each instruction book for the particular instrument being used. Even when the power is turned off, contact with the scope connections, when the instrument is removed from the case, can cause severe shock or burns from the high-voltage charge that may be stored in the capacitors. As a standard practice one should *always remove the power cord and discharge the capacitors* before attempting to touch any of the internal connections.

10-7. BASIC OPERATION OF OSCILLOSCOPE (FAMILIARIZATION TECHNIQUES)

To a person operating the oscilloscope for the first time, the array of a dozen or so controls understandably gives a feeling of great complexity, with accompanying fears of either dire results or frustrating nonperformance, if any of the numerous controls should be set improperly. Let it be said at the very outset that *these fears can be overcome with ridiculous ease after even the slightest acquaintance with actual operation.*

There will be three stages in the familiarization process:

Stage 1: To produce a centered spot
Stage 2: To produce suitable vertical height and horizontal width
Stage 3: To produce the desired wave-form

For each of the three stages a separate diagram is given (Figs. 10-9 to 10-11), showing in solid lines the controls actually involved, while the controls not necessary for that stage are indicated by dotted lines on these generalized diagrams. The controls of this generalized oscilloscope are typical of those on popular service-type oscilloscopes [which are mainly of the free-running (or recurrent) sweep type, as opposed to the triggered-sweep type; the latter is discussed later as more suitable for laboratory and educational use]. The general example chosen for the diagram here is fine for the purpose of familiarization because of the convenient arrangement of the controls: four controls for the CR tube arranged around its face (shown solid in Fig. 10-9); two controls for the vertical (or y-axis) deflection, enclosed by the curving line at the left; two controls for the horizontal (or x-axis) deflection at the right; and finally four controls in between for the sweeping functions.

Stage 1: Obtaining a Centered Spot

The spot is originally positioned at the center of the screen by the four controls surrounding the CR tube face (Fig. 10-9). None of the controls beneath the CR tube is operational at this time. However, to ensure that the nonoperational controls are not in a position to disturb the operation of obtaining a centered spot, a preliminary precaution is taken that nothing is connected to the V input posts at the left and that the y-amp control next to these posts is turned to its minimum position (completely counterclockwise). Likewise, the H input posts at the right remain unconnected and the x-amp control next to them is also turned completely counterclockwise. With the power switch turned on, the spot is centered on the screen with the following four controls:

1. Turn INTENSITY control practically completely clockwise.
2. Move the spot up and down with the y-POSITION control, until the spot is halfway up.
3. Move the spot left and right with the x-POSITION control until the spot is midway and centered on the screen.
4. Adjust the FOCUS control until the smallest size of spot is clearly seen.

(**Caution Note:** The usual caution against leaving the bright spot motionless, to prevent burning of the screen, may safely be ignored for the few minutes it takes to make the initial adjustments. If, however, the spot is to remain motionless for more than a few minutes, the screen will still be safe if the intensity control is used to make the spot barely visible.)

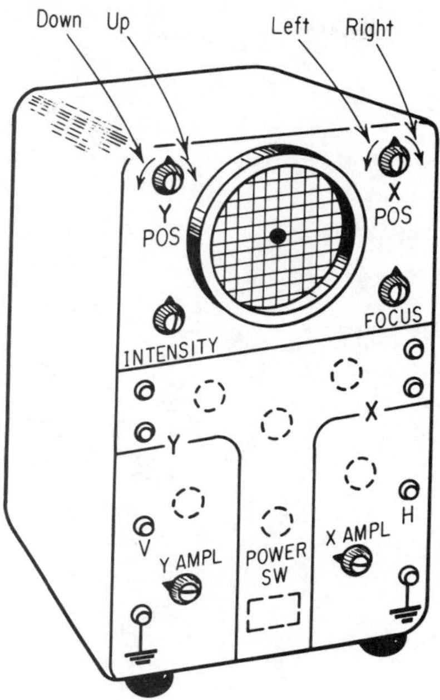

Figure 10-9. Panel controls for obtaining a centered spot on screen.

Stage 2: Obtaining Suitable Vertical Height and Horizontal Width

For this stage (Fig. 10-10), a 60-Hz test signal (usually provided as a test terminal on the scope) is connected to the ungrounded *V* input terminal. (A length of wire hung from this ungrounded *V* input terminal will also serve to supply a 60-Hz signal by stray ac pickup, if a test signal post is not available.) The two *H* terminal posts on the right are left unconnected. Since a satisfactory centered spot has already been obtained, the four controls around the face of the CR tube are no longer needed for this and the following familiarization steps. Now we work only with the two vertical controls below the CR tube on the left to obtain suitable height of the *y* trace, after which the two horizontal (*H*) controls on the right are used to obtain suitable width of the *x* trace:

1. Click the COARSE *y*-ATTENUATOR control so that the switch point selects an intermediate amount of *y* amplification (picking the point labeled 10 gives next-to-the-greatest amount of amplification, since point 1 denotes the least attenuation and point 1000 denotes the greatest attenuation); simultaneously turn up the continuous *y*-AMPLITUDE control until a vertical trace about 2–3 in. high is obtained on the 5-in. scope screen. Note the number (from 0 to 100) on the *y*-AMPLITUDE control, and return it to zero, to return to the centered-spot

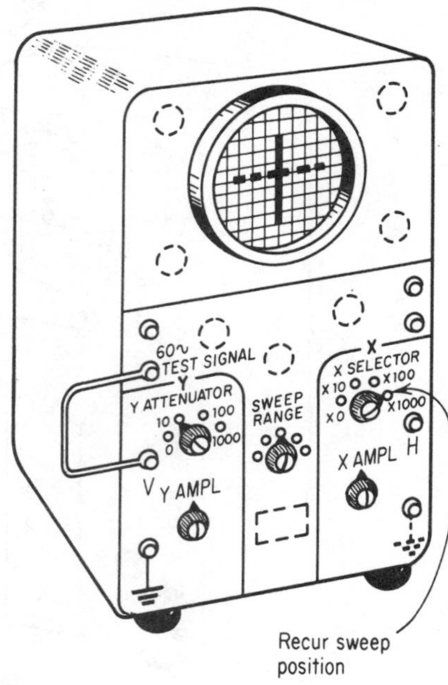

Figure 10-10. Panel controls for obtaining suitable vertical height and horizontal width of screen pattern.

condition. (During all this time, the x-AMPLITUDE continuous control at the right has remained at zero.)

2. Starting again with the centered-spot condition, the horizontal sweep is started by clicking the x-SELECTOR switch to its sweep position. (The x-selector switch is variously labeled x or *H amplifier* and allows choice of connecting the *H amplifier* to either the H-input post or to the sweep oscillator; see Fig. 10-3.) Simultaneously, advance the x-AMPLITUDE continuous control until the spot beocmes a horizontal trace about 4-in. wide on the 5-in. scope face.

3. Control the speed of sweeping with the SWEEP RANGE switch. Note that when a switch point in the counterclockwise position is selected, the sweeping speed is so slow that the eye can detect the flicker as the horizontal trace sweeps slowly to the right and abruptly flies back to the left. As the SWEEP RANGE switch is clicked progressively to the right (clockwise), the rate of sweeping increases until the flicker is no longer visible and a straight horizontal trace is obtained. [Note also that the horizontal trace is obtained only at the SWEEP SWITCH position of the x-SELECTOR switch. Any other position would serve to amplify a signal that might be connected to the H input posts, but since these posts are unconnected, no resulting horizontal deflection is obtained. Thus, a sweep trace is obtained only at the SWEEP position (which does not require any input to the H posts).]

Stage 3: Obtaining Desired Wave-Form

If we now leave the x-AMPLITUDE continuous control at the right in the position found in step 2 (which gave a suitable 4-in. horizontal trace), and return the y-AMPLITUDE continuous control to the number previously noted in step 1 to produce a 2–3-in. vertical trace, we have all the settings required to display the 60-cycle sine-wave form, except for *synchronization.* It can therefore be expected that the result of the steps up to this point will be some unstable moving pattern, generally unrecognizable, unless we just happened to be lucky in the random settings of the three center controls (not yet used) for synchronizing the sweep. These three controls, together with the SWEEP RANGE switch directly below them, are the only tricky settings of all the assorted knobs and controls. The types of trace obtained will differ widely for the same 60-cycle input signal, depending on the rate of sweeping selected and its synchronization with the incoming signal. Therefore, a little faith and patience are needed at this point, until the wild patterns are brought under control to produce the desired stationary wave-form. We will aim at obtaining the wave-form shown in Fig. 10-11, where two cycles of the 60-Hz wave-form are displayed. From what has been said in the previous discussion, this aim will be attained when the sweep oscillator is set close to one-half of this 60-Hz frequency, or at 30 Hz, and when simultaneously a suitable amount of internal synchronizing signal is applied. To achieve this result,

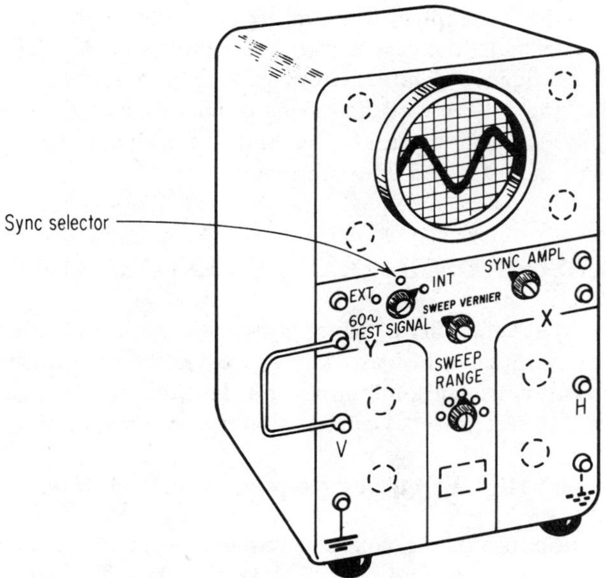

Figure 10-11. Panel controls for obtaining desired wave-form.

we must pay reasonably close attention to the four controls shown dark in the figure.

1. Click the SYNC selector switch to the INT position for internal synchronization.
2. Click the SWEEP RANGE switch at the bottom to the position marked 50. (This will allow sawtooth sweep frequencies between 10 and 50 cycles to be generated, depending on the setting of the SWEEP VERNIER continuous control above it.)
3. Set the SWEEP VERNIER continuous control to a position where two cycles of the incoming signal appear in an almost (but not quite) stationary pattern. (Do not be disturbed by the complexity of the patterns that result while trying to arrive at the two-cycle condition—usually the pattern looks most complicated just before it settles down to a recognizable two-cycle wave-form.)
4. Turn the SYNC AMP control at the right until the two-cycle pattern locks into a stationary display.
5. To ensure that too much SYNC AMP has not been applied, with the attendant danger of disturbing the wave-form, back off on this control a bit, and set the SWEEP VERNIER closer to the desired 30-cycle rate, as evidenced by the condition where the pattern locks in, with the smallest practical amount of the SYNC AMP control.

With the aid of the above dozen steps, the humblest beginner should be able to produce just as excellent a display as any experienced professional without, it is hoped, too much trial and trouble. Although these twelve simple steps by no means exhaust the very versatile capabilities of the oscilloscope, the understanding and confidence obtained from producing a simple wave-form as desired should go a long way toward mastering other capabilities of the scope, such as intensity (Z) modulation, external sync, and phase and frequency comparison by Lissajous figures, as discussed later sections.

10-8. APPRAISING OSCILLOSCOPE PERFORMANCE

The great versatility of which the oscilloscope is capable has resulted in the intensive development of many kinds of advanced instruments. Before discussing these, we shall consider some important desirable features that go beyond the general-purpose (service-type) oscilloscope previously described.

Vertical High-Frequency Response and Rise Time

The response of the medium-frequency general-purpose scopes is generally satisfactory up to around 4 or 5 MHz. The high-frequency scopes provide adequate performance up to around 10 MHz (with some considerably higher); they have a proportionately better rise time. Rise time is the more important specification for "faster" scopes, while the frequency response of the "slower" scopes is generally expressed in terms of the passband or bandwidth. The *rise time* t_r is related to the *bandwidth BW* by the constant-product relation:

$$t_r(BW) = 0.35$$

at optimum transient response. Thus, in the representative case discussed in a later section, a 15-MHz bandwidth offers a nominal rise time of 0.35/15 MHz or 0.023 μsec (or 23 nsec). Factors larger than 0.35 probably indicate overshoot in excess of 2%.

Ideally, scopes should have a vertical system capable of rising in about one-fifth the time that the fastest observed step rises. In that case, the rise-time measurement can be depended upon within a 2% error. However, *for comparing the rise time of two signals*, as is the more frequent requirement, scopes having a *rise time equal* to the rise time of signals applied to them are usually adequate.

Alternating-Current and Direct-Current Coupling

Most oscilloscopes provide dc coupling (in both vertical- and horizontal-deflection systems) in addition to the regular ac coupling. This, in many instances, is a very desirable feature for measuring the action of dc levels, as, for example, in transistor analysis. Even when the dc-coupling arrangement is used for displaying ac signals, this feature provides undistorted and unattenuated response to low-frequency signals and is also useful when examining low-frequency response to fast signals.

Triggered Sweep

Together with the higher passband, the provision for the highly useful *triggered sweep* is one of the outstanding features of practically all advanced scopes. The sweep generator for producing constant-speed horizontal deflection of the beam is present in all scopes. In the simpler basic models these generators run continuously (recurrent sweep), and the control and calibration of the sweep is based on the *repetition frequency*. In order to present a stable display, the repetition frequency of the sweep must be forced to run in synchronization with the input signal on the vertical plates. This is done in the simpler models by carefully adjusting the free-running sweep frequency to a value close to the exact signal frequency (or some submultiple of the exact frequency), and then, depending on the internal sync signal (a sample derived from the input), locking the sweep generator into exact step. This method causes quite some difficulty with low-level input signals, since the sample from the incoming signal then becomes too small to achieve positive synchronization. More importantly, this method is severely limited to displaying recurrent signals of *constant frequency and amplitude*. Hence, when attempting to display voice or music signals from a microphone, the pattern keeps falling in and out of sync as the frequency and amplitude of the microphone output varies, resulting in a constantly shifting unstable pattern.

The limitations inherent in the recurrent (or free-running) sweep are largely overcome by the use of the triggered (or driven) sweep of the advanced scopes. Here, the input signal is caused to generate substantial pulses that "drive" or "trigger" the sweep, thus ensuring that the sweep (if generated at all) must necessarily be in step with the trigger that drives it. As a result, the display remains stable in spite of variations in the frequency or amplitude of the input signal. For most signals (except for the most extreme ones) the display will be stable without any manipulation of the triggering controls, thus providing an "*automatic mode*" *of triggering* in these scopes. In such cases, the sweep can be calibrated in terms of a direct unit of time—for example, 100 μsec/cm—and thus provide an *accurately known time base* for computing the period or frequency of the input signal.

The difference in action between the recurrent type of sweep—which must be readjusted for each new change in frequency—and the triggered sweep—which automatically adjusts itself to the frequency at which it is driven—arises from the different kind of relaxation circuit used for generating the sweep in each case. In the recurrent type of sweep, the generator is free running (or astable) and will synchronize with the internal sync signal only when the free-running frequency has been carefully adjusted to be very close to the frequency at which the locking action is desired. On the other hand, in the triggered type of sweep the triggers exert definite control of the repetition frequency of the sweep within wide limits, and hence the sweep frequency is *forced* to agree with the input signal and thus provide a stable pattern within these wide limits.

Triggered-Sweep Scope for Technical and Educational Use

It is to be expected that the advantages of the triggered-sweep type of oscilloscope can be found in the advanced forms of laboratory oscilloscopes but generally also at a much advanced cost. As a rough estimate, the most common progression is from a service-type scope (in the $200 range) to a much more versatile laboratory type (from close to $1000 and up). However, it is possible to retain the triggered-sweep feature in an *in-between price range* (around $300) by selecting a scope with specifications that are somewhat limited but still quite satisfactory for most *technical and educational* uses. Though perhaps much less widely promoted, it is nevertheless particularly suitable for scientific and teaching personnel, who are not primarily engaged in the electronic field. (To cite a particular example, science teachers in secondary schools would greatly welcome the ability of the triggered-sweep feature in that it can automatically follow the rapidly varying frequencies of speech and music in a display demonstration of properties of sound.)

An example of such a relatively inexpensive triggered-sweep oscilloscope is illustrated in Fig. 10-12. The model shown [*Tektronix Telequipment model S51B* and even more so the specifically educational model (*S51E*)] has a minimum number of control knobs for simplicity of operation. Despite this simplification,

Figure 10-12. Triggered-sweep oscilloscope, relatively inexpensive type, suitable for technical and educational use. (*Tektronix Telequipment model S51B.*)

its performance is entirely adequate for most scope functions, with the possible exception in cases where very high sensitivity is required. The most favorable areas of application can be evaluated by comparing its specifications given below with those of a more advanced laboratory type discussed in Sec. 10-9.

Specifications [***Telequipment Model S51B*** (***Fig. 10-12***)]

Deflection Sensitivity: 100 mV/cm.

Bandwidth: Direct current −3 MHz.

Calibrated V/cm: Nine positions, frequency-compensated.

Time-Base Range: 1 μsec/cm–100 msec/cm in six calibrated steps and variable between steps.

Trigger: INT, EXT, AUTO, or TRIG level.

CRT Screen: 8 × 10 cm at 3-kV accelerating potential, with intensity (Z) modulation capability.

The circuit arrangement in an oscilloscope employing a triggered sweep, such as a Schmitt trigger circuit, is discussed further in Sec. 10-9.

10-9. REPRESENTATIVE EXAMPLE OF TRIGGERED-SWEEP LABORATORY OSCILLOSCOPE

The oscilloscope pictured in Fig. 10-13 (*Hewlett-Packard model 1202A*) is a laboratory, high-sensitivity, medium-frequency type, possessing the highly useful feature of a triggered sweep. When set to the auto position for universal automatic triggering, the sweep will automatically synchronize with almost any input, without manual adjustment of the fine sweep time control. As a result, with all periodic signals (except the very weakest ones), once the initial triggering adjustments are made, the scope pattern will remain locked in a stable form, while the input signal varies in frequency over the full rated range of scope response. The sawtooth sweep generator may also be operated in a free-running position, in which case, of course, the sweep time controls must be carefully set to synchronize exactly with the given frequency of the input signal.

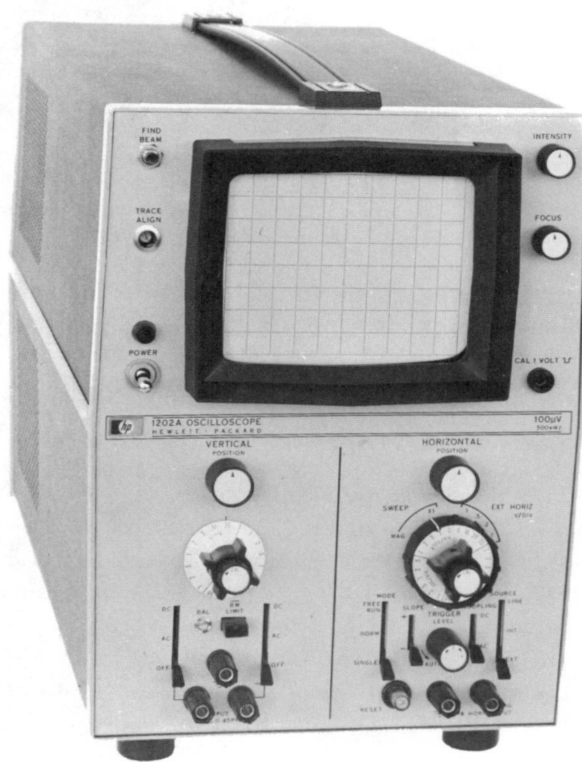

Figure 10-13. Triggered-sweep type of laboratory oscilloscope solid-state version, medium frequency, 500 kHz. (*Hewlett-Packard model 1202A.*)

Schmitt Trigger Action

The electronic action of a triggered sweep that ensures a sweep frequency, always in step with a changing signal frequency, is obtained from the basic action of a pulse circuit known as a Schmitt-trigger circuit (shown in the tube version for clarity in Fig. 10-14). It is a two-tube bistable arrangement, i.e., a circuit having two stable states, with a common cathode resistor (R_k) as the coupling element.

In the circuit diagram (Fig. 10-14), assume that tube B is conducting, with its grid G_2 receiving a positive voltage from the voltage-divider action of resistors R_1 and R_2. The resulting voltage across cathode resistor R_k is high enough to cut off tube A, whose grid G_1 is free to receive the incoming signal. The input signal may be a sine wave or some complex distorted wave, shown for two cycles at the left of the diagram. As the first cycle of the incoming signal rises above a minimum

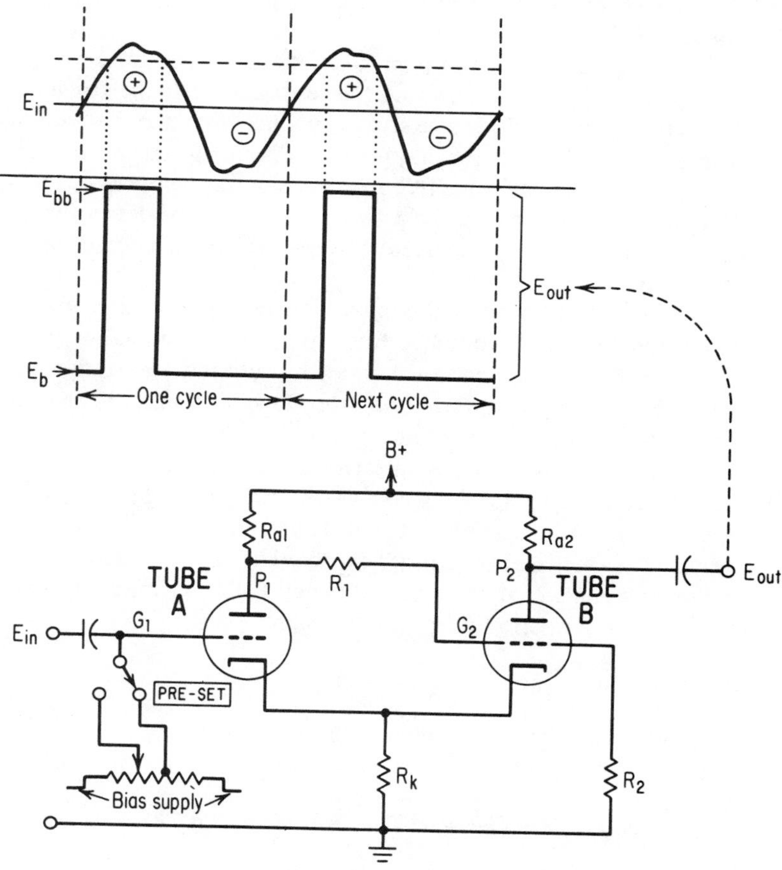

Figure 10-14. Schmitt-trigger circuit; the input wave-form is shown at the upper left with the trigger output directly beneath it. (*Hewlett-Packard model 130A oscilloscope.*)

positive value, tube A goes into conduction, lowering the plate voltage P_1, and consequently lowering the positive bias on G_2. As a result of this regenerative action, tube B switches to its nonconducting state (since its cathode is still at a high positive potential), and the output voltage at P_2 rises abruptly. The P_2 output remains at the high level until the first cycle of the input falls below the dashed line, causing tube A to revert to its nonconducting state, at which time tube B is forced into conduction. The resulting conduction current in tube B abruptly lowers the plate voltage of P_2, causing the output voltage to fall practically vertically, and completing one cycle. The circuit remains in this stable state, until the point in the second cycle of the input signal is reached, high enough to cause the process to repeat itself. Thus, the output of the Schmitt trigger consists of high-amplitude rectangular-pulse triggers, in step with the input signal. These triggers produce very stable and definite synchronization of the sawtooth generating circuit, in spite of the fact the input signal might have been comparatively weak and ragged in shape.

The input circuit of the grid G_1 of tube A contains a potentiometer voltage-divider arrangement for setting the level at which the triggering takes place. At its extreme counterclockwise setting, it clicks into a preset condition that provides optimum automatic triggering for most input signals; at the other end of its rotation (extreme clockwise), a free-running condition prevails, at which time a positive feedback arrangement derived from the sawtooth allows the sweep generator to free-run, producing a recurrent sweep—which, of course, must be synchronized manually.

The greatly simplified explanation of trigger action given here does not include the details of counteracting the hysteresis inherent in the Schmitt-trigger circuit, nor does it mention the various alternative arrangements for setting the level and polarity of internal or external sync sources. The specific details of the sawtooth generating arrangement, which takes the form of a feedback (or Miller) integrator to provide better linearity of the sweep, are also omitted here, in favor of the detailed explanations given later in Chapter 11 on signal generators and in the instrument instruction manual. It will suffice here to say in summary that the triggered sweep makes it practical to view stable screen patterns obtained from periodic signals that shift in frequency (such as microphone outputs) and also from many nonperiodic or transient signals.

Specification Features

The other pertinent features of this laboratory type of oscilloscope include the following specifications.

Specifications (Hewlett-Packard, model 1202A)

Deflection Sensitivity: V: 0.1 mV/cm–20 V/cm
H: 0.1 mV/cm–1 V/cm.

Bandwidth: Direct-current-coupled, dc–500 kHz (rise time = 0.7 μsec); ac-coupled, 2 Hz–500 kHz.

Input Coupling: Single-ended $RC = 1\ \text{M}\Omega$ shunted by 45 pF; differential: common-mode rejection = 100 dB.

Time-Base Range: From 1 μsec/division to 5 sec/division (21 positions) calibrated, and variable between ranges.

Trigger Level and Slope: INT, EXT, or AUTO, with variable selection on each, including single-sweep and free-run positions.

CRT Screen: 8 × 10-cm monoaccelerator at 3 kV, with intensity (Z) modulation and pushbutton beam finder.

10-10. SPECIAL FEATURES IN OSCILLOSCOPES

Many other special advanced features are available in commercial oscilloscopes, but space limitation here permits us only a glance at the more important ones.

Plug-In Arrangements for Special Characteristics

Manufacturers of advanced-type oscilloscopes tend more and more to supply desired special features in the form of a number of different arrangements that can be plugged into the basic cathode-ray-tube unit. This development is well illustrated by the widely used *Tektronix type 547* oscilloscope, which forms a basic unit into which one can plug any one of a number of different vertical amplifier arrangements. Thus, by acquiring the basic plug-in-type oscilloscope, with a single plug-in vertical amplifier (let us say of the *wide-band type*), subsequent purchases of different plug-in types can be used to enable the original instrument to perform special functions such as *dual-trace display*, *high-gain amplification* of low-level signals, *differential amplifications*, *fast-rise pulse amplification*, *spectrum analysis*, and the like.

The Basic Plug-In-Type Oscilloscope (dc, 60 MHz)

A general-purpose dual-beam oscilloscope with plug-ins is illustrated in Fig. 10-15. The mainframe is designed to accommodate two vertical plug-ins and one horizontal plug-in. The mainframe includes the CRT, power supply, two separate vertical and horizontal deflection systems, and auxiliary circuits. The model shown includes optional digital readout of plug-in scale factors on the CRT.

To the right of the CRT are the display controls and power switch. Other controls are located on the plug-ins. The horizontal time-base plug-in is located on the lower right with two vertical plug-ins to the left of it. The horizontal plug-in shown provides two independent time bases with sweep rates of 50 nsec/div to 2 sec/div in 25 calibrated steps. A ×10 magnifier provides sweep rates up to 5 nsec/div. A variety of vertical plug-ins are available for various applications with up to four traces per plug-in.

In addition to the standard oscilloscope functions, this mainframe can accept various accessory plug-ins, such as a semiconductor curve tracer and a spectrum analyzer.

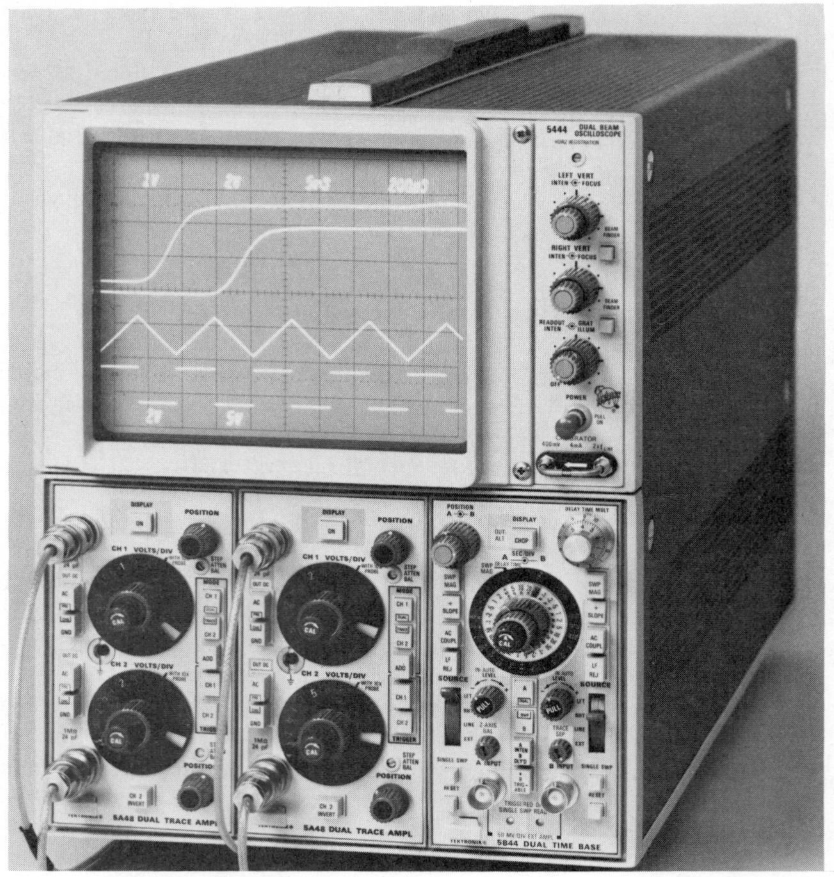

Figure 10-15. Laboratory-type high-frequency scope, showing plug-ins. [(*Tektronix 5444 mainframe with two 5A4B dual-trace vertical plug-ins and a 5B44 dual-time-base horizontal plug-in.* (Courtesy Tektronix, Inc.)]

Sampling Oscillography

As signal frequencies keep getting higher, extending far beyond 50 MHz into the UHF range around 500 MHz (or 0.5 GHz) and beyond, the conventional vertical amplifier and horizontal sweep methods become less and less able to follow the wave-form of the exceedingly fast wave. The method for displaying such superhigh frequencies on an oscilloscope is then modified to a *sampling method* shown in Fig. 10-16. Here the function of the sweep is only to trace a single dot of the signal wave (rather than a continuous curve). By means of a progressive delay, the sampling pulse is made to sample successive points on the signal input [as shown in Fig. 10-16(a)], until the entire wave is traced by the successive dots. This process

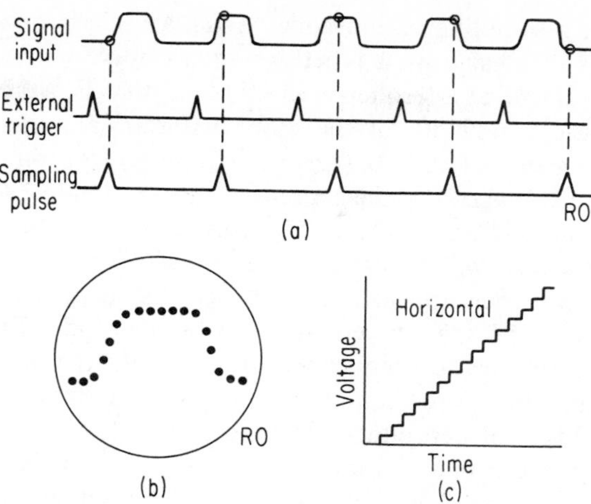

Figure 10-16. Principle of sampling oscilloscope: (a) successive sampling pulses occur progressively after the trigger derived from the signal input; (b) appearance of succession of sampling dots; (c) staircase output for advancing the sampling pulse. (*Hewlett-Packard Application Note No. 36.*)

is then repeated over and over again enough times to form the screen image, as shown in Fig. 10-16(b). Thus the samples provide the vertical deflection and are plotted simultaneously with a staricase signal [Fig. 10-16(c)] that provides the horizontal deflection.

The highly simplified presentation of Fig. 10-16(a) shows that the outline of the signal wave can be taken with as little as five sample pulses. A better outline of the wave is shown in Fig. 10-16(b), with the use of a greater number of samples. If we increase the number of samples taken to display a signal cycle, the display will obviously begin to appear as a continuous curve, rather than a succession of dots.

The result of the sampling action can be profitably compared to the perhaps more familiar action of the *stroboscope*, in stopping rotary motion by means of pulsed flashes of light. If, instead of making the spinning wheel appear to stop its rotation, we instead time the flashes slightly out of exact synchronization, to catch a different point successively on the rotating wheel, we will then see the appearance of a slowly rotating wheel as a representation of the rapidly spinning one. In an analogous manner, the sampling oscilloscope will display a single cycle of, say, a 1000-MHz (or 1-GHz) signal wave, as if it were handling a much lower frequency, well within its sweep-speed capability.

As a clarification of the sampling process,[2] let us, as an example, take the

[2]Condensed from "Viewing Signals in the Nanosecond Domain," by EID Staff Survey, *Electronic Instrument Digest*, Sept.–Oct. 1965.

case of a sampling oscilloscope having a basic bandwidth of only 1 MHz, in the process of measuring a repetitive square wave (period = 100 nsec). This implies measurement of a frequency of 10 MHz, that is, a frequency 10 times as great as the basic bandwidth of the scope without sampling. Suppose that we select a sampling rate of 50 kHz (period = 20 μsec) to produce a "strobe" pulse 0.1 nsec wide and delayed (by the staircase generator) by 0.2 nsec.

The display on the CRT screen will appear as a series of dots, more than 150 of which will be used (in our example) to display the 50-nsec pulse. These pulses will then appear every 20 μsec (based on our 50-kHz repetition rate), so that the CRT may sweep that slowly. The individual output pulses (which we have simplistically been calling dots) would then be somewhat less than 1 μsec wide and therefore require only a moderate bandwith (within the basic 1-MHz capability) for reproduction. In practice, the inevitable smoothing out of the display results in a continuous line, at low sweep rates.

The electronic considerations of the actual sampling process at such very high frequencies is, of course, much more complicated by the fact that the time interval for the sampling act must necessarily have a certain minimum duration, and this time, together with the charging time constant of the sampling circuit, places limits on the possible rise time or bandwidth that the oscilloscope can resolve. Recent advances in the use of high-speed-switching by conventional and step-recovery diodes have enhanced the capability of sampling oscilloscopes. (In the case of the *Hewlett-Packard sampling oscilloscope*, a dual-trace unit is available with a bandwidth of dc–1 GHz; newer models of various manufacturers, beyond this capability, are forthcoming.)

Another approach to displaying signals far into the microwave field is offered by the *traveling-wave oscilloscope*. This approach has the added capability of displaying a single transient, although it suffers by comparison with the sampling oscilloscope, since the latter is superior in the areas of large display size, high sensitivity (3 mV/cm or better for the sampling type), excellent trace contrast and high input impedance.

In summary, the sampling oscilloscope is able to respond and store rapid bits of information and present them in a continuous display. And it is this ability that enables the sampling oscilloscope to side-step the usual limitations inherent in conventional high-frequency oscilloscopes: limited sensitivity and bandwidth and small display size. The sampling technique immediately translates the input signal into a lower-frequency domain, where conventional low-frequency circuitry is then capable of producing a highly effective oscilloscope display.

Storage and Null Readout Scope

Another system for obtaining a digital readout of analog data is used in the *Analab type 1220* mainframe with *type 700 null readout* plug-in, which also features a *storage tube* (Fig. 10-17). Direct reading of amplitude on the *A* and *B* dual channels is provided from a 1% accurate null dial, in ranges from 100 μV to 20 V/cm,

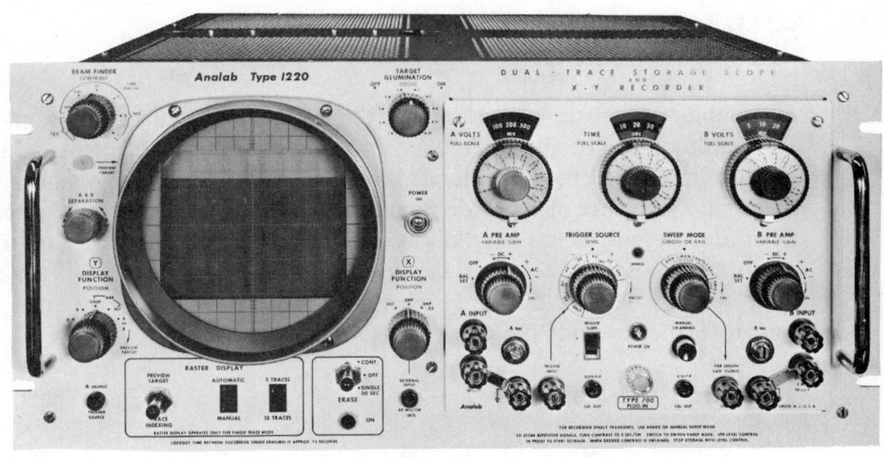

Figure 10-17. Storage oscilloscope, with a dual trace and null readout features provided by plug-in amplifier; the scope itself provides for preview target (2 cm high), storage target (7 cm high), and X–Y recording. (*Analab type 1220 scope with type 700 plug-in.*)

without interpolation. The time base is also provided with null readout for the ranges from 5 sec to 1 μsec/cm. It also measures relative phase angle directly in degrees.

The storage tube makes it possible to store repetitive signals from very low rates up to 100 kHz and single transients to several kilohertz. Images can be observed on a "preview target," 2 cm high, as in an ordinary scope and then transferred to the "storage target," 7 cm high. Images of high contrast *can remain stored* for up to several months, if desired. Erase time is nominally 30 sec.

Among its many possible applications, it can provide readout for low-repetition-rate analog computers, record *x–y* plots (as in antenna patterns or Smith charts), observe and store transients (such as pulse stimuli in neurological studies), and in general provide leisurely examination of any wave-form stored on the scope.

A CONTINUOUS ERASE mode provides a running record in which present and previous traces are continually visible. For example, in monitoring the electrocardiogram wave-forms of a patient during an operation, a five-line presentation could be chosen, so that when the traces are finishing the fifth line, the first line of traces has been erased and is ready to receive the tracing corresponding to a sixth line, and so on. This allows a highly useful comparison of present output, on a given line, with some previous output that is going to be erased but is still stored on some preceding line. Many interesting applications are suggested by possible combinations of the long-time storage and the relatively short-time erase capabilities of such an instrument.

10-11. AUXILIARY EQUIPMENT FOR OSCILLOSCOPES

The ability to observe and analyze electrical signals and their wave-forms makes the cathode-ray oscilloscope one of the most useful instruments in the electronic industry. Its usefulness is extended by its ability to monitor nonelectrical quantities as well, by the use of the wide variety of *special transducers* that supply the related electrical signal, as has been discussed in a previous section on that topic. In addition to such special features as *plug-in units* and *generators for frequency sampling and delay*, which have already been described, other accessory or auxiliary equipment further extends the usefulness of the oscilloscope.

Marker Generators

The marker generator provides visual marks on the scope trace to signify precise time-interval or frequency references. Such markers are available to allow a selection of time intervals from 1 μsec all the way up to 5 sec. These are particularly useful in such pulse measurements as rise time and pulse duration. Other marker generators providing an accurately known frequency, are discussed in Chapter 11 under the heading "Sweep Generators."

Electronic Switch

The electronic switch is a very useful accessory for displaying two signals simultaneously on single-trace scopes. The electronic switch is essentially a square-wave generator. The two signals that are to be viewed are each connected to a separate grid of a dual-section amplifier tube, operating with cutoff bias. The square wave is applied in such a fashion that it alternately unblanks each tube section, thus presenting each signal alternately to the vertical amplifier of the scope. When the switching rate (determined by the square-wave frequency) is high enough, each signal appears continuous, even though it is actually being periodically interrupted. Switching rates as high as 100 kHz are available.

The electronic switch is a useful accessory for the less expensive single-beam scopes that do not have a plug-in facility. Where dual-trace plug-ins are provided, the electronic switching action is already incorporated in the plug-in unit and so does not require the accessory for dual presentation. In some cases, however, the electronic switch can be used with a dual-trace unit to expand the capability for displaying four wave-forms simultaneously.

Signal Averaging to Extract Signals from Noise

As a special-purpose digital addition to an oscilloscope, a *signal averager* has a powerful potential for extracting repetitive signals from noise. By means of a digital implementation of a sampling and storing process (a signal averager such as the *Hewlett-Packard model 5480A*[3]), the signal-to-noise ratio, as displayed on the scope screen, can be improved by factors of well over hundreds of times (the S/N ratio improves as $\sqrt{m}$, where m is the number of repetitions). This method, as well as other means for contending with noise, is further discussed in Chapter 19.

Oscilloscope Probes

The probe performs the very important function of introducing the test circuit into the testing instrument, without altering, loading, or otherwise disturbing the test circuit.

Although probes may be given many different names, they fall into three principal types:[4]

1. The direct probe (or shielded test cable)
2. The circuit-isolation (or voltage-divider) probe
3. The detector (or demodulator) probe

Direct Probe

The simplest probe form is the direct probe in the form of a shielded coaxial cable. It avoids the stray pickup that can be quite troublesome when low-level signals are being examined (with correspondingly high-gain settings on the scope). It is effective in relatively low-impedance or low-frequency circuits, where cable capacitance is not important, and in such cases allows the maximum sensitivity of the scope to be used. However, in using the shielded probe, the shunt capacitance of the probe and cable is added to the input impedance and capacity of the scope and acts to lower the scope's response to high-impedance and high-frequency circuits being tested.

Isolation Probes

To avoid the undesired circuit loading by the combination of shielded probe and scope (or even by the input impedance of the scope itself), the isolation probe is used to decrease the input capacitance and increase the input resistance of the oscilloscope. In its common form of *low-capacitance probe* it attenuates the scope's input capacitance (usually by a 10: 1 ratio), as shown in Fig. 10-18(a). By presenting a greatly reduced input capacitance (and correspondingly increased impedance) to the circuit under test, it necessarily attenuates the input signal by the same ratio. In a *cathode-follower probe*, only the very low input capacitance of the cathode follower itself is presented to the test circuit. Some attenuation in voltage is produced by the cathode-follower isolation, but it is of minor significance. In cases where it is desired to measure signals having voltages high enough to damage severely

[3]C. R. Trimble, "What Is Signal Averaging?" *Hewlett-Packard Journal*, Apr. 1968.

[4]D. Herdt, "Expanding Oscilloscope Use with Accessories," *Industrial Electronic Engineering*, Sept. 1961.

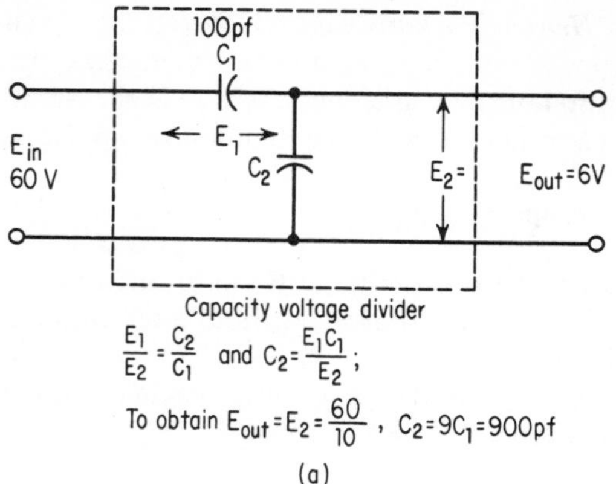

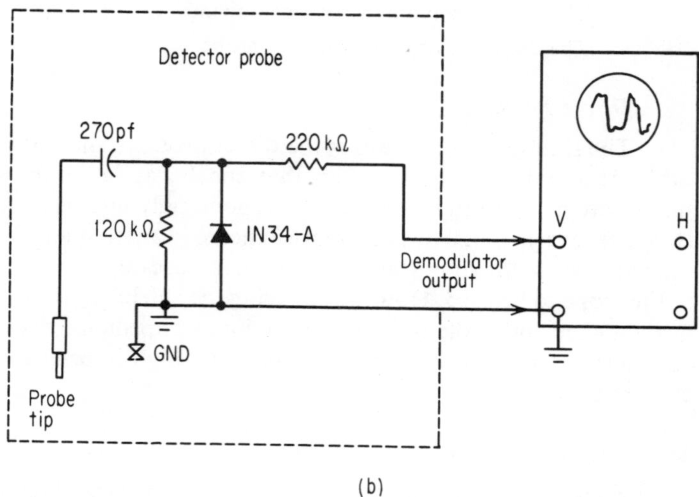

Figure 10-18. Oscilloscope probes: (a) capacity voltage-divider probe for 10: 1 attenuation; (b) shunt-rectifier type of semiconductor detector (or demodulator) probe. [From A. A. Ghirardi and R. G. Middleton, *How to Use Test Probes* (New York: Hayden–Rider, 1964).]

the input circuitry of the scope, attenuation of the signal is the desired objective, and a *high-voltage probe* is used. Such a probe consists of a resistive or capacitive voltage-divider circuit, which allows only a small portion of the input signal to reach the scope but accurately preserves the wave-form being measured.

When isolation probes are used to measure the magnitude of test voltages, the attenuation factor by which the probe divides the input signal voltage must obviously be known as a constant factor to preserve the measurement calibration.

Detector or Demodulator Probe

In analyzing the response to modulated signals used in comunication, as in AM, FM, and TV receivers, the detector probe functions to separate the lower-frequency-modulation component from the high-frequency carrier. By rectifying and bypassing action, the amplitude of the modulation envelope (which is proportional to the response of the receiver to the much-higher-frequency carrier signal) is displayed on the scope. This allows a scope capable of audio-frequency response to perform signal-tracing tests on communication signals in the range of hundreds of megahertz—a range that would otherwise be far beyond the capabilities of practically all scopes, except the very highly specialized ones.

The detector or demodulating circuit of the probe is shown in Fig. 10-18(b). After undergoing the rectifying action of the diode D and the RF bypassing action of the capacitors, the audio signal corresponding to the original modulation is passed on to the output terminals. The resulting display of the audio signal can then be used effectively as an indication of proper alignment of each tuned circuit and for further visual analysis of the overall tuned-circuit response, as would be done in a visual display of bandpass circuits.

10-12. SUMMARY OF OSCILLOSCOPE APPLICATIONS

In view of the extensive uses of the versatile oscilloscope, we cannot attempt a complete listing of all the applications here. The *major application fields*, however, may be briefly summarized as follows:

1. *For Measurement Instrumentation*
   - (a) Measurements of voltage (dc and peak-to-peak)
   - (b) Amplifier gain
   - (c) Frequency and phase relation
   - (d) Rise time and duration of pulse
   - (e) Modulation percentage
2. *For Visual Indication and Analysis*
   - (a) Distortion indication (and also square-wave testing, Chapter 12)
   - (b) Null indication (in bridge and potentiometer comparison instruments)
   - (c) Periodic motion studies (as in the automobile ignition system)
   - (d) Timing comparisons (such as pulse trains in telemetry and digital techniques) and time-domain reflectometry (TDR)
   - (e) Response of tuned circuits (and also sweep techniques, Chapter 11)

In addition, some mention should also be made of developments currently taking place in the higher-frequency field, such as *distributed amplifiers and traveling-wave oscilloscopes*. These types are capable of handling frequencies from 100 MHz up, as exemplified by the *Tektronix 581* type of traveling-wave oscilloscope, which has a rise time of 3.5 nsec. There is also much activity in the development of *memory and storage-tube types of oscilloscopes*, of which only one example was given, the *Analab storage oscilloscope type 1220*. Further development of storage and digital displays is taking place as part of the progress in computer data processing and telemetry applications. *Sampling techniques* also extend the frequency range.

The overall significance of the oscilloscope can well be summarized by starting with the basic fact that all branches of electrical and electronic industries have one thing in common: all are concerned with changing voltage and current.

"With an oscilloscope, one can look at rapid changes in voltage and current better than in any way yet devised. The role the oscilloscope plays extends from medicine to missiles. Any changing phenomena transducible to voltage or current can be studied with an oscilloscope; and the nature of change is a common concern in every walk of life."[5]

QUESTIONS

Q10-1. (a) Explain the formation of a small spot of light on the screen of an oscilloscope when both vertical- and horizontal-deflecting circuits are inoperative;
(b) Why is the spot of light concentrated at one point instead of spread out over the entire screen?
(c) Why is it not advisable to allow this small spot to remain on the screen without deflection for any extended period of time?

Q10-2. Explain why a sine-wave voltage applied to the vertical terminals causes a vertical line to appear, if the horizontal circuits are inoperative.

Q10-3. If a linear-rising sawtooth wave is applied to the vertical terminals, what would be the effect on the resulting screen pattern, if a sine-wave voltage were to be used for the horizontal-deflection circuit?

Q10-4. What type of screen pattern is to be expected if:
(a) A linear-rising sawtooth wave is applied simultaneously to both vertical- and horizontal-deflection circuits?
(b) A sine-wave voltage is applied simultaneously to both vertical- and horizontal-deflection circuits?

Q10-5. Compare the merits of a relaxation type of sweep circuit for horizontal deflection using:
(a) A unijunction transistor.
(b) A bipolar junction transistor.

Q10-6. Explain why it is preferable to use an oscilloscope rather than a VTVM to check the following ac voltages:

[5] J. Mulvey of Tektronix, "Factors Affecting the Validity of Oscilloscope Measurements," *Electrical Design News*, Nov. 1960.

(a) Output from an AF amplifiers.
(b) Ripple from a well-regulated dc power supply.
(c) A pulse-type output, as commonly found in television receivers.

Q10-7. Explain how a scope is calibrated to read directly in volts.

Q10-8. Explain why the output of a photocell exposed to sunlight cannot be measured in the ordinary manner on an inexpensive general-purpose scope.

Q10-9. Suggest two methods for making it possible to measure the photocell output of Question Q10-8.

Q10-10. Mention the types of applications for which the use of a triggered (or driven) sweep is decidedly superior to the more common free-running (or recurrent) type of sweep.

Q10-11. In a scope using a triggered sweep:
(a) Explain why the baseline disappears when the signal to the vertical input is reduced to practically zero.
(b) State how the baseline can be restored under the no-signal condition of part (a).

Q10-12. Explain the result of using *too small* a setting of the SYNC AMP control in a general-purpose oscilloscope.

Q10-13. Explain the basis for the precaution usually given (when using the nontriggered) or free-running type of sweep against using *too large* a setting for the SYNC AMP control.

Q10-14. Discuss the relation between the wide-band and the high-sensitivity capabilities of a given oscilloscope instrument.

Q10-15. (a) Discuss the relation between the wide-band capability and fast-rise-time feature in a particular oscilloscope.
(b) What property (or properties) of the oscilloscope governs its ability to reproduce a square wave most faithfully?

Q10-16. What characteristics of an amplifier are revealed when an oscilloscope is used to examine the amplifier output resulting from a square-wave input?

Q10-17. State three types of plug-in units that may be used to advantage in the type of oscilloscope designed to accept such plug-in units.

Q10-18. Discuss two types of accessory oscilloscope probes and their uses.

PROBLEMS

P10-1. In the screen-pattern oscillograms shown in Fig. P10-1, a *sine-wave signal of unknown frequency is connected to the vertical input terminals of the scope;* at the same time, a 60-Hz voltage is connected to the horizontal input (with the horizontal selector switch set so that the horizontal (or x) input actuates the deflection plates and the sawtooth sweep is inoperative). State the *frequency* of the unknown signal for each case shown in the diagram.

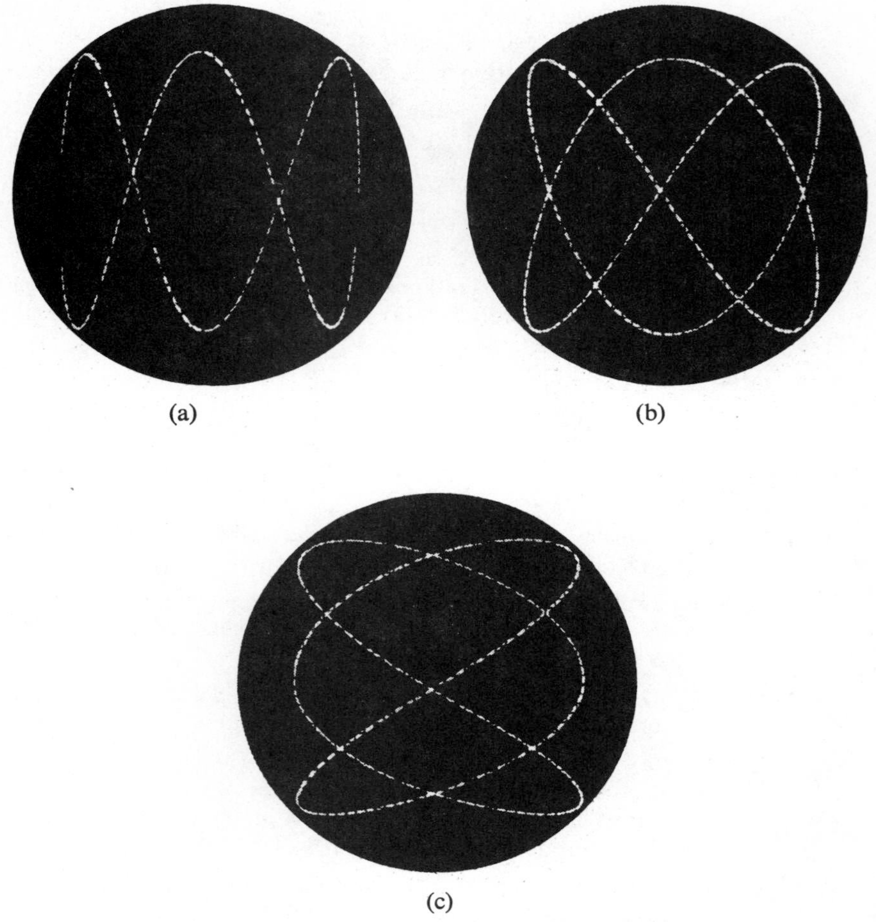

Figure P10-1. Oscilloscope patterns for determining unknown frequencies connected to the vertical plates, while a known frequency (such as 60 Hertz) is connected to the horizontal plates. [From J. H. Ruiter, *Modern Oscilloscopes and Their Uses* (New York: Holt, Rinehart and Winston, 1949).]

P10-2. In the oscilloscope patterns shown in Fig. P10-2, the signal connected to the vertical plates has the *same frequency* in both parts of the figure. If the input to the horizontal plate is a 60-Hz voltage:
(a) State the unknown frequency.
(b) Explain what causes the difference in the appearance of the two patterns.

P10-3. The sketches shown in Fig. P10-3 give the values for the y intercept and y maximum for two cases where voltages of the *same frequency but of different phase* are connected to the V and H plates of the scope. Find the phase difference in each case.

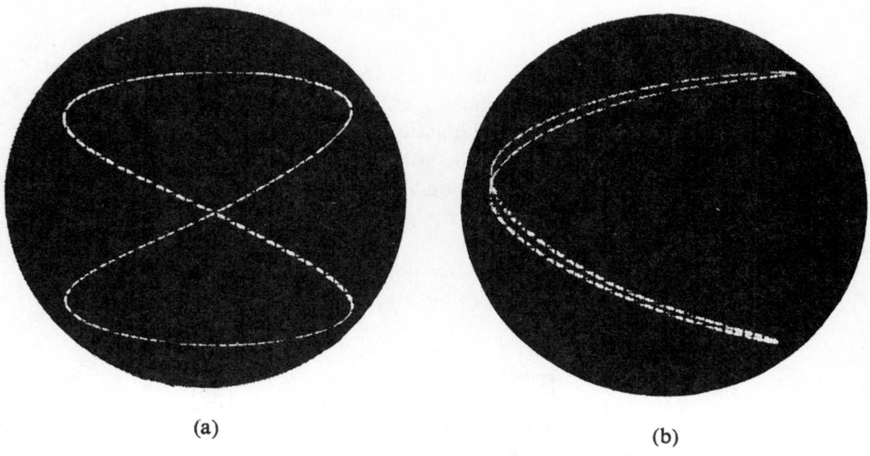

Figure P10-2. Screen patterns produced with 60-Hertz horizontal input; the signal voltage applied to the vertical input has the same frequency in (a) and (b).

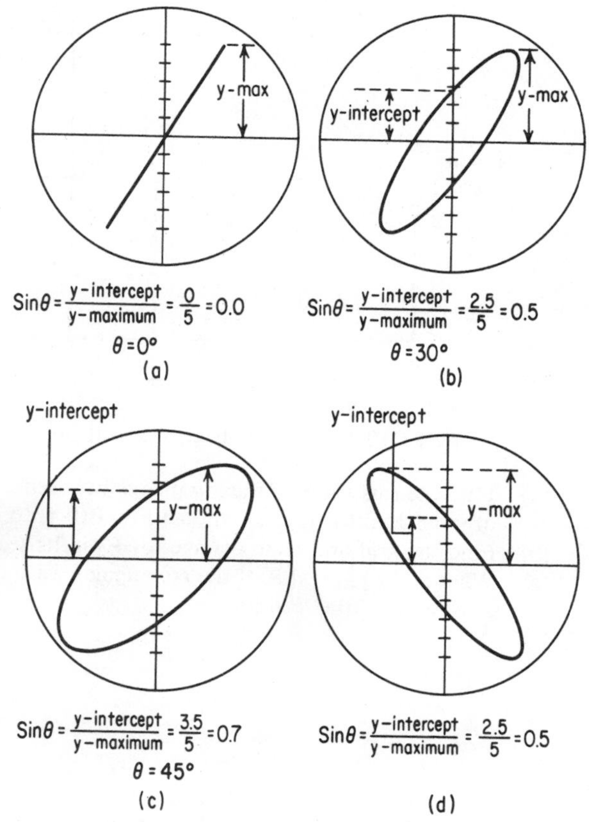

Figure P10-3. Lissajous figures for measuring phase difference between two voltages at the same frequency.

P10-4. A Schmitt trigger circuit is shown in Fig. P10-4 in its quiescent (no-input-signal) condition, with tube V_1 cut off (cutoff voltage approx. -9 V) and tube V_2 conducting. Trace the action that occurs as the input sine-wave voltage rises to a value higher than the voltage across cathode resistor V_{R_k} and then falls below this positive value, producing a pulse or trigger output at E_{out}. [The commutating capacitor (50 pF) for the purpose of producing a faster rise time of the pulse is optional.]

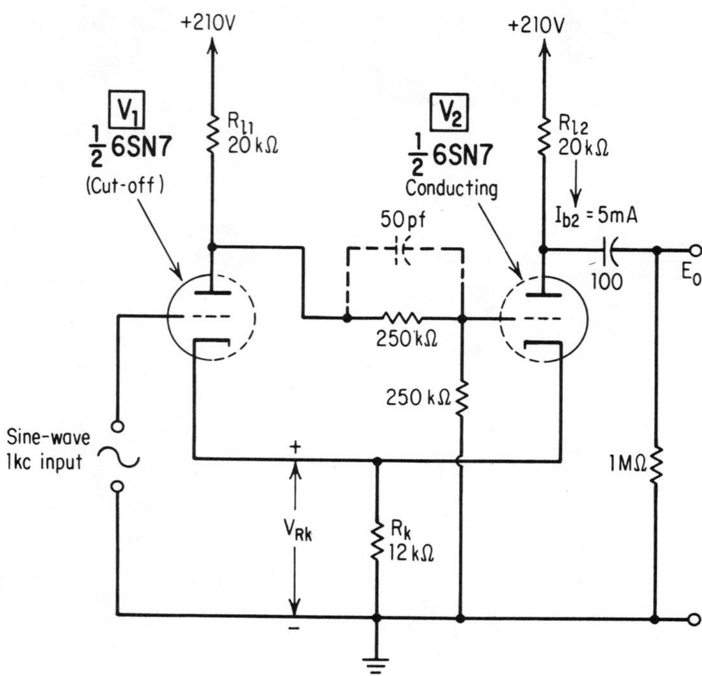

Figure P10-4. Schmitt trigger circuit (in quiescent state.)

P10-5. (a) Find the rms value of the input sine wave required to initiate the transition of the Schmitt-trigger circuit (of Fig. P10-4) to its second stable state.
(b) Find the grid and plate voltages of V_2 in its second stable state.
(c) What is the p–p value of the rectangular wave generated at the plate of V_2 as a result of the transition?

11

Signal Generators

11-1. TYPES OF GENERATED SIGNALS

Many different kinds of signals, at both audio and radio frequencies, are required at various times in an instrumentation system. In most cases, the particular signal required by the instrument is internally generated by a *self-contained oscillator*. The oscillator circuit commonly appears in a *fixed-frequency form* as, for example, when it provides a 1000-Hz excitation signal for an ac bridge. In other cases, such as in a Q meter, the self-contained oscillator appears in the form of a *variable-frequency arrangement* for covering Q measurements over a wide range of frequencies from a few hundred kilohertz into the megahertz range.

In contrast with the self-contained oscillators that generate only the specific signals required by the instrument in question, the class of generators that are available as *separate instruments to provide signals for general test purposes are usually designated as signal generators*. These AF and RF generators are designed to provide extensive and continuous coverage over as wide a range of frequency as practical. The type of signal provided by the signal generator may be assumed to be a continuous sine-wave signal, unless the generator is specifically called a *square wave*, *pulse*, or *function generator*. In RF signal generators, additional provision is generally made to modulate the continuous-wave (CW) signal to provide a modulated RF signal (MCW).

The frequency ranges covered by signal generators often overlap the strict definition limits of the frequency bands. It is useful, however, to retain the terminology given in Table 11-1, which defines the *frequency-band* limits. In the discussions that follow, it will be seen that the frequency coverage of a particular signal generator will often encompass more than just a single one of the arbitrary bands given in the table.

TABLE 11-1. Frequency-Band Limits

Abbreviation	*Band*	*Approximate Range*
AF	Audio frequencies	20 Hz–20 kHz
RF	Radio frequencies	Generally above 30 kHz
VLF	Very low frequencies	15–100 kHz
LF	Low frequencies	100–500 kHz
BDCST	Broadcast frequencies	0.5–1.5 MHz
VIDEO	Video frequencies	Direct current—5 MHz
HF	High frequencies	1.5–30 MHz
VHF	Very high frequencies	30–300 MHz
UHF	Ultrahigh frequencies	300–3000 MHz
MICROWAVE	Microwave frequencies	Beyond 3000 MHz (= 3 kMHz or 3 GHz or shorter than 100 mm*)

*The highest-frequency or shortest-wavelength RF waves (in the millimeter range) merge with the longest-wavelength infrared (IR) waves (fractions of a millimeter).

There is also a difference in how wide a frequency range can be expected to be covered by instruments of the same designation, but in a different price class within that designation. The less expensive *service type of AF generator*, for example, commonly covers from 20 Hz to 200 kHz, which is far beyond the AF range. In the more advanced *laboratory types of AF generators* the frequency range extends quite a bit further, as illustrated by two such models in Fig. 11-1. The *Hewlett-Packard* tube model shown in part (a), *model 200CD*, covers from 5 Hz to 600 kHz. The *Marconi model 1370A*, shown in (b), is a wide-range oscillator, generating both sine and square waves, and covers a range of 10 Hz up to 10 MHz for sine waves (and to 100 kHz for square waves). This exceptional range of 1,000,000–1 in frequency for sine waves is obtained from a two-stage *RC* oscillator circuit basically similar to the one used in the oscillator of (a) but with added tubes and greater circuit complexity.

11-2. TYPES OF OSCILLATORS

An oscillator is a circuit that continuously generates a periodic time-varying wave-form of some kind. Typical generated wave-forms are sine, square, triangular, sawtooth, and trapezoidal waves. The most common wave-form for use in testing and adjusting analog circuits and systems is the sine wave. Circuits that generate

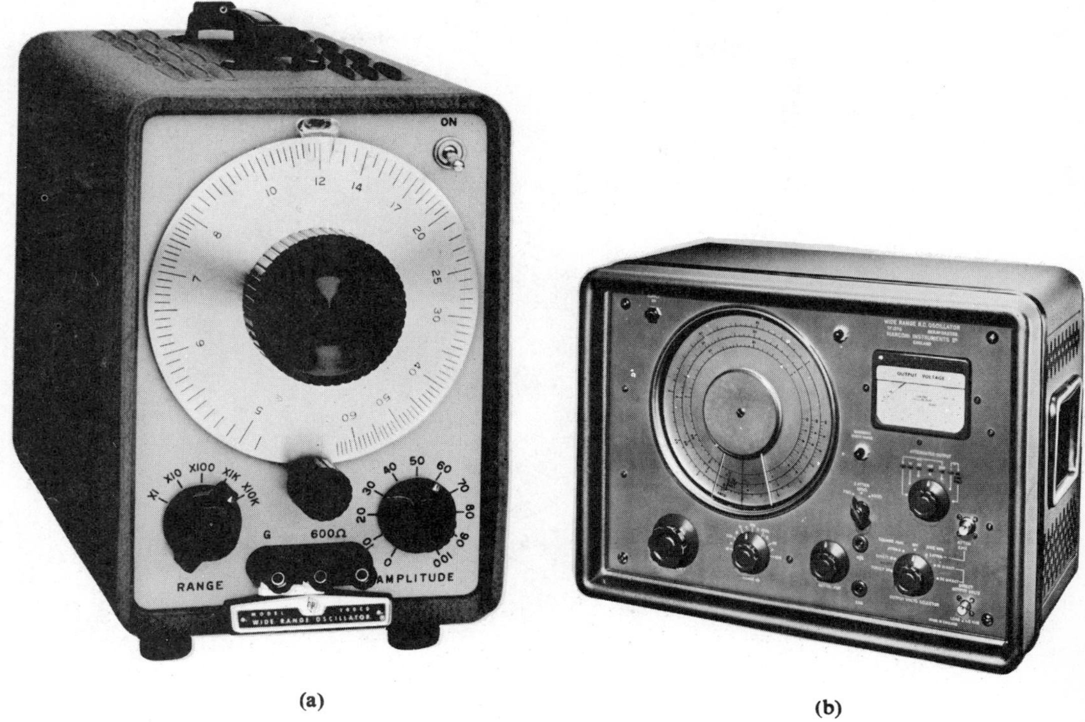

(a) (b)

Figure 11-1. Laboratory-type audio-frequency generators: (a) *Hewlett-Packard model 200CD*; (b) AF generator of both sine and square waves (*Marconi sine- and square-wave generator, type TF1307A*).

sine waves are called sinusoidal oscillators. Circuits that generate other types of wave-forms are referred to as nonsinusoidal oscillators, multivibrators, relaxation oscillators, and pulse generators.

The necessary and sufficient condition for the production of a continuous sinusoidal wave-form is *Barkhausen's criterion.* A closed-loop system is required with a loop gain of exactly 1 and a phase shift of exactly zero degrees. In general, one cannot design an electronic circuit that meets these requirements without some form of automatic gain control. The simplest way to get around this problem is to design a circuit with a loop gain slightly greater than 1 and then allow some nonlinearity in the circuit to limit the gain after oscillations have reached sufficient amplitude. The nonlinearity may come about, for example, through saturation or cutoff of an amplifier on the peaks of the sine wave. This technique naturally introduces some distortion and may not be acceptable for applications in which a nearly pure sine wave is required, as in measuring distortion in high-fidelity audio equipment.

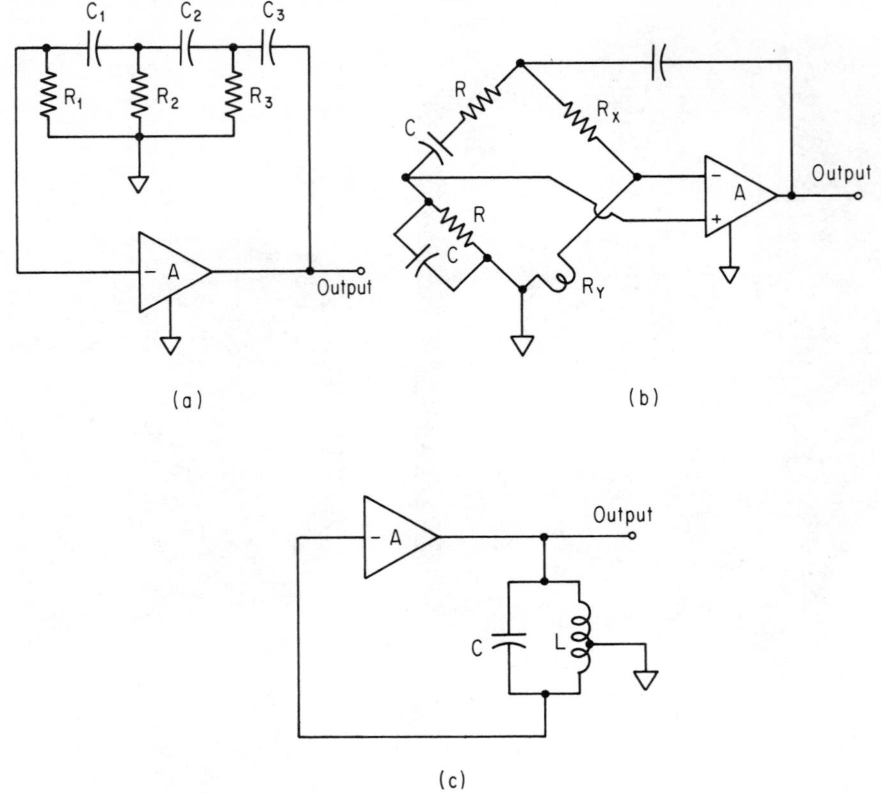

Figure 11-2. Three popular oscillator circuits: (a) phase-shift oscillator; (b) Wien-bridge oscillator; (c) Hartley oscillator.

Three common types of oscillators are shown in Fig. 11-2. Each has advantages and disadvantages, depending on the application. The phase-shift oscillator of Fig. 11-2(a) has the advantages of simplicity and low cost. It is used to generate sine waves in the audio and ultrasonic ranges when a certain amount of distortion can be tolerated. When very-low-distortion sine waves in this frequency range are desired, the Wien-bridge oscillator of Fig. 11-2(b) is preferred. For higher frequencies, ranging from the ultrasonic to the UHF range, an *LC* oscillator such as the Hartley oscillator of Fig. 11-2(c) is normally used. The operation of each of these circuits will now be described.

The phase-shift oscillator makes use of three *RC* sections in the feedback path of an inverting amplifier to satisfy Barkhausen's criterion for oscillation. From elementary network theory we know that each *RC* section is capable of producing a maximum phase shift of 90°. Therefore, three cascaded *RC* sections could produce a maximum phase shift of 270°. At some frequency, the phase shift from output to input will be exactly 180°. The phase shift through the inverting amplifier brings the total loop phase to 360° (which is the same as 0°), satisfying the zero-loop-phase requirement for oscillation.

In addition to being phase-shifted in going through the *RC* network, the signal is attenuated. The amplifier, therefore, must have sufficient gain to overcome this attenuation and bring the overall loop gain up to something greater than 1. Under these conditions, the circuit will begin to oscillate when power is applied, and the amplitude of the oscillations will gradually build up until the amplifier begins to saturate or cut off on the peaks of the wave. The greater the gain of the amplifier, the harder it will be driven into saturation or cutoff, and the more distorted the output will become. For this reason, the amplifier should be designed for a voltage gain only slightly in excess of the minimum required for oscillation.

For the special case in which all three resistors are equal, $R_1 = R_2 = R_3 = R$, and all capacitors are equal, $C_1 = C_2 = C_3 = C$, the frequency of oscillation is given by

$$f_0 = \frac{1}{2\pi\sqrt{6}\,RC}$$

Under these conditions the magnitude of the voltage gain of the amplifier must be equal to or greater than 29:

$$A \geq 29$$

To vary the frequency of the phase-shift oscillator, any or all of the components of the *RC* feedback network can be varied. We might, for example, make a variable-frequency oscillator by making the three resistors the three elements of a triple-ganged potentiometer.

The Wien-bridge oscillator of Fig. 11-2(b) is often used for audio signal generators in which a very-low-distortion output is required. A differential amplifier is used to amplify the feedback signals from the two sides of a bridge network. The signal at the left-hand side of the bridge is in phase with the output signal at only one frequency, which is determined by R and C. This frequency is

$$f_0 = \frac{1}{2\pi RC}$$

Negative feedback is supplied by a resistive voltage divider consisting of R_x and R_y. Resistor R_y is a temperature-dependent resistor with a positive temperature coefficient, typically a tungsten-filament lamp. As the oscillations build up, the filament of the lamp begins to heat up and its resistance R_y increases. This increases the negative feedback, which reduces the loop gain. Thus the lamp acts as an automatic gain control which stabilizes the amplitude of oscillations at a voltage level that prevents the amplifier from going into saturation or cutoff. The thermal time constant of the lamp must be very long compared to the period of the oscillations, a condition that is easy to achieve at normal audio frequencies.

In modern, solid-state signal generators, the tungsten-filament lamp might be replaced by a voltage-variable resistor such as a field-effect transistor. In this case, a portion of the output voltage is rectified and low-pass-filtered to provide the control voltage for the gate of the FET. Regardless of the method of automatic gain control used, the principle of operation is the same as described above. Wien-bridge oscillators are capable of delivering sine-wave outputs with harmonic distortion of only a small fraction of 1%.

The Hartley oscillator of Fig. 11-2(c) is useful for generating radio-frequency signals. Feedback is provided through the use of a tapped inductor. The signal at the input to the amplifier will be 180° out of phase with the output at a frequency given by

$$f_0 = \frac{1}{2\pi\sqrt{2MLC}}$$

where M is the mutual inductance between the two sections of the inductor on either side of the tap.

The frequency of the oscillator can be varied by making the capacitor C an air-variable capacitor. Band switching may then be accomplished by switching inductors.

The oscillators presented here are only three of many possible configurations.

11-3. REPRESENTATIVE AUDIO-FREQUENCY OSCILLATORS (LABORATORY TYPE)

A widely used tube version of the Wien-bridge type of *RC* oscillator was previously illustrated in Fig. 11-1(a) for the *Hewlett-Packard AF signal generator model 200CD*, covering the range 5 Hz–600 kHz. This range, it will be noted, includes signals in the subsonic, audio, and ultrasonic bands. This very wide range is obtained in five overlapping decade bands, the first band covering 5–60 Hz, and the last 50–600 kHz. At all frequencies, an output as great as 20 V rms can be obtained with no load; when delivering a signal to a 600-Ω load, to match the 600-Ω internal impedance of the oscillator, the voltage across the load would naturally be one-half of the open-circuit voltage, or 10 V. The power obtained in such a matched load would thus be E^2/R or $10 \times 10\ \text{V}/600\ \Omega = \frac{1}{6}$ W or 167 mW. Although this figure does not look impressive from the power viewpoint, it should be remembered that when used, as it is primarily intended to be, for test signal purposes, it represents a comparatively large amount of available signal voltage under ordinary test conditions. Of fundamental importance is the fact that for any given setting of the amplitude control, as the frequency is varied, the output signal is stable and retains its undistorted wave-form within the closely *specified tolerances* given below:

Specifications

Calibration Accuracy: ±2% under normal conditions.

Frequency Response: Within ±1 dB (of a 1000-Hz reference) over the entire frequency range.

Frequency Stability: Negligible shift in output frequency for ±10% line-voltage variations.

Distortion: Less than $\frac{1}{2}$% below 500 kHz (less than 1% above 500 kHz), independent of load impedance.

Balanced Output: May be obtained (at maximum output) with better than 1% balance; or may be operated single-ended (with low side grounded), at an internal impedance of 600 Ω, for any portion of output attenuator.

Before going on to discussion of the newer solid-state versions of this basic Wien-bridge type of *RC* oscillator, it is interesting to note that popularity of the original tube version formed the basis of the now greatly expanded Hewlett-Packard Co. Also, the improved capabilities of the solid-state versions (illustrated in Fig. 11-3 are retained in the transistor versions, thus retaining the flexibility of the *RC* type of positive feedback and the stability of the negative feedback, as explained in the preceding section.

The solid-state version of the *Hewlett-Packard AF signal generator* (*model 204C*) is illustrated in Fig. 11-3(a). Here, the frequency range covered is extended (from the 600-kHz range of the tube model 200CD) up to 1.2 MHz in six overlapping ranges. The output voltage, however, becomes less: 2.5 V rms into 600 Ω (as opposed to 10 V rms into 600 Ω for the tube version). An added feature is a low-distortion mode, which improves the flatness of frequency response at the low frequencies (5–100 Hz) over the normal mode. This low-distortion mode allows a ±1% flatness to extend all the way down to the lowest frequnecy of 5 Hz, whereas in the normal mode, the low-frequency region between 5 and 100 Hz is flat to only +5% and −1%.

As another added feature in the solid-state version, provision for *synchronization* has been added. This assures that the stable signals over the entire frequency band having a dial accuracy of ±3% can be synchronized with a more accurate external source. The locking range for an external signal, such as a crystal or tuning-fork reference, is ±1% per volt of SYNC, up to a maximum of 5 V rms (or 7 V peak). Thus, at a maximum SYNC input of 7 V peak, the oscillator will be locked to the reference signal accuracy within a tuning range of ±7% of the frequency indicated on the dial.

An example of a solid-state AF signal generator that also incorporates the synchronization feature is illustrated in Fig. 11-3(b), for the *General Radio model 1310A*. As will be seen from the specifications for this model given below, the lock-in range for synchronization here is appreciably larger (up to ±30% for maximum SYNC input of 10 V rms) and, for that reason, the discussion of this feature will refer to this particular model.

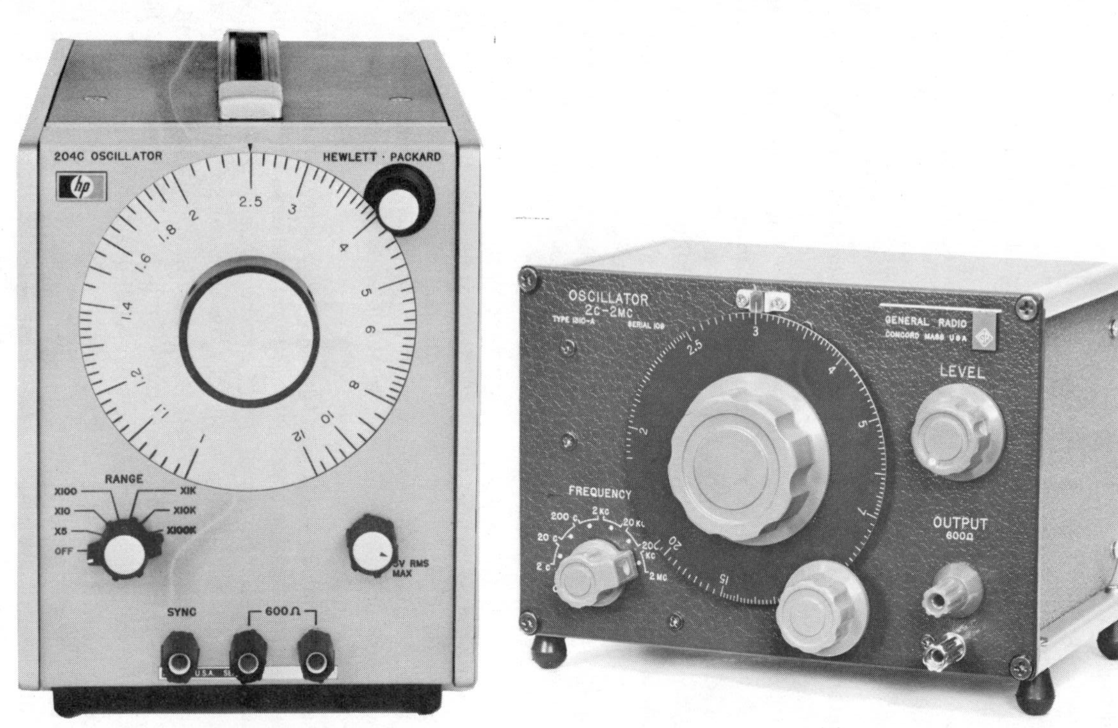

(a) (b)

Figure 11-3. Two solid-state *RC*-type AF signal generators: (a) *Hewlett-Packard model 204C*; (b) *General Radio model 1310A*.

Specifications

Frequency Range: 2 Hz to 2 MHz (six ranges).

Accuracy: ±2% of setting.

Output: 20 V into open circuit; 10 V (160 mW) into 600 Ω.

Distortion: <0.25%.

Synchronization: At input connector, lock range ±3%/V rms input, up to 10 V. At output connector, constant-amplitude, high-impedance (27 kΩ) output to drive counter or oscilloscope.

The synchronization function is accomplished by injecting the external SYNC source through the auxiliary connector into the active *RC* oscillator circuit. Then, when the oscillator is tuned within a certain range of this SYNC signal, normal oscillations cease, and the oscillator appears to be "locked in," and oscillates stably at the injected-signal frequency. At this time, the oscillator is not oscillating in the conventional sense, but is, in fact, producing a frequency-selective regeneration of the input.

The range of frequencies over which this locking takes place depends (linearly) on the amplitude of the injected SYNC signal and is ±3% for each volt of SYNC signal in this model. As an example of this ±3% lock-range suppose that a reference oscillator (at, say, 1 kHz) is available, having an output in the neighborhood of 1 V, and accurate to 0.1%. Without synchronization, and setting the dial of the oscillator with ordinary care to the 1-kHz mark, we would be assured only of the basic ±2% accuracy of the oscillator, so that we could only depend on the oscillator output at any amplitude to be somewhere between 980 and 1020 Hz. Then, by plugging the 1-V output of the reference oscillator into the synchronizing jack (and allowing about 1 sec for the time constant of the synchronizing mechanism) the 1-kHz output of the oscillator would become dependable within the ±0.1% tolerance of the reference oscillator, thus producing an improvement of 20 times over the original ±2% oscillator tolerance.

Another application of the synchronizing jack is its use in suppressing harmonics. It should also be noted that the same jack can also be used as an output connector to provide a constant-amplitude synchronizing signal for an oscilloscope or counter.

11-4. SQUARE-WAVE GENERATION

Combination Sine- and Square-Wave Generators

The production of square waves can be derived conveniently from the output of a sine-wave generator, and therefore, it is quite often a commercial practice to offer both sine- and square-wave outputs in a single model. This is especially true in instruments where a highly refined rise time of the square wave is not a primary consideration.

The process of shaping a sine wave to produce square-wave output can be accomplished in a number of ways, of which the following three may be mentioned. For low output levels, a simple diode *clipping* circuit may be used; for higher output levels, either an *overdriven amplifier* or, for better wave-form, the triggering of a bistable circuit can be used. The bistable circuit form, which can be triggered by a sine wave, is known as a *Schmitt-trigger* circuit.

Clipping Circuit

Approximate square waves may be produced by shaping an input sine wave with a diode clipping circuit. To produce a single-peak amplitude of, say, 6 V with an ordinary diode [Fig. 11-4(a)], a biasing battery would be used so that the input wave would rise (sinusoidally) to a 6-V value before the diode conducted, to clip that half of the wave and prevent any further rise; a similar diode and 6-V biasing battery would be used to do the same, in the reverse polarity sense, for the other half of the wave. A simpler method for achieving this same clipping action using only two Zener diodes, connected back to back, without the use of any batteries, is shown in Fig. 11-4(d). Here, there is no diode conduction on either half of the sine-wave input cycle, until the voltage has risen up to the breakdown voltage of the Zener diode (6 V in this case). As a result, an approximate square-wave output of 12 V peak-to-peak is produced from the 36-V peak-to-peak sine-wave input of the example, as shown in the wave-form diagrams, Fig. 11-4(b) and (c).

Although diode clipping action has the virtue of simplicity, the output of the passive circuit is small, and the rising and falling portions of the output square wave are necessarily slightly slanted, rather than vertical, as they would be in a true square wave. In the example shown (Fig. 11-4), the rise time of the output wave would depend on the time required for the sine wave to reach one-third of its peak. Calculated for the 250 μsec of the first quarter cycle of 90° (for the 1000-Hz input of the example), the fraction of this time required for the angle whose sine is 0.33 becomes 19.5°/90° or 0.216; this fraction of the 250-μsec time gives approximately 54 μsec for the total time required for the input wave to reach the 6-V output value, or twice that amount (108 μsec) for the peak-to-peak value. Although such a rise time might be acceptable for certain applications where only an approximate square wave is required, it must be compared with rise times of only 1 μsec and less, for cases where the squareness of the wave and the corresponding fast rise time are important.

Comparator Circuit

Square waves can be generated from other periodic wave-forms (such as sine waves) by driving a high-gain amplifier between saturation and cutoff. An operational amplifier, because of its very high gain, can be used for this purpose. In this application, the op amp is operated without any negative feedback as shown

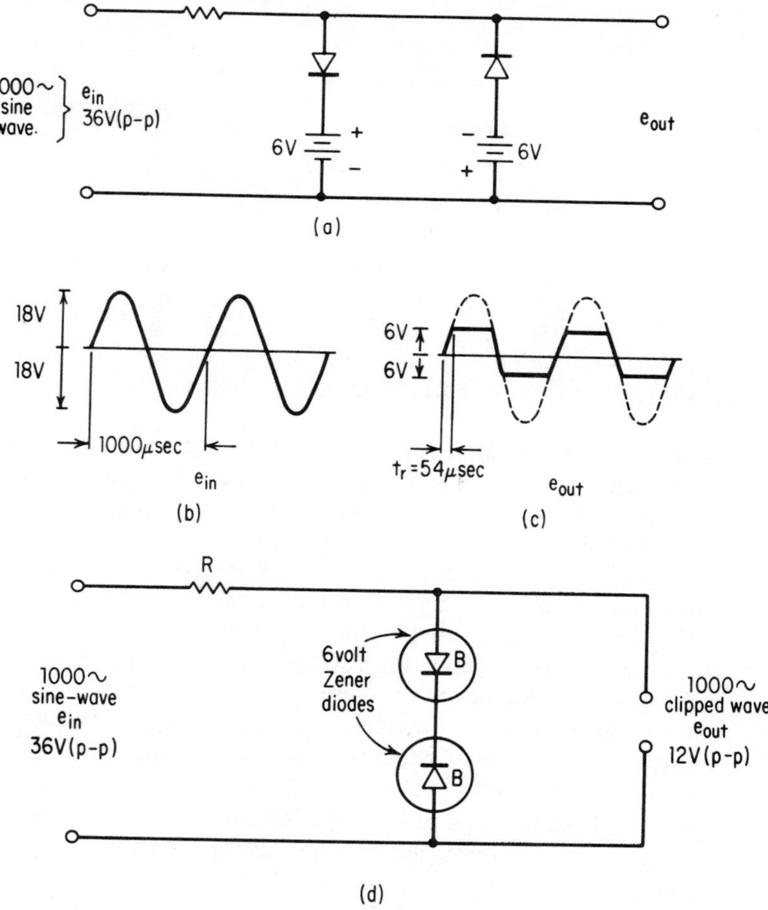

Figure 11-4. Clipping circuit for producing approximate square waves from a sine-wave input: (a) circuit using diodes and biasing batteries; (b) sine-wave input e_{in}; (c) approximate square-wave output; (d) circuit using back to back Zener diodes and no batteries.

in Fig. 11-5. A very small positive input voltage will cause the output of the op amp to saturate at a voltage very nearly equal to the positive supply while a small negative input will cause it to saturate in the opposite direction.

An integrated circuit called a comparator is very much like an operational amplifier except that it is specifically designed for this kind of open-loop operation and has a much higher slew rate than a typical op amp. Thus the square waves generated with a comparator will have shorter rise and fall times than those generated with an op amp.

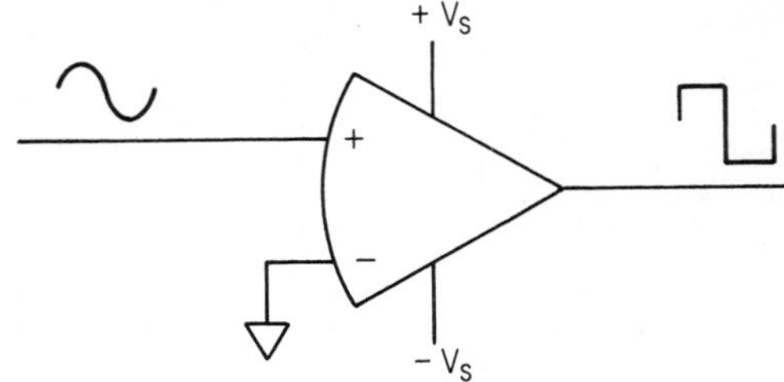

Figure 11-5. Sine-square wave conversion using an operational amplifier.

11-5. COMMERCIAL SQUARE-WAVE GENERATORS

When it is important to have a square wave with characteristics that closely approach the ideal wave-form, the wave-shaping circuits of commercial generators are quite a bit more complex than the relatively simple circuits already mentioned.

The necessity for an excellent square wave-form can be appreciated in the highly effective method of *square-wave testing of amplifier response*. The interpretation of some of the response patterns obtained on an oscilloscope when a square-wave signal is used as the amplifier input is shown in Fig. 11-6. The need for a horizontal flat top in the test signal can be seen from the fact that an amplifier with poor low-frequency response will produce a tilt in its output, as in part (b) of the figure, which approaches the differentiated wave-form of (c), as a limit of exaggerated poor low-frequency response. Similarly, an amplifier with poor high-frequency response is evidenced by an output as in (d), showing rounding of the rising and falling steps of the input square wave until, as a limit in (e), the exaggerated poor high-frequency response approaches a triangular wave-form, showing integration of the input square wave. Presence of oscillation in the circuit would be evidenced by the "ringing" shown in (f). It can be seen that this highly effective method of judging amplifier response at a glance depends greatly on the perfection of the original square-wave signal.

11-6. RADIO-FREQUENCY SIGNAL GENERATORS

In the category of RF signal generators, many models are required to cover the wide expanse in frequency range from 30 kHz to 3000 MHz, in combination with specific requirements for *modulating* the RF signal and often additional requirements for an output closely *calibrated in amplitude as well as in frequency*. In order to emphasize the distinction between the many models available in the RF field, the discussion will center first around a service-type RF generator, commonly employed for testing communication receivers. This will be followed by a laboratory

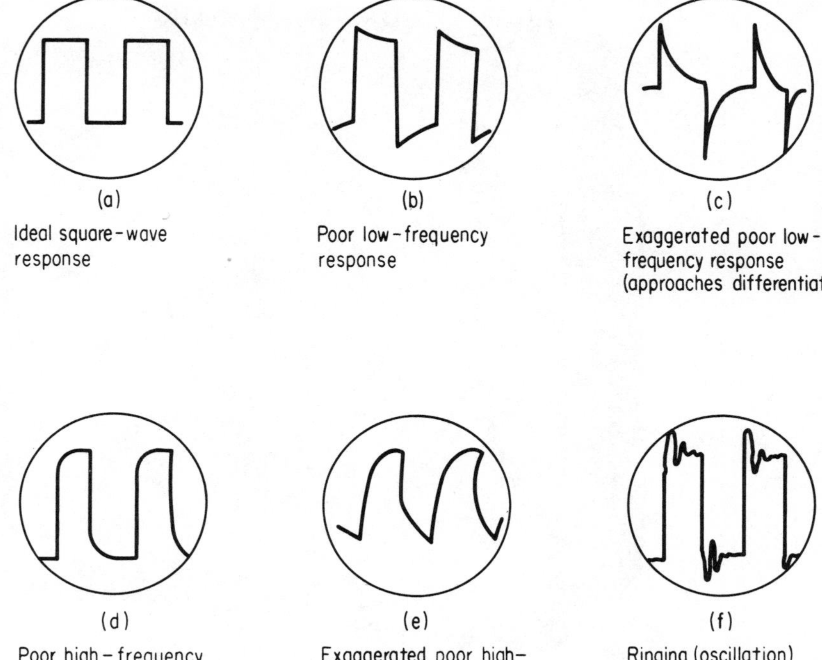

Figure 11-6. Amplifier response in square-wave testing: (a) ideal square-wave response; (b) poor low-frequency response; (c) exaggerated poor low-frequency response (approaches differentiation); (d) poor high-frequency response; (e) exaggerated poor high-frequency response (approaches integration); (f) ringing (oscillation).

model, generally called a "microvolter," that provides an RF signal that is not only at a known frequency but whose amplitude can be accurately controlled and read directly in microvolts.

11-7. REPRESENTATIVE SERVICE-TYPE RADIO-FREQUENCY SIGNAL GENERATOR

A representative model of the service-type RF signal generator is illustrated in Fig. 11-7. It provides three types of signals generally used in receiver testing: a variable-frequency RF signal (also called continuous-wave, abbreviated CW), modulated RF signals in the same frequency range as the unmodulated signals, and a fixed-frequency AF signal.

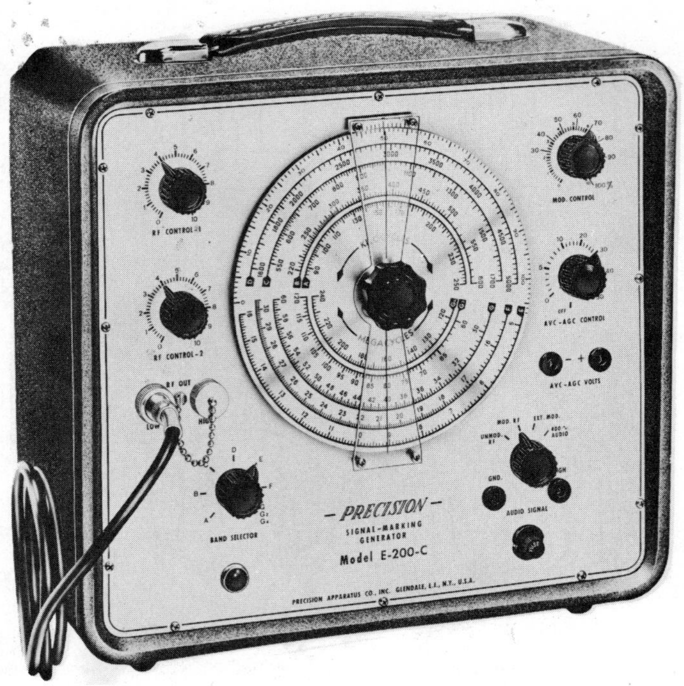

Figure 11-7. A representative RF signal generator of the service type, *Precision model E-200C.* (Courtesy of Precision Apparatus, Division of Dynascan.)

The *RF signal*, in eight bands, provides continuous coverage on fundamentals from 88 kHz through 110 MHz and extended output is available on second and fourth harmonics through 440 MHz, as follows:

Band	*Coverage*
A	88–230 kHz
B	220–600 kHz
C	550–1700 kHz
D	1.60–5.0 MHz
E	5.5–15.5 MHz
F	15–29 MHz
G	29–55 MHz
H	55–110 MHz
H_2 (2nd harmonic)	110–220 MHz
H_4 (4th harmonic)	220–440 MHz

Band *H* is arranged to provide strong harmonics of the same order of stability as the original band, so that the dial is calibrated directly for the second harmonic as band H_2: 110–220 MHz. The 220–440-MHz range is read on band H_4. The accuracy of calibration is kept to $\pm 1\%$ on all bands, under widely varying conditions.

Modulation of the RF signal is selected as either unmodulated output (i.e., continuous-wave), 400-Hz modulated output (with variable modulation from 0 to 100%), or externally modulated output. It is usual practice to modulate the signal 30% and to display the detected 400-Hz note on an output meter or oscilloscope. Modulation at greater percentages, up to 100%, is desirable when aligning receivers that are greatly out of line.

Audio-frequency output is provided as a 400-Hz sine-wave audio signal, independently controlled and available as a separate output, which is continuously controllable from zero up to almost 30-V peak AF voltage. The use of a cathode-follower output for the AF oscillator section ensures independence of the 400-Hz signal.

11-8. LABORATORY-TYPE RADIO-FREQUENCY GENERATORS

Generator Requirements at Radio Frequencies

A radio-frequency generator suitable for laboratory applications must not only be able to generate frequencies from around 100 kHz up to the neighborhood of 30 MHz, but must also be exceptionally *stable as to the frequency and amplitude* of the output signal. While it is very easy to get a triode tube or transistor to oscillate in this megacycle range (and sometimes exasperating to try to keep it from oscillating), it is much more difficult to keep the frequency (and amplitude) constant in spite of slight changes that occur in normal operating conditions, with regard to ambient temperature, supply voltage, and loading conditions. Pentodes employing electron-coupled oscillator circuits are generally used for greater stability. The field-effect transistor possesses practically equal advantages for the solid-state versions.

Frequency Stability for "Standard" Signal Generators

Where a $\pm 1\%$ change in a nominal output frequency of 1000 cycles (or ± 10 cycles) might easily be tolerated for an AF signal, the same change of $\pm 1\%$ in a 10-MHz signal would cause a shift of 100,000 cycles that might easily detune a high-Q tuned circuit. However, the problem of maintaining and checking the frequency stability of high-frequency circuits is greatly simplified by the use of *crystal oscillator and checking circuits.* The crystal oscillator, fortunately, is inherently very stable and can easily be arranged to provide constant frequencies within much better than 0.01% (or 1 part in 10,000) and, when used in crystal ovens, can be made—without great difficulty—to furnish accuracies of 1 part in

1,000,000 (±0.0001%). For most laboratory applications, *direct calibration of the variable-frequency dial to around 1%* is generally sufficient, with the added precaution of checking this frequency against a crystal calibrator, whenever greater precision is desired.

Amplitude Stability

The generator used to supply the kind of RF signal suitable as a reference in the laboratory must possess other qualities in addition to generating a reliably known frequency. The so-called "standard" signal generator must also provide that the signal be *accurately calibrated in amplitude* and be able to be modulated to a *known percentage of modulation.*

The requirement for providing a known amplitude, calibrated in microvolts, is met by the incorporation of a low-impedance variable attenuator, monitored by a meter generally labeled "carrier microvolts." A low-impedance characteristic is necessary to avoid large changes in output as the generator is fed into various loads. The output of the generator is therefore provided by a coaxial cable terminated in a relatively low resistance, almost always less than 100 Ω. The impedance seen by the load, which is this resistor in parallel with the attenuator, is usually a much lower value (typically around 5 Ω). In this manner, a low output impedance is obtained at any setting of the attenuator, which can vary the output from a value as low as a few microvolts, up to values well beyond 100,000 μV (generally 1 or 2 V). The total power output is not a very important factor, generally being in the milliwatt range, up to 1 or 2 W.

The means for monitoring the carrier level and also the percentage modulation are discussed in greater detail in the representative example of a "standard signal" or "microvolter" generator that follows.

11-9. REPRESENTATIVE "STANDARD SIGNAL" RADIO-FREQUENCY GENERATOR ("MICROVOLTER")

The example of a laboratory-type generator illustrated in Fig. 11-8(a) is the *Measurements "Standard" Signal Generator* (*model 180*).

The block diagram of the arrangement used in the *Measurements model 180* instrument is shown in Fig. 11-9. The significant specifications follow.

Specifications

Carrier Frequency Range: From 2 to 400 MHz, in six bands (*model 180R* extends to 475 MHz).

Dial Accuracy: Individually calibrated direct-reading to an accuracy of ±0.5%.

Modulation: AM continuously variable from 0 to 30%, monitored by panel meter; internal modulation of 400, 1000, and 10,000 Hz, plus external sine-wave or pulse modulation.

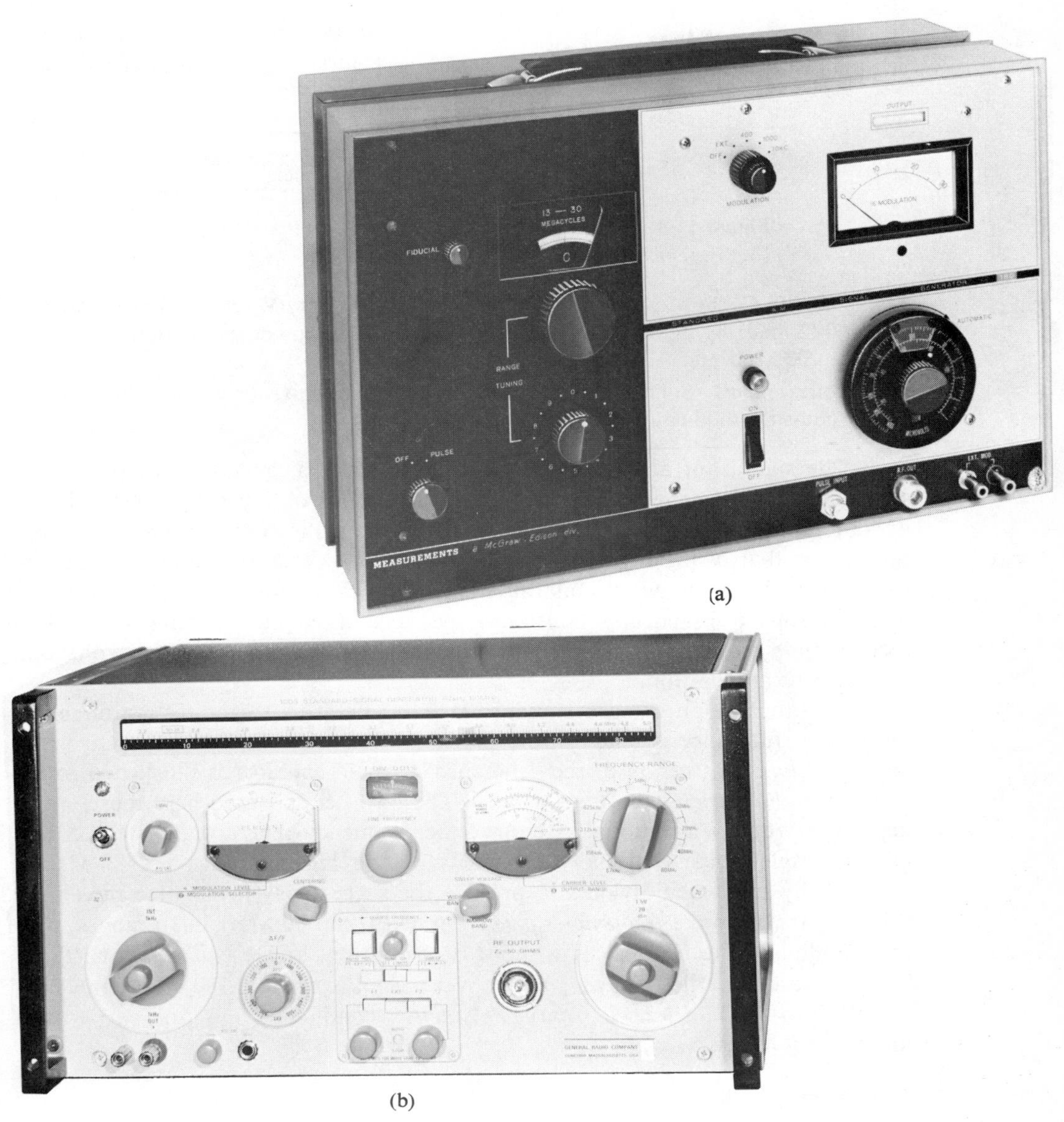

Figure 11-8. "Standard" RF signal generators: (a) *Measurements model 180*; (b) *General Radio* (solid-state version) *model 1103*.

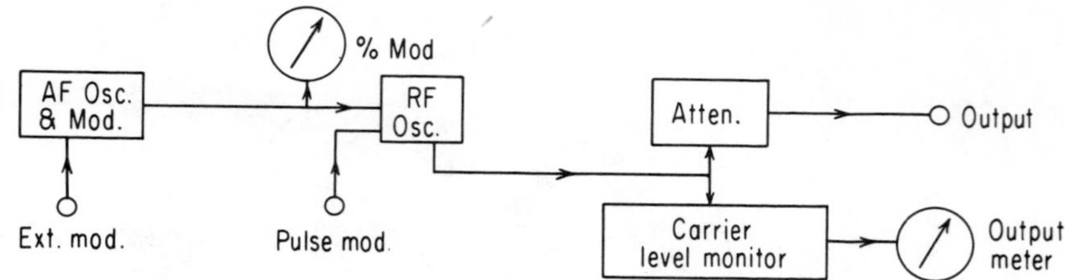

Figure 11-9. Block diagram of "standard" signal generator. (*Measurements model 180.*)

Output: Calibrated output from 0.01 up to 100,000 μV (as developed across a 50-Ω load) by mutual-inductance type of piston attenuator; accuracy $\pm 10\%$ up to 200 MHz and $\pm 15\%$ at higher frequencies.

Power Supply: Solid-state (115-V input), supplying Nuvistor carrier oscillator and solid-state modulator.

The significant ability of the standard signal RF generator to produce a signal of known frequency and required amplitude is obtained by careful design of amplifying and isolating circuits, in addition to the primary circuits, whose function is that of producing a stable oscillation. As seen in the block diagram, the RF signal and the AF modulating signal are each monitored as stable voltages, working into a practically constant-impedance attenuator arrangement, with rectifier-type meters monitoring both the RF carrier output in microvolts and the per cent modulation.

The application of such standard signal generators to the testing of radio-receiver performance is taken up in Chapter 15.

A *solid-state version* of the "standard" signal generator is illustrated in the *General Radio model 1003* [Fig. 11-8(b)]. The novel circuit used here represents a departure from conventional circuitry; the frequency-generating system covers only a single range [optimally designed for the highest range (34–80 MHz)], followed by frequency dividers to provide the successively lower nine ranges. In this way, it covers the *frequency range* from 67 kHz to 80 MHz in 10 ranges, with the stability of the top ranges without deterioration, and a *calibration accuracy* of $\frac{1}{4}\%$ is attained. The *RF signal output* over the entire range is 180 mW, equivalent to 6 V into 50 Ω. The *two front-panel meters* continually monitor carrier and modulation levels.

Frequency Synthesizer

In testing very sharply tuned circuits, where the stability and resolution are measured to small fractions of a single cycle while testing in megahertz ranges, the *frequency synthesizer* offers an advanced testing instrument that goes beyond even the ± 1 digit of the extremely accurate *digital frequency counters.* Each of these instruments is discussed separately; the frequency counter in Sec. 17-1, and the frequency synthesizer in Chapter 15.

11-10. SPECIAL-PURPOSE SIGNAL GENERATORS

The generators of sine- and square-wave RF outputs in the megahertz range have been discussed so far from the standpoint of either an unmodulated or an amplitude-modulated (AM) signal. Other types that are valuable for special proposes include *frequency-modulated (FM) generators*, *sweep generators*, *pulse generators*, and *function generators*.

FM Generators

For the testing of FM receivers, a frequency-modulated signal is obviously necessary. The FM generator is basically similar to the AM generator shown in the previous example, with the main exception that the audio modulation is caused to change the center frequency of the oscillator rather than its amplitude. This is generally done by a reactance-tube arrangement, and the resulting percentage modulation determines the frequency deviation, either side of center (typically, ±75 kHz at 100% modulation for broadcast FM transmissions). To cover other test requirements, the amount of frequency deviation is made variable so that both greater and smaller deviations are available.

The center frequency required, of course, depends upon the equipment to be tested. In the case of broadcast FM, the range of frequencies (88–108 MHz) is much higher than for broadcast AM ($\frac{1}{2}$–$1\frac{1}{2}$ MHz), and the center frequencies of the RF oscillators are correspondingly higher in the FM signal generators. A laboratory-type FM generator is illustrated in Fig. 11-10 (*Marconi FM/AM generator type 1006B/6*).

Figure 11-10. FM/AM signal generator (laboratory type). (*Marconi type 1006B/6.*)

Marconi FM Generator

The *Marconi* FM signal generator is designed to cover frequency deviations of ± 400 kHz of carriers whose center frequencies range from 10 to 470 MHz. It can therefore simulate signals for the FM broadcast band (which has center frequencies in the 88–108-MHz range), with 100% frequency deviation at ± 75 kHz; it likewise covers the FM sound in television receivers, where the center frequency can be between 56 and 216 MHz (for TV channels 2–13), with 100% frequency deviation at ± 25 kHz. Internal modulation frequencies of 1 kHz and 5 kHz are available, and variable amounts of FM deviation, up to ± 100 kHz (on ranges 10–50 MHz), and to ± 400 kHz (on ranges 50–270 MHz), decreasing to ± 300 kHz (on the final range of 270–470 MHz). Meters monitor both carrier level and FM deviation. Crystal calibration is provided at 1 and 10 MHz.

Additional coverage at the higher frequencies of FM telemetry is also available in a separate model. The source impedance of both models is standardized at 50 Ω.

Sweep Generators

In order to obtain bandwidth measurements of RF amplifiers, the conventional unmodulated signal can be used in a step-by-step process of changing the frequency by small amounts either side of resonance and monitoring the response. By the use of a *sweep generator*, however, where the resonant frequency is periodically swept by a given amount on either side, it is possible to obtain a visual representation of the bandwidth response curve on the screen of an oscilloscope that is synchronized with the periodic variation of frequency. This method is widely used in testing the bandpass circuits of television and radar receivers, both of which have very wide bandpass characteristics.

A versatile sweep generator, having the desirable characteristics of solid-state, electronic sweep, wide-sweep, and marker generators, is illustrated in Fig. 11-11 (*Kay Electric model 154C*). Over the frequency range of 50 kHz–110 MHz, it is capable of continuously variable sweep width from 50 kHz to the full 110 MHz wide. The RF output level of 1 V into 50 Ω, $\pm\frac{1}{4}$ dB, is metered. The sweep may be disabled (for continuous wave CW) or set at line-lock, manual, or external modulation, with sweep rates from 0.01 to 1000 sweeps/sec.

The *sweeping action* takes place in a voltage-controlled oscillator (VCO), which is mixed with an unswept one to produce a *beat frequency* that is amplified by a broadband RF amplifier. The sweeping action is produced by a sawtooth voltage that is impressed across the voltage-variable capacitors in the oscillator tank circuit, thus changing the output frequency linearly with respect to time, and then snapping back to the original unswept frequency. The automatic-gain circuits (AGC) keep the sweep output constant throughout the RF range.

Calibrated amounts of attenuation are introduced by the row of toggle switches seen at the bottom right, while the center frequency is displayed digitally

Figure 11-11. Laboratory-type sweep generator: solid-state, with sweep-width capability up to 110 MHz and provision for pulse and "birdie" marker plug-ins. (*Kay Electric model 154C.*)

by the multi-turn dial at the top right. The left-hand side of the instrument takes the *marker plug-ins*. Pulse or "birdie" markers are available for specified frequencies.

Application of the sweep generator to produce an oscilloscope display of bandwidth and frequency-response curves is discussed in Chapter 15.

Pulse Generators

The discussion of signal generators thus far has emphasized the generation of sine waves in the various modulation forms of unmodulated (CW), amplitude-modulated (AM), and frequency-modulated (FM). Pulse modulation (PM) is another important field, especially in the areas of data transmission and telemetry; these areas are referred to in Chapter 19 (see, in particular, Sec. 19-8).

Function Generators

Instruments for generating a wide variety of signals, called *function generators*, are quite useful for certain testing operations in the laboratory.

As an example, the *Hewlett-Packard model 3310A* function generator provides a variety of waves, including *square waves, ramps, triangles, and pulses,* in addition to the conventional sine waves. The repetition rate is selectable from 0.0005 Hz to 5 MHz and thus offers a range that covers subaudio, audio, ultrasonic, and video frequencies.

While sine waves are the basic signal for frequency-response tests, the square waves and pulses find use for *transient response;* triangles and ramps are also useful test signals where interest is centered on *rate of change* (*or slewing rate*), and are also useful for driving recorders and sweeping oscillators.

QUESTIONS

Q11-1. Compare the method for generating a 400-Hz signal voltage using rotating machinery with the method using an electronic oscillator for:
(a) Low-power output.
(b) High-power output.

Q11-2. How are the following properties related to the ability of an electron device to generate a signal by oscillation?
(a) Type of feedback.
(b) Amount of amplification.
(c) Negative-resistance property.

Q11-3. A transistor audio-frequency oscillator uses the primary of a transformer in its collector circuit and the secondary in the base circuit. What is the effect on oscillation of reversing:
(a) The primary terminals?
(b) The secondary terminals?

Q11-4. Explain the method used for varying the frequency in a beat-frequency audio oscillator.

Q11-5. Explain the method used for varying the frequency in an *RC* (Wien-bridge) audio oscillator.

Q11-6. In an RF signal generator, why is a combination of a Hartley oscillator and a crystal oscillator to be preferred to the use of either type alone?

Q11-7. Describe three important kinds of output that should be provided in a service-type RF signal generator.

Q11-8. State the important aspects in which a laboratory "standard signal" type of RF generator differs from a service type of RF signal generator.

Q11-9. Describe the "squaring" process and the type of wave-form obtained in producing a square-wave output from a sine-wave input, by each of the following methods:
(a) Clipping circuit.
(b) Overdriven amplifier circuit.

Q11-10. How is the output of a square-wave generator useful:
(a) In testing low-frequency response of an amplifier?
(b) In testing amplifier high-frequency response?
(c) State two other uses of a square-wave generator.

Q11-11. Describe how the output of an AM signal generator differs in its essential characteristics from that of an FM signal generator.

Q11-12. State the deviation ($\pm$ kHz from the center frequency) required for testing:
(a) Broadcast FM receivers.
(b) Television sound receivers.

Q11-13. State the principle by which sweeping action is produced in a sweep generator, when it is accomplished:
(a) By mechanical action.
(b) By electronic action.

PROBLEMS

P11-1. (a) Sketch, by block diagram, the setup you would use to calibrate the frequency of a signal generator by means of an oscilloscope.
(b) Describe the procedure for this calibration.

P11-2. (a) Draw a similar block diagram for calibrating a signal generator with the aid of a crystal oscillator and receiver.
(b) Describe the procedure for this calibration.

P11-3. Using a block diagram, describe the procedure for obtaining a visual display on an oscilloscope of the bandwidth response of an RF amplifier (455 kHz) with the use of a sweep generator (but no marker generator).

P11-4. If a marker generator is to be added to the sweep generator setup for Problem P11-3, describe the new setup and additional procedure, in order to obtain three marker points on the visual response curve.

P11-5. Draw a schematic circuit for producing positive peaked pulses having a repetition rate of 1000 pulses/sec from a 1-kHz sine wave. (*Suggestion:* The simplest circuit would require only two Zener diodes together with a diode rectifier, a capacitor, and some resistors.)

12

Component Test Methods

12-1. TRANSISTOR "QUICK CHECK"

In tackling the problem of transistor testing, one is immediately confronted by the tremendous number of bipolar types—over 5000 different type numbers—and an additional large number of the junction FETs and MOSFETs. As an initial approach, a "quick check" for a given bipolar transistor is useful, if only to establish whether it is of the p–n–p or n–p–n type, and to see if it is in operating condition—as opposed to a shorted or open transistor. In contrast to the tube situation, where the defect is commonly found to arise from the gradual weakening of the tube's ability to emit electrons, the transistor rarely shows such a progressive weakening of collector current as a common fault. Since the transistor does not depend upon a heated cathode for thermionic emission, it is more likely to go bad by means of a complete failure caused by a short- or open-circuit condition. Hence it is quite often enough to determine only that the transistor still retains its ability to amplify, without necessarily obtaining an exact measure of how much that amplification is. Such a "quick-check" method is described next.

Transistor Check with Ohmmeter

A very simple method for obtaining a quick check of the relative dc amplifying ability of a transistor can be devised around the use of an ohmmeter.[1] The circuit

[1]S. D. Prensky, "Multimeter Transistor Checker," *Radio-Electronics*, Aug. 1956.

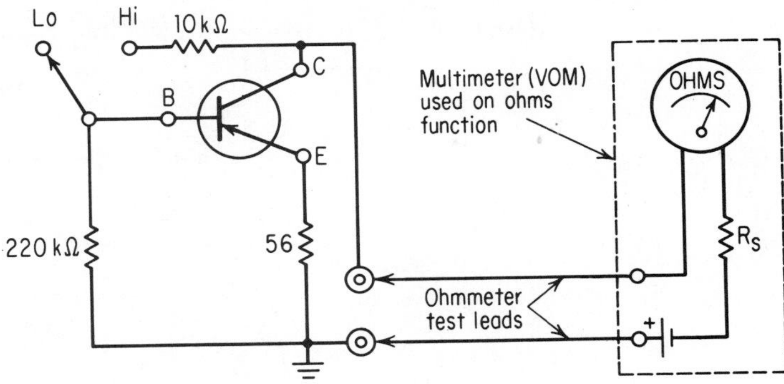

Figure 12-1. Transistor check with multimeter, using self-contained battery and resistor of multimeter on its ohms function to check triode operation of transistor. (From S. D. Prensky, "Multimeter Transistor Checker," *Radio-Electronics*, Aug. 1956.)

in Fig. 12-1 shows that only three fixed resistors and a switch are needed, when connected to an ohmmeter, to accomplish this quick check. While an ohmmeter could be used simply to test the forward and back resistance between two of the transistor elements taken at a time (as if it were a double diode with a common base), the method being described here does not test each diode separately but instead tests the whole unit at once. This check thus involves the amplifying ability of the transistor in addition to its rectifying properties.

The quick-check circuit in Fig. 12-1 takes advantage of the circuitry already present in a conventional ohmmeter, which contains a 1.5-V cell for voltage supply, a sensitive meter, and a series limiting resistor R_s, which protects both the meter and the transistor under test. The test leads of this ohmmeter combination are applied to the collector and ground terminals of the unknown transistor arranged in its common-emitter configuration. By means of the switch, a reading on the meter is first obtained, under the condition of zero bias on the base, which is returned to ground through the 220-kΩ resistor. This collector-current reading (taken as a percentage of full scale, rather than in ohms) is called the LO reading. When the switch is thrown to its HI position, the base is biased by allowing the base current to flow through the 10-kΩ resistor, and a new value of collector current, called the HI reading, is obtained. The net change between the HI and LO readings then gives a comparative picture of dc current-amplification action in the transistor.

In its commercial form, the quick-check type for indicating defective transistors is also designated as a "go–no go" circuit (an example is the *Sencore go–no go transistor/diode checker model TR115*). The functional circuit of such a *checker* of the ability of a transistor to produce dc gain (or dc beta) is shown in Fig. 12-2(a), which operates in a manner very similar to the quick-check circuit of Fig. 12-1. It is important to note here that this is a qualitative check, rather

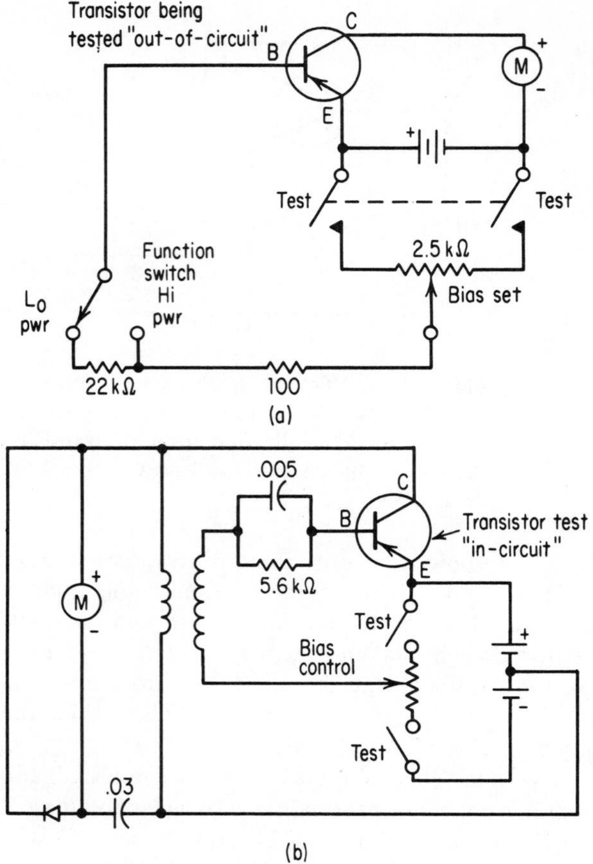

Figure 12-2. Functional circuit of transistor check: (a) dc gain test out of circuit; (b) in-circuit test of transistor in oscillating circuit.

than a quantitative *measure* of the h_{fe} (or ac beta) parameter that is generally given in a transistor specification. It is effective, however, as a "fast and dirty" way of spotting shorts, opens, and excessive leakage of defective transistors (and also diodes).

In-Circuit Transistor Checking

A method that is able to check transistors already wired into a circuit, without removing them from the circuit, is a welcome type of test instrument, especially with transistors that are often difficult to remove or subject to easy damage from a hot soldering iron.

One method for accomplishing an in-circuit rough check is by means of an oscillating-circuit transistor test, shown functionally in simplified form in Fig. 12-2(b). The transistor being checked is left in its original circuit in the equipment, such as a radio receiver, without being disconnected in any way, and the equipment is deenergized. When the test leads of the test instrument are properly clipped onto the emitter, base, and collector terminals of the transistor being tested, the transistor becomes part of an audio-frequency oscillator circuit of the tester, as shown in the functional circuit diagram. In this circuit, positive feedback from the collector circuit in the primary of the audio transformer is inductively coupled back into the secondary winding connected to the base circuit of the transistor. With the bias control set on an arbitrary set point and the test pushbutton depressed, the transistor in question will produce an audio-frequency oscillation if it is capable of producing even a small amount of ac gain. This is true even when the original circuit in which the transistor is still wired contains shunting elements, as long as shunting impedance across any of the elements is greater than about 150 Ω. Since this condition is satisfied in the vast majority of cases, it becomes possible to rough-check the transistor in this way, without disturbing its original connections.

The indication of audio-frequency oscillation is obtained by coupling the resulting audio output through capacitor C to the rectifier–meter combination. This shunt rectifier circuit produces an indication on the dc meter showing that AF output is being produced, without any interpretation being attached to the amount of the indication. Thus, although no quantitative measure is obtained, the fact that the transistor is capable of AF oscillation does avoid the nuisance of disconnecting the transistor. From the evidence of AF oscillation, it may generally be presumed that the transistor is, at least, in an operating condition.

It is interesting to note that this scheme of testing for ability to oscillate forms the basis of so-called "meterless" testing, when the meter indication is replaced by a headphone or a loudspeaker. This same idea, now being used in a few commercial transistor checkers, was previously used in the past for a quick check of tubes, without requiring a meter.[2]

12-2. TRANSITOR TESTERS AND ANALYZERS (CURVE TRACERS)

The term transistor "tester" (or analyzer) has been reserved, in this text, for instruments giving quantitative measurements of transistor test parameters. As a minimum, the tester should be able to provide direct-reading values for at least the following two important measurements: (1) a value for *the forward gain* in the common-emitter configuration (h_{fe} for ac gain or alternately h_{FE} for dc gain), commonly referred to as the ac or dc beta (β), and (2) a value for collector-to-base

[2]S. D. Prensky, "Meterless Tube Checker," *Radio Craft* (now *Radio Electronics*), 1930.

leakage current, with the emitter open (I_{cbo}, or sometimes simply I_{co}). The latter measurement of collector-to-base reverse current is generally regarded as the most significant one in the aging process of a transistor. It is a comparatively difficult measurement to make and interpret accurately not only because of the small currents involved (generally only a few microamperes), but also because of the extreme temperature sensitivity of the measurement. The simple act of holding the transistor between one's fingers while making the measurement, for example, can increase the value of the reading by as much as 50%. When made carefully, however, with due regard for temperature restrictions, the reverse-current I_{co} measurement together with the value for the current gain β forms a combination that is ample for determining the operating condition of a transistor in most maintenance applications. These two tests, accordingly, are the basis for most service-type testers. More extensive tests of transistor parameters are discussed under laboratory-type transistor testers, generally called analyzers or curve tracers.

In the transistor development process, as the number of conventional and unipolar transistor types have proliferated, the situation regarding transistor testing has also changed. The requirements for obtaining significant quantitative measurements for such a host of different types have made it quite impractical for any given tester to be capable of handling all of the numerous types. These limitations have resulted in a pronounced trend toward *in-circuit testing* of transistors for service work.

On the other hand, the laboratory testing of transistors has tended toward the *curve-tracer* instrument, where the transistor characteristics are displayed on an oscilloscope screen and form the basis of interpreting the desired transistor parameters.

In-Circuit Transistor Testing

The in-circuit semiconductor tester illustrated in Fig. 12-3(a) (*AEL model 259*) measures diodes, transistors, and FETs both in circuit and out of circuit (Table 12-1). In taking advantage of the great convenience in measuring a transistor (conventional or FET) that is already wired into a circuit, the quantitative value obtained (for ac beta, for example) will depend in some degree on the components to which the transistor is already connected in the circuit. This is taken into account by the specification that as long as the emitter-to-base load of the transistor is more than 50 Ω, the ac beta (or h_{fe}) value obtained on the 1–100 range will be accurate within ±10%. Similarly, in testing an FET in circuit, the value of transconductance within ±10% calls for a minimum drain-to-source load of 3 kΩ and a gate-to-source load of at least 100 Ω. Since these conditions can be expected to be present a good part of the time, this method tends to make up in convenience what it might lose in the precision of the results.

The *Sencore Transistor and FET Tester*, shown in Fig. 12-3(b), also provides in- or out-of-circuit tests for both low- and high-power bipolar transistors and

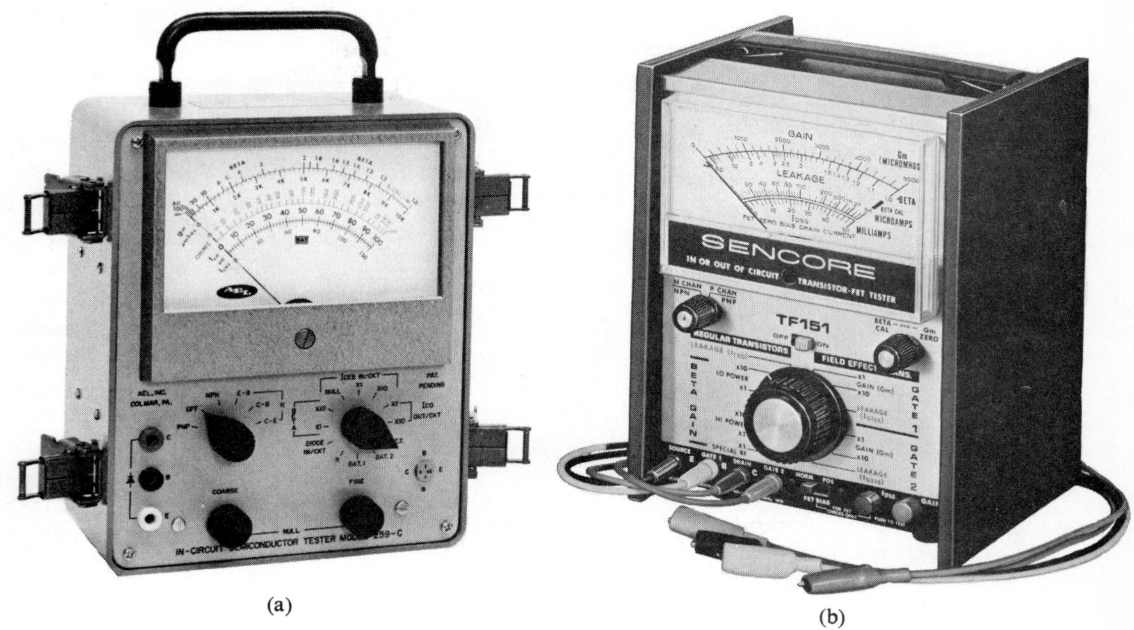

(a) (b)

Figure 12-3. In-circuit transistor testers measure both out of circuit and in circuit, and also cover both bipolar transistors and FETs: (a) *American Electronics Labs model 259C*; (b) *Sencore model TF151.*

for FETs (including the dual-gate variety). The reference book provided with the instruction manual lists as many as 12,000 types in all—a dramatic indication of how the solid-state devices have "just growed like Topsy."

Transistor Curve-Tracer Oscilloscope Display

Transistor characteristic curves (such as the family of output characteristics in the form of a plot of collector current I_C vs. collector voltage V_{CE}, with base current I_B as a running parameter, are particularly useful in evaluating a given transistor. Curves of this nature can be obtained as a continuous screen display by means of an oscilloscope curve tracer, such as the one illustrated in Fig. 12-4 (*Tektronix type 576*). By means of the various controls, dynamic characteristics of a wide variety of transistor types can be displayed (including FETs, SCRs, unijunctions, and others) to produce a family of curves similar to those shown in Chapter 6.

Steps of current or voltage in staircase form are selected by the corresponding controls and applied to the transistor under test producing a family of a number of curves (4 to 10, as desired), with calibrated deflection factors for each of the

TABLE 12-1. Specifications for In-Circuit Semiconductor Tester Model 259

Out of Circuit

Beta (h_{fe}, 1 kHz); $I_c = 1.0$ mA; $V_{CB} = 1$ V (max.)

Range	Beta	Accuracy
×1	1–100	±5%
×10	10–1000	±5%

I_{CBO}; $V_{CB} = 6$ V; $I_E = 0$

Range	I_{CBO}	Accuracy*
×1	0–100 μA	±3%
×10	0–1 mA	±3%

I_R, $V_R = 6$ V

Range	I_R	Accuracy*
×1	0–100 μA	±3%
×10	0–1 mA	±3%

Field-Effect Transistors
Transconductance (g_m, 1 kHz)

Range	g_m	Accuracy*
×1	0–10,000 μ℧	±5%

In Circuit

Beta (h_{fe}, 1 kHz); $I_c = 1.0$ mA; $V_{CB} = 1$ V (max.)

Range	g_m	*E–B* Load (min.)	Accuracy
×1	1–100	50 Ω	±10%
×10	10–1000	500 Ω	±10%

Field-effect Transistors
Transconductance (g_m, 1 kHz)

Range	g_m	Drain-to-source load	Gate-to-source load	Accuracy
×1	0–10,000 μ℧	3000 Ω (min.)	100 Ω (min.)	±10%

*Of full-scale reading.
Courtesy of American Electronics Labs (AEL).

variables. In this way, by either photographing or tracing, each transistor can be characterized under working conditions, and also if desired, compared to a reference (or standard) transistor of that type by manual switching.

The curves are displayed on a rectangular 6.5-in. CRT, providing 10- × 12-cm divisions for a high-resolution display.

A fiber-optics readout is automatically displayed at the right of the screen displaying the curves. The optical fibers transmit light from lamps to the front-panel display, where characters are formed. The lamps are driven from an I_C decoder computer that makes a "beta-per-division" computation corresponding to the other three parameters, as selected. The setting for these other three variable parameters are also simultaneously displayed on the front-panel display, showing the following calibrated values: VERT (usually for reading I_C in microamperes or milliamperes per division), HORIZ (usually for reading V_{CE} in millivolts or volts per division) and PER STEP values (usually base drive in microamperes or milliamperes per division) as a running parameter.

The versatile capability of such a curve tracer, (with the aid of a plug-in test fixture), furnishes a very convenient method for obtaining an immediate display of the most important characteristics of practically all semiconductor devices.

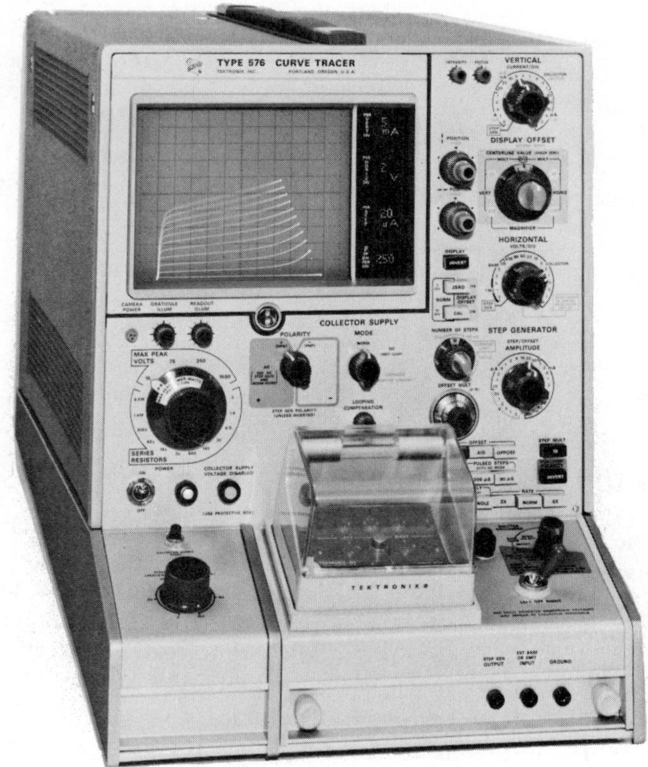

Figure 12-4. Curve tracer provides a CRT screen display of characteristic curves for transistors (or other semiconductors) under test. A fiber-optics display panel (at the right of the CRT screen) simultaneously displays the calibrated values being used in the test (see the text). (*Tektronix Curve Tracer model 576.*)

12-3. PRACTICAL METHODS FOR TESTING COMPONENTS

The development of instrumentation for investigating components of electronic circuits is similarly marked by a division of instruments into two classes: (1) the *basic measuring group* of instruments (discussed in previous chapters), and (2) the group of *practical test instruments*. The emphasis in this chapter is on the practical test instruments (and the methods employed in using them) to test circuit components (R, L, and C) and their properties (U, Z, and Q).

The basic measurement of resistance R by the dc bridge was considered in Chapter 4, and the measurement of inductance L, capacitance C, and impedance Z was given in Chapter 5. The practical method for checking resistance by an ohmmeter also appeared previously, using the nonelectronic form of the volt–ohm–milliammeter (VOM) in Chapter 2 and employing the electronic form (as one function included in the electronic voltmeter or VTVM) in Chapter 7. This chapter will enlarge on other practical instruments for testing properties of components as follows:

1. R and C bridge in a service-type test instrument.
2. Q meters for determining quality factor and associated L and C in the laboratory.
3. Direct-reading impedance testing for polar impedance, giving a direct-reading value of the magnitude and phase of an impedance (Z-angle).

12-4. RC CHECKER (SERVICE TYPE)

The ac bridge circuit, as adapted for checking resistance and capacitance in maintenance applications, is the basis of the RC checker, illustrated in Fig. 12-5. The operation of the bridge circuit is simplified by the use of the voltage-divider potentiometer in place of a slide-wire and the calibrated scale on the voltage-divider dial, so that both R and C can be read off directly from the dial markings. The use of such a test-bridge circuit sacrifices some of the high precision possessed by laboratory measuring bridges, in favor of a practical instrument at reasonable cost. It still retains sufficient accuracy for the tolerances usually encountered in maintenence work.

Figure 12-5. Resistance–capacitance checker of the ac-bridge type. (*Heathkit model C-3*, Heath Company, subsidiary of Daystrom, Inc.)

The balance indicator is a tuning-eye tube usde to indicate the null condition at balance, when its shadow angle is wide open. In addition to its function of checking resistance and capacitance values over wide ranges, it also includes provisions for measuring the leakage properties of electrolytic capacitors under operating conditions with the proper polarizing voltage applied and, in addition, provides for measuring the power factor of such capacitors, as related to this leakage resistance.

The ranges covered by a typical RC checker, such as the one illustrated (which happens to be supplied in kit form), are as follows:

Capacity: Four range: 10–5000 pF (0.005 μF), 0.001–0.5 μF, 0.1–50 μF, 20–1000 μF.

Dc Leakage Test: In milliamperes.

Polarizing Voltages: Five ranges: 25, 150, 250, 350, and 400 V dc.

Power Factor: 0–45%.

Resistance: Two ranges: 100 Ω to 50 kΩ, 10 kΩ to 5 MΩ.

Ac Bridge Circuit: Powered by 60-Hz line voltage from secondary of power transformer.

Indicator: Tuning-eye tube null indicator.

Simplified circuits for each function are shown in Fig. 12-6.

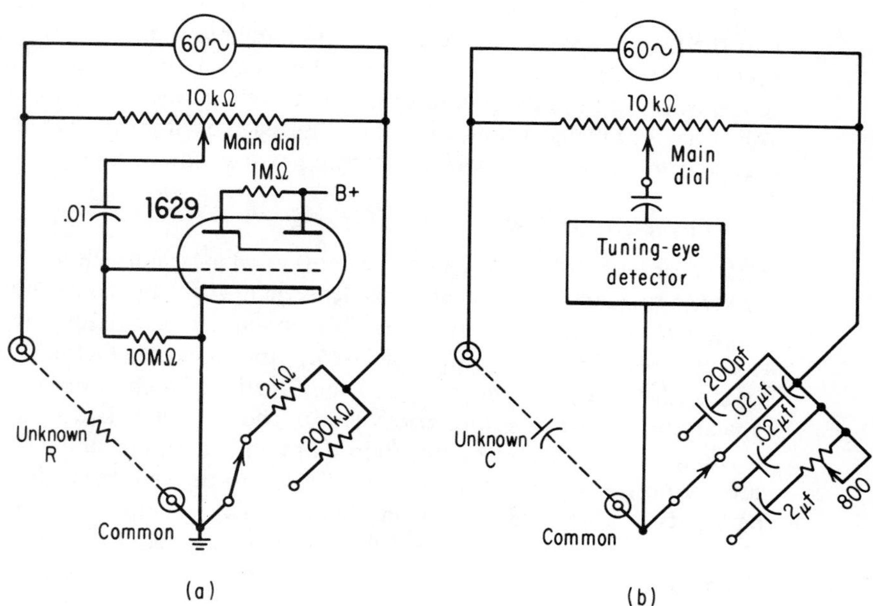

Figure 12-6. RC checker functional circuits: (a) for resistance R; (b) for capacitance C.

Resistance and Capacitance Tests

The functional circuit for testing *resistance* is shown in Fig. 12-6(a). The bridge circuit is powered by a 60-Hz line voltage and produces an off-balance ac output voltage at the grid of the tuning-eye detector while the circuit is in an unbalanced condition. This unbalanced voltage at the grid of the tuning-eye tube is rectified by the diode action between its grid and cathode. This rectified voltage is dc-amplified by triode action, causing the eye opening to decrease its shadow angle and to approach a closed, overlapping appearance as a limiting condition of greatest unbalance. When the circuit is balanced, the null-output condition is indicated by the widest opening of the eye, at which time the value of resistance is read off directly from the appropriate scale on the main dial.

Capacitance Test

The circuit for testing an unknown capacitor is shown in functional form in Fig. 12-6(b). The method for obtaining the capacity value is similar to the bridge-balancing method used for determining resistance, for the various types of non-electrolytic capacitors. In the electrolytic types, additional tests are made for leakage and power factor.

Leakage and Power-Factor Tests

With the proper polarizing voltage applied to the electrolytic capacitor under test and the function switch set to the LEAKAGE function, any leakage current that is present produces a dc voltage at the grid of the tuning-eye tube. This produces a tendency for the eye to close in proportion to the amount of leakage present. The values are so arranged that a partially shorted capacitor (or any leakage greatly in excess of tolerance) causes an overlapping of the shadow angle.

Power-Factor Test

In order to obtain a balance condition when testing electrolytic capacitors, the series-resistance equivalent of the leakage in such capacitors must be balanced out by a physical resistance in series with the standard capacitor. This is introduced as a variable-resistance control of 800 Ω, labeled power factor [Fig. 12-6(b)], in series with the 2-μF standard, when the function switch is turned to its highest-capacity position for testing electrolytics. The correct balance point is obtained by simultaneous settings of the main dial and the power-factor control, and the value of power factor can be read off directly in per cent from this control, while the capacity value is obtained from the main dial.

The power factor is a merit figure for the capacitor, depending on the size of the series resistance R_s. It is defined as the cosine of the phase angle θ:

$$\text{power factor} = \cos\theta$$

where

$$\theta = \tan^{-1}\left(\frac{X_c}{R_s}\right) \quad \text{or} \quad \frac{1}{2\pi fCR_s}$$

Electrolytic capacitors show power factors up to around 20%, compared to nonelectrolytic capacitors whose power factor is close to 0% and can usually be considered negligible.

12-5. THE *Q* METER

The determination of the quality factor Q is the means most widely used in the laboratory for testing RF coils and other inductors. The quality factor of a coil is defined as the ratio of its reactance to its resistance at the test frequency, or

$$Q = \frac{\omega_0 L}{R}$$

where $\omega_0 = 2\pi f_0$ (or 2π times the resonant frequency) and R represents the effective resistance in series with the inductance of the coil. This effective resistance is practically never measured directly, since, in addition to the dc resistance of the wire, it also includes all other resistance losses, such as the skin-effect and eddy-current losses. The effective resistance R also depends to a large degree on the core material and on the manner in which a given coil is wound. Since the value of R represents the sum of all these real losses, its value varies with frequency in a highly complex manner. The effective resistance R is therefore determined indirectly by a measurement of Q.

A knowledge of the Q of the coil is also very useful in resonant circuits. In the usual parallel-resonant circuits, the Q determines the resonant rise in impedance, i.e., the factor by which the inductive reactance is multiplied to give the equivalent resistive impedance R_0 at resonance, or

$$R_0 = \omega_0 LQ$$

As a helpful comparison, it will be recalled that in the series-resonant case, shown in Fig. 12-7, the Q determines the resonant rise in voltage, i.e., the factor by which the input voltage is multiplied to give the voltage across the reactive element. With a 12-V input, for example, at the resonant frequency $f_0 = 1$ mHz, the positive inductance reactance $X_L = 1$ kΩ is cancelled by the negative capacitive reactance, $X_C = 1$ kΩ, leaving the 10 Ω of effective resistance to determine the value of the current $I = 12$ V/10 Ω $= 1.2$ A. The magnitude of the voltage drop IZ across the capacitor, indicated by the high-impedance ac voltmeter V, is 1.2 A × 1 kΩ or 1200 V. By using the value of the coil $Q = \omega_0 L/R$ or 1 kΩ/10 $= 100$,

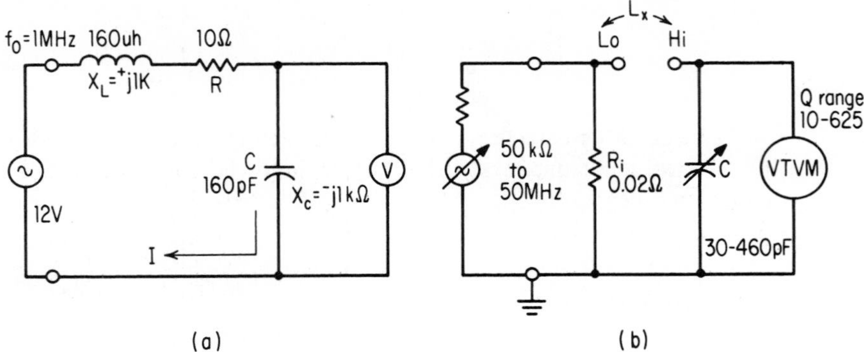

Figure 12-7. Principle of Q meter: (a) circuit to illustrate resonant rise of voltage in a series circuit; (b) circuit of a Q meter.

the same result is obtained by multiplying the input of 12 V by $Q = 100$, to obtain the resonant voltage across the capacitor (or inductor) as equal to 12×100 or 1200 V.

Principle of Q Meter

The resonant rise of voltage across the capacitor in a series circuit is the basic principle used in the Q meter. Essentially, the series-resonant circuit of Fig. 12-7(a) is modified to form a Q meter by connecting a self-contained vacuum-tube voltmeter V across capacitor C. This voltmeter reading as a measure of the voltage magnification can be calibrated directly in terms of Q. This requires that the input voltage be across a very low input impedance R_i and monitored to ensure that it remains at a constant value for each measurement. The self-contained oscillator supplying the input voltage is calibrated for the required frequency ranges, and the capacitor dial is also marked in pico-farads of capacity. The capacity scale can be used to determine the inductance value of a coil connected across the LO and HI terminals of L_x, by noting the capacity required to resonate the coil at the known frequency.

Commercial Q Meters

Two commercial forms of Q meters are illustrated in Fig. 12-8.

Boonton Q Meter

The *Boonton Q Meter type 260-A* in Fig. 12-8(a) covers the frequency range of 50 kHz–50 MHz. (Another *Boonton model, type 190-A*, covers the range 20–260 MHz.) The Q meter illustrated consists of a self-contained, continuously variable oscillator, whose controlled and measured output is applied in series

(a)

(b)

Figure 12-8. Commercial Q meters: (a) *Boonton type 260-A*; (b) *Marconi Q meter TF 1245*, with *oscillator TF 1246*.

with the series-tuned resonant circuit. The reactive voltage across the internal variable capacitor of the tuned circuit is measured by the self-contained VTVM, in terms of the desired Q.

When measuring coils, the unknown coil is connected externally and becomes the inductor in the series-resonant circuit. The change in effective circuit Q_e compared to the original coil Q, can be measured by connecting low impedances in series with the coil. Similarly high impedances connected in parallel with the capacitor will also show the effect of circuit components on the effective circuit Q_e. A thermocouple measuring system (protected against overload) is used to set the oscillator voltage, and the indication is displayed on the "multiply-Q-by" dial. In this way the values obtained on the Q dial (0–250) either can be made to be direct reading (when the multiplier dial is set to 1) or may be multiplied by constant factors of 1.5, 2, or 2.5, depending the setting of the multiplier dial. A separate low-Q scale is provided for Q values of 0–10. Also a delta Q (ΔQ) scale is provided for reading very small changes in Q resulting from a small variation of the test-circuit parameters. The specified accuracies of this model are given as $\pm 1\%$ for frequency, $\pm 5\%$ for direct-reading Q (up to 250) on the indicating meter up to 30 MHz ($\pm 10\%$ at 50 MHz), the main capacitance dial calibration, 30–460 pF direct reading, is accurate to 1% or 1.0 pF, whichever is greater. For the higher frequencies, the *type 190A Q* meter provides a frequency range of 20-260 MHz.

Marconi Q Meter

The version of the Q meter illustrated in Fig. 12-8(b) shows the Q-meter portion (*Marconi type TF 1245*) linked with a separate oscillator (*type TF 1246*) for the frequency range of 40 kHz–50 MHz. The output impedance of the oscillator is of the correct value of low impedance (0.5 Ω) to feed into the Q-meter circuit, which divides the oscillator output down to provide the required 0.02-Ω driving voltage. For higher frequencies, another oscillator type (*TF 1247*) is linked to the basis Q meter, which then becomes capable of measuring Q at frequencies from 20–300 MHz in the very-high-frequency (VHF) range. The output impedance of this oscillator is also 0.5 Ω, designed to feed directly into the basis Q-meter circuit. For frequencies lower than 40 kHz, any laboratory-type audio generator can be linked with the basic Q meter by means of a coupling transformer, provided as an accessory, that transforms the usual 600-Ω output impedance of the AF generator down to the 0.5-Ω impedance required to feed the basic Q meter.

Functional Arrangement for Wide-Range *Q* Meter

The functional arrangement that makes it possible to cover Q measurements over the wide range 1 kHz–300 MHz in the Marconi Q meter is shown in Fig. 12-9. There are two test circuits in the Q meter proper: the top one (HI-I) for measurements from 1 kHz–50 MHz, and the bottom one (HI-II) used from 20 up to 300 MHz. The separate oscillator, usually linked to the top test circuit, supplies

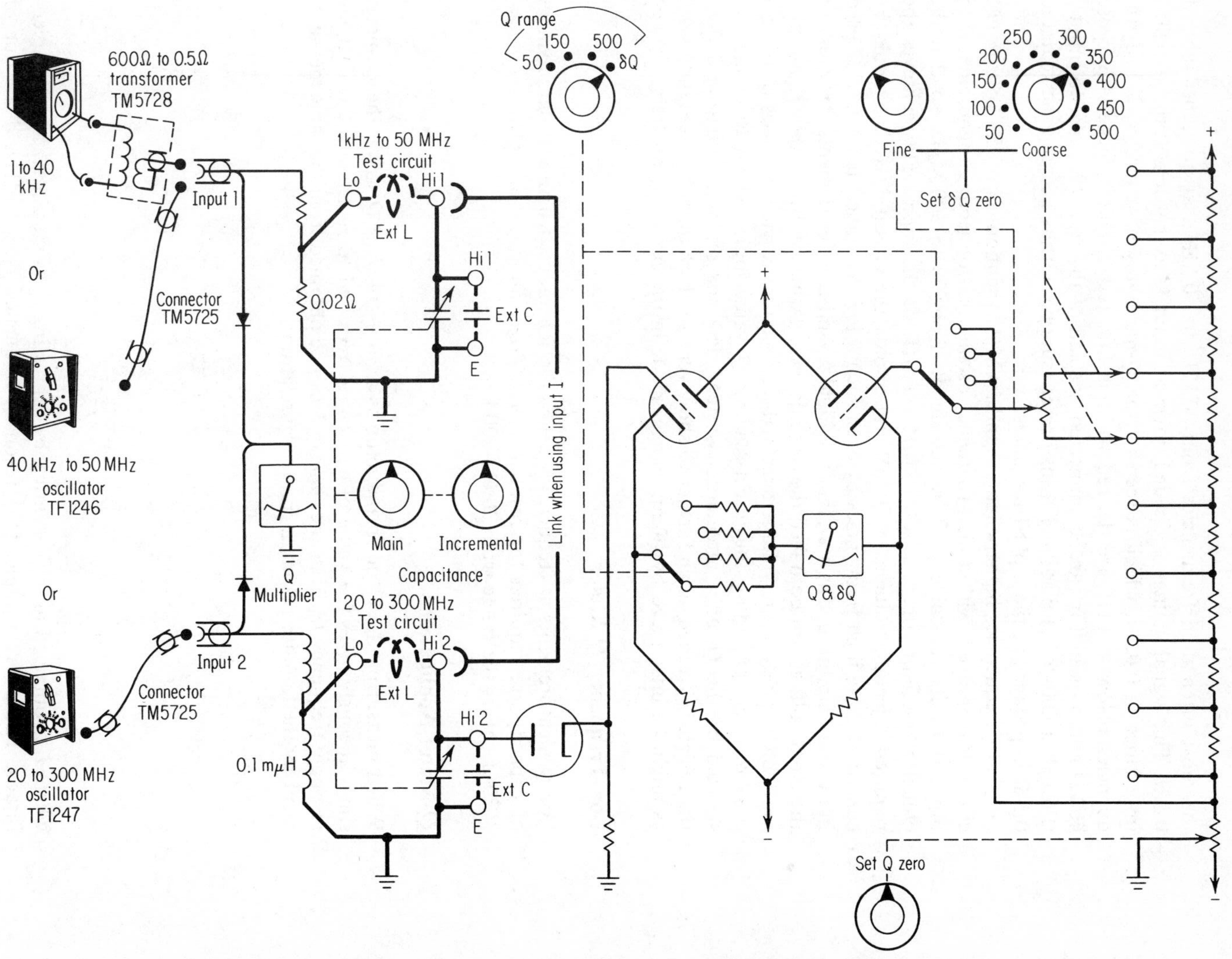

Figure 12-9. Functional arrangement in *Marconi Q meter*.

the most often used frequencies from 40 kHz to 50 MHz, at an impedance of 0.5 Ω. This signal voltage is divided down at input I, so that it is at the standard impedance of 0.02 Ω, which is seen by the Q-meter resonant circuit. When frequencies below 40 kHz are desired for this first test circuit, a general-purpose 600-Ω audio oscillator can be connected, instead of the *type TF 1246*, to input I through a 600–0.5-Ω matching transformer (supplied), thus keeping the required 0.5-Ω impedance at the input I terminals.

For frequencies higher than around 50 MHz the input II of the bottom test circuit is employed, with a high-frequency oscillator. Connecting this oscillator to input II by its coaxial cable supplies signals from 20 to 300 MHz, so that the Q-meter resonant circuit sees 0.1 mμH (or nH) at the input. This achieves the required injection voltage of 20 mV but avoids the circuit loss that would be caused at these high frequencies by even such a low resistance as the 0.02 Ω used in the test I circuit. In this way, the individual crystal-rectifier elements that monitor the input voltage can both feed the same dc meter, which indicates the Q multiplying factor. The output voltage across the resonating capacitor passes through a tube rectifier, providing the direct-reading Q value indicated on the meter, as selected by the Q-range attenuation system. The last set of attenuators divides the selected Q range down further to show very small changes in Q (ΔQ) of ± 25, as might be introduced by slight changes made in the circuit parameters.

Low-Frequency Q Meter

For Q measurements at frequencies below the RF range, where very large values of capacity are required for resonance, the *Freed low-frequency Q meter model 1030-A* covers the range from 20 Hz–200 kHz.

Q-Meter Applications

When measuring the Q of an inductor as a *direct-reading Q value*, the Q meter simultaneously gives the value of capacity C required to resonate the coil at the resonant frequency f_0. Thus the *value of L* can be obtained from the same measurement by the familiar resonant-frequency relation:

$$f_0 = \frac{1}{2\pi\sqrt{LC}}$$

or

$$L = \frac{25{,}400}{f^2C}$$

where L is obtained in microhenries, by expressing f_0 in megahertz and C in picofarads. Thus, for example, at a resonant frequency $f_0 = 1$ MHz, if a value of

$C = 160$ pF is required for resonance,

$$L\,(\mu\text{H}) = \frac{25{,}400}{1 \times 10^{12}(160 \times 10^{-12})} = 160\ \mu\text{H}$$

as a close approximation.

Measuring Self-Capacitance C_0

It is often necessary to know the distributed *self-capacitance* C_0 of the coil under test. A rough idea of its value can be obtained by finding the *self-resonant frequency* of the coil when no external capacity is connected across it. The Q meter can be used to obtain a better value for the self-capacitance C_0 by a quick and simple method, known as the *frequency-doubling method.* In this method, the Q of the coil is determined with the variable capacity of the Q meter set at its maximum (call this C_1 at, say, 500 pF), and the frequency at which this resonance occurs is noted as f_1. Next, the oscillator is set to twice this frequency ($f_2 = 2f_1$) and resonance is restored by a new setting of the variable capacitor at some lower value, noted as C_2. The self-capacitance (C_0) of the inductor under test is then given by[3]

$$C_0 = \frac{C_1 - 4C_2}{3}$$

Finding True Inductance L_x

The value of self-capacitance thus found can be used to find the *true inductance* L_x as contrasted with the indicated inductance L_{ind}. This correction is applied, when necessary, by the expression for true inductance L_x:

$$L_x = L_{\text{ind}}\left(\frac{C_{\text{ind}}}{C_{\text{ind}} + C_0}\right)$$

This correction is small when working at ordinary frequencies, but the effect becomes greater at the higher frequencies. Thus, if we assume that an unknown coil has a comparatively high distributed self-capacitance (C_0) of 2 pF, we would find, on measurement at 5 MHz, that it resonated with an indicated capacitance (C_{ind}) of 30 pF, giving an indicated inductance (L_{ind}) of about 34 μH. Applying

[3]This expression is derived by Terman and Pettit (see Appendix B), as follows: since the effective capacitance at f_1 is $C_1 + C_0$ and the effective capacitance required for the double frequency $C_2 + C_0$ is four times as great,

$$C_1 + C_0 = 4(C_2 + C_0)$$

$$C_1 - 4C_2 = 4C_0 - C_0 = 3C_0$$

then

$$C_0 = \frac{C_1 - 4C_2}{3}$$

the correction for the true inductance L_x, we find

$$L_x = 34\ \mu\text{H}\left(\frac{30\ \text{pF}}{30 + 2\ \text{pF}}\right) = 34(0.938) = 32\ \text{pF}$$

For the extreme assumption made for the self-capacitance C_0 in this example, the true inductance value is seen to be below the indicated one by over 5%. In most practical cases, however, the difference is considerably less than this and is ordinarily considered negligible.

Measuring Unknown Capacity C

The Q meter may also be used to determine the *capacity value* C of an unknown capacitor. With a coil of the proper range connected across the L terminals, the circuit is resonated at a given frequency—first with, and then without, the unknown capacitor C connected across the calibrated variable capacitor of the Q meter. The difference between the two readings then gives the unknown value of C. For values of the unknown C larger than the 500 pF available in the variable capacitor of the Q meter, a series method is used, for which a special series test jig (obtainable as an accessory) is employed.

Finding Bandwidth of Tuned Circuit ($BW_{3\ \text{dB}}$)

The *3-dB bandwidth* of a tuned circuit can easily be obtained on the Q meter as another way of obtaining Q, from the relation

$$BW_{3\ \text{dB}} = \frac{f_0}{Q}$$

where the *3-dB* bandwidth ($BW_{3\ \text{dB}}$) represents the difference in the two frequencies on either side of resonance, at which the output (the Q value in this case) has dropped to 70.7% of its value at resonance.

The Q meter is applicable to many other tuned-circuit and impedance tests. The *characteristic impedance* Z_0 of a transmission line, for example, can be determined by open- and short-circuit measurements made on the transmission line by the Q meter.

12-6. PRACTICAL INDUCTOR CHECKING

There is a dearth of practical methods for speedy, convenient testing of inductors by means of a *direct-reading inductive-impedance check*, as a substitute for more lengthy, indirect-reading methods of *measuring coil impedances* in the laboratory by the conventional bridge method. In *resistance testing*, the ohmmeter method is an accepted substitute for the bridge, whenever a rapid, convenient resistance check is desired; and in the case of *capacitance testing*, a number of rapid methods are available in service instruments as previously described. But not so in the case of service-type instruments for rapid checking of inductors; here the field is very sparse indeed. As a result, the testing of inductors is very seldom undertaken in maintenance work (other than the making of a continuity or dc resistance test of a winding—which begs the question). In the main, inductors are still measured in the laboratory by comparatively laborious methods and are practically never checked in the field for either inductance L or impedance Z values.

The reason underlying the scarcity of direct-reading check methods for inductive impedance Z is that the impedance of a coil depends basically on the resistance R_L inevitably associated with its reactance X_L. Accordingly, a simple ac test for coil Z, similar to the dc method of ohmmeter testing, does not yield the same satisfactory information in the case of the coil, as it does in the case of a resistor having only resistance, or in the case of a capacitor, where the value of impedance Z is made up of practically only reactance X_c, with generally negligible contribution of the resistance. (Where there is appreciable resistance associated with the reactance X_c, as with electrolytic capacitors, we are generally interested only in the extent of the leakage resistance, to make sure that the capacitor has not broken down.)

The prospect for obtaining a rapid direct-reading impedance check on an inductor is not altogether hopeless, however, if the problem is approached with an understanding of the inherent difficulty involved. If allowance is made for this difficulty by accepting a reasonably lower tolerance for an inductor check wherever practical (or by suitable interpretation of the meaning of the check result), the convenience of rapid inductor checking can still be retained by a number of methods, even though they have not become as popular or as well known as the simpler resistance or capacitor checking methods.

In cases where only the *magnitude of the coil impedance* $|Z|$ is required (without the need for knowing the exact ratio of X_L to R or the value of Q), simple checking methods that are used for capacitive impedance apply equally well for checking inductive impedance. Thus, among the measurement methods for determining the impedance discussed in Chapter 4, the equal-deflection method given in that chapter can be employed as a practical test method for a rapid check on the overall impedance magnitude $|Z|$ of a coil. Although this determination of the impedance magnitude does not provide any knowledge of the separate reactive and resistance factors that make it up, or of the ratio between them that determines the Q (equal to X_L/R), a knowledge of the overall magnitude is often sufficient. For example, when testing the impedance variation of a speaker voice coil at different frequencies, a graph of the changes in magnitude (without the phase-angle variations) is valuable. In many other instances, as in the case of testing the primary and secondary impedances of a transformer, the value of Q is often unimportant; here also, a test method giving only the impedance magnitude (usually at 1000 cycles) provides quite sufficient information for test purposes. In such cases, lacking a commercial instrument, a do-it-yourself test using, say, the equal-deflection method employing only general-purpose instruments, could be used to advantage.

12-7. DIRECT-READING VECTOR IMPEDANCE TEST

When it is required to have some knowledge of the reactive and resistive factors, in addition to obtaining a direct-reading value of the impedance magnitude, a test method for determining the *vector impedance* may be employed. This method determines the impedance in polar form; that is, it gives the magnitude $|Z|$ and phase angle ϕ of the impedance being tested, rather than the individual resistance and reactance in the rectangular form of $R + jX$.

The test circuit is shown in Fig. 12-10(a). Using resistors of equal value for R_{AB} and R_{BC}, the voltage drops E_{AB} and E_{BC} will be equal, and each will be half of the input voltage E_{in}, or E_{AC}. Since the same current I_1 flows through the variable

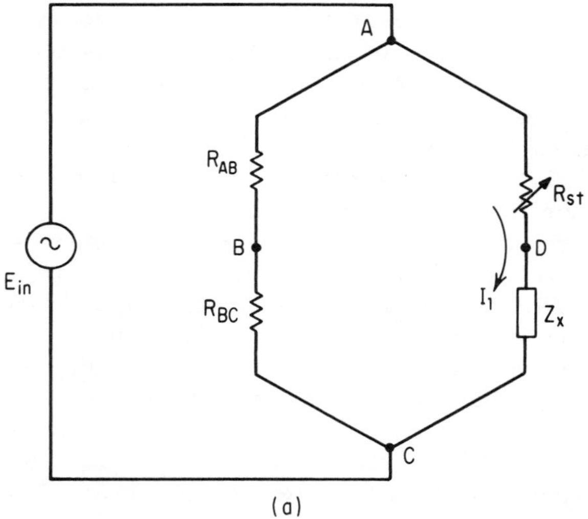

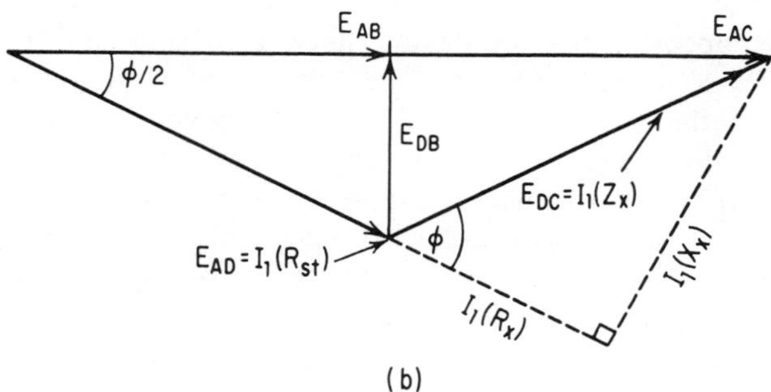

Figure 12-10. Vector (polar) impedance method for inductor checking: (a) circuit diagram; (b) vector diagram.

standard resistor R_{st} and the unknown impedance Z_x in series with it, the magnitude of Z_x can be obtained by the equal-deflection method by obtaining equal voltage drops E_{AD} and E_{DC} and reading the value of the calibrated standard R_{st} required to produce this condition.

An estimate of the phase angle ϕ of the impedance Z_x can be obtained by interpreting the voltage reading E_{DB} under these circumstances. From the geometry of the vector diagram shown in Fig. 12-10(b), the angle between E_{AB} and E_{AD} is half of the phase angle ϕ, since E_{AD} has been made equal to E_{DC}. Therefore,

$$\frac{\phi}{2} = \arctan \frac{E_{DB}}{E_{AB}}$$

Since E_{AB} is known as half of the known input voltage E_{in}, the voltmeter reading of E_{DB} can be interpreted in terms of $\phi/2$ (and therefore of the phase angle ϕ of the unknown impedance Z_x.

12-8. COMMERCIAL VECTOR-IMPEDANCE METER

An example of a commercial instrument that provides a direct measure of the polar impedance (Z-angle) of inductors and capacitors is illustrated in Fig. 12-11. (The model illustrated is the *high-frequency Hewlett-Packard, model 4815A*, for frequencies above 500 kHz and up to 108 MHz; the similar *low-frequency model*

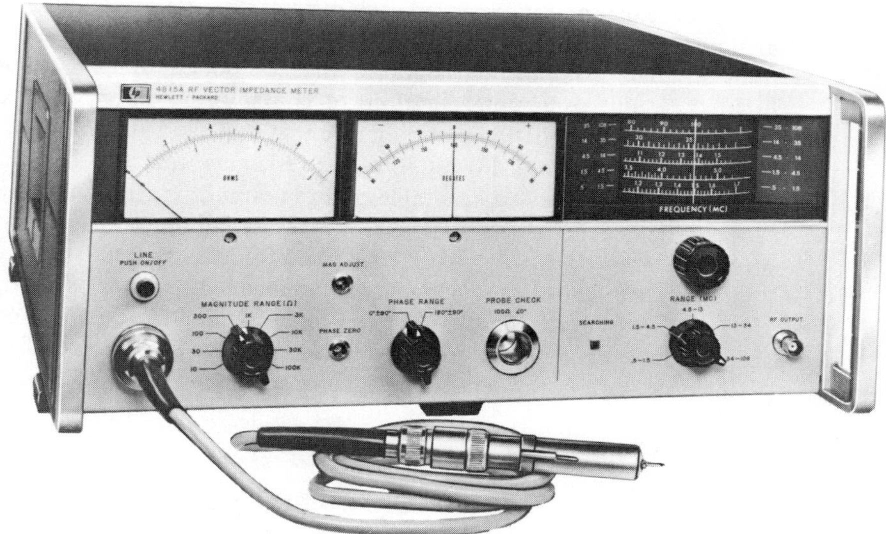

Figure 12-11. Vector-impedance meter for direct measurement of polar impedance (*Z*-angle). (*Hewlett-Packard model 4815A*.)

4800A is described below for component measurement at frequencies from 5 Hz to 500 kHz, which includes the standard component-measuring frequency of 1000 Hz, or 1 kHz. The measurement of the two-terminal impedance will yield the magnitude $|Z|$ up to 10 MΩ and the phase angle ϕ to ± 90, at the frequency of interest. Analog outputs are available for X–Y recording of these *Z*-angle values, as the frequency is mechanically swept over the full range.

By setting the frequency dial to the calibrated setting of 15.92 (and its decade multiples up to $\times\ 10^4 = 159.2$ kHz), direct readings can be made of inductance (and hence the apparent Q) and of capacitance (and its associated D). In this way, a convenient check can easily be made to determine quickly the component and circuit values of reactive elements.

12-9. OSCILLATING-CIRCUIT TEST FOR COMPONENTS

In testing individual components (resistors, coils, or capacitors) it is usually necessary, as a general rule, to remove the component from the circuit (or in some cases, at least to disconnect one end). It would, however, be a great convenience if it were feasible to test the component *in-circuit*, i.e., without unsoldering any of the circuit connections—especially so, in the case of the densely packed situation that is true of the widely used printed circuits. Such in-circuit testing capability can often be achieved by the use of an oscillating-circuit type of test.

Based on the general principle that an *LC* circuit will oscillate only if the transistor, coil, and capacitor are each in proper operative condition, this in-circuit method can be effective as a quick-check way of spotting open or shorted conditions in these components. Because the component remains connected in its original circuit, some consideration, of course, must be given to elements that might shunt (or load) the component under test, and this imposes a limitation, usually specified as the lowest shunting resistance that can be tolerated for the test. Thus, in the previous example of an in-circuit transistor tester [Fig. 12-2(b)], it was specified that the resistance load across emitter to base must be 50 Ω (or more) in order to ensure $\pm 10\%$ accuracy for the in-circuit beta check. Similar limitations are stated for in-circuit measurements of coils or capacitors.

Since, for quick-check purposes, only a qualitative indication is required, the indicator can be as simple as a tuning-eye tube as evidence of oscillation and yet be able rapidly to identify an open or short condition of a coil or capacitor.

QUESTIONS

Q12-1. Discuss two main reasons to justify the statement that there is less need, in general, for testing transistors than tubes.

Q12-2. Under what special circumstances will a transistor fail more readily than a tube?

Q12-3. State the special provisions needed in a transistor tested to provide tests for:
(a) Power transistors,
(b) Switching transistors.

Q12-4. What principle may be employed to permit in-circuit testing of transistors?

Q12-5. State how the ratio of forward current to reverse current can be used as a basis for testing semiconductor diodes.

Q12-6. (a) In transistor testing, explain the difference between values for I_{cbo} (or I_{co}) and I_{ceo}.
(b) Explain why it is important to obtain a good measurement of the value for I_{co} in transistor testing.

Q12-7. If only an ohmmeter is available for transistor testing, show how it is possible to obtain an indication of the amplifying ability of the transistor under test, in addition to obtaining readings merely of resistance values.

Q12-8. When a 20-μF electrolytic capacitor is tested on a capacity bridge, its capacity reads 0.3 μF. Account for the probable defect in the capacitor, assuming that the test result is valid.

Q12-9. When an electrolytic capacitor is tested for leakage resistance with an ohmmeter, the resistance indication initially shows a very low resistance value, which quickly increases in a steady manner to a high value. Explain.

Q12-10. State why it is much more difficult, in general, to find the inductance and impedance of a coil, than it is to find the capacity and impedance of a capacitor on service-type testers.

Q12-11. (a) Explain why different values should be expected for the Q of a coil when measuring the Q at various frequencies.
(b) Account for the fact that the Q of a coil may be found to be substantially constant over a range of hundreds of kilohertz (as in the broadcast band).

Q12-12. What principle is used to make possible an in-circuit check of capacity, without disconnecting any leads, even though the capacitor is shunted by resistors in the circuit?

Q12-13. State in what respect the reading obtained for an unknown coil on a polar-impedance measurement differs from that obtained by a bridge measurement.

PROBLEMS

P12-1. A 5-H choke coil measured on an ohmmeter is found to have a resistance R_{dc} of 190 Ω. If it is found to have a $Q = 5$ at 60 Hz:
(a) Find its equivalent resistance R_L.
(b) Explain the difference between the values for R_{dc} and R_L.

P12-2. Show by diagram how an unknown resistance of 30 MΩ can be measured with a VTVM having only voltage scales and an input impedance of 10 MΩ.

P12-3. An unknown coil, shown in the Fig. P12-3, is found on an R–X meter to have a parallel reactance X_p of 75 Ω and a parallel resistance R_p of 100 Ω, measured at 1 kHz. Find the equivalent series reactance X_s and series resistance R_s at that frequency.

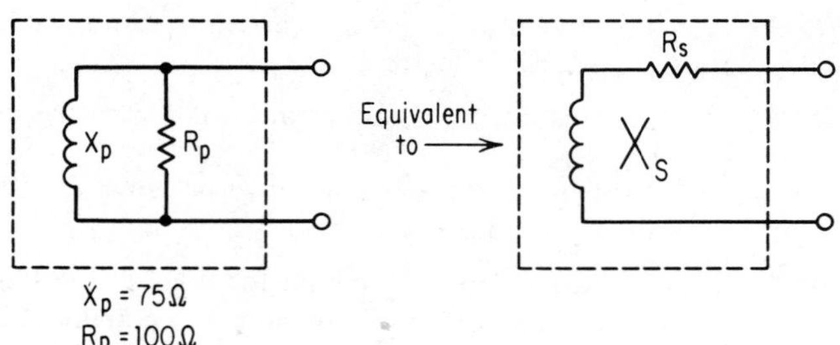

Figure P12-3

P12-4. In the Q-meter circuit shown in Fig. P12-4, find the full-scale range of the VTVM required to indicate a full-scale Q range of 150.

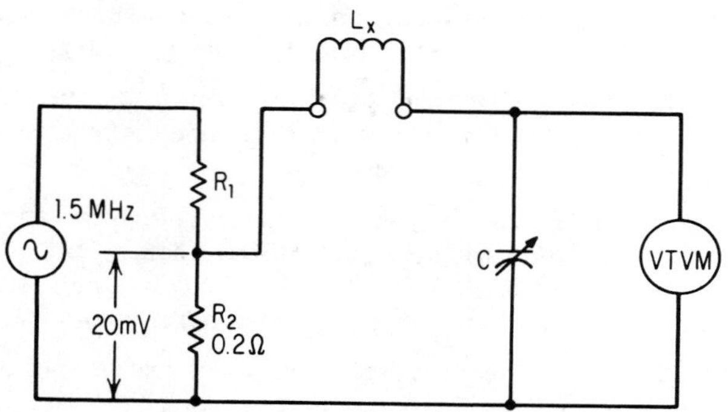

Figure P12-4

P12-5. In the Q-meter circuit for Problem P12-4, the values at resonance are as follows:

$$\text{coil } L_x = 320\ \mu\text{H}$$
$$f_x = 1.5\ \text{MHz}$$

(a) Find the value of C required for resonance.
(b) What value of Q is indicated when the VTVM reading is 1.2 V?

13

Integrated Circuits

13-1. GROWTH OF INTEGRATED CIRCUITS

The impressive development of integrated circuits (ICs) is a clear indication of the great potential possessed by these units for *simplifying the design* of electronic circuits, while at the same time producing a unit that has *greater reliability* and a *smaller cost* than can be achieved with separate, discrete elements. This powerful triple advantage (simplicity, reliability, and low cost) does not even take into account the additional feature of greatly reduced size, which was one of the important original factors in the integrated-circuit (or microminiature) development.

The early experiments for fabricating multiple transistors from a single silicon slice (or chip) were aimed at their use in digital computers that employed the semiconductor diodes and transistors in the thousands. The successful outcome of this intense experimentation—resulting in the integrated circuits—was indeed a technological breakthrough. On a single silicon chip, no larger than a decent-sized dot, an IC is made to provide more than a dozen transistors, already internally interconnected with another dozen or so of associated resistors; the whole unit (with its 8, or 10, or 12 external leads) may be packaged in a single T0-5 can, of a size that previously housed just a single transistor. Extending this *medium-size integration* (*MSI*) to *large-size integration* (*LSI*), results in a highly compact unit containing literally hundreds of transistors, forming a very versatile *building block*, for use as a basic amplifier.[1]

The appearance of the basic chip and its internal interconnections is shown in Fig. 13-1, greatly enlarged from its actual microscopic size. Most of the space

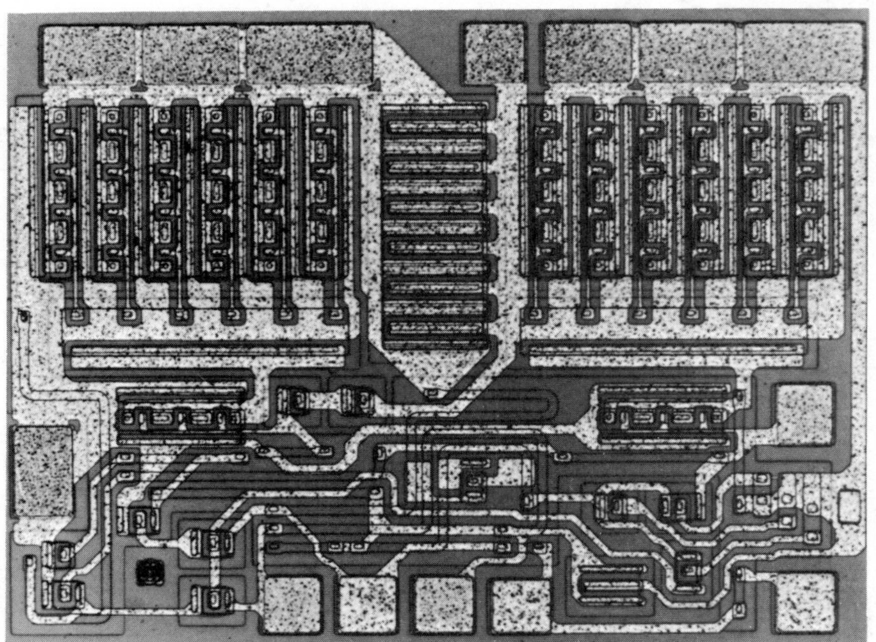

Figure 13-1. A monolithic integrated-circuit chip, greatly enlarged—actual size much less than $\frac{1}{4}$ in. on longest side; this chip is packaged in a 10-lead TO-5 can, to form a 1-W IC power amplifier. (*MC-1554G*, Motorola Semiconductor Products.)

on this minute chip can be seen to be occupied by the zigzag resistors. (The advanced solid-state fabrication methods for accomplishing this new breed of semiconductor, including the refined techniques of diffusion, masking, bonding, and the like, must be left as a separate story.) Externally, the multilead IC takes one of the three popular packaged forms shown in Fig. 13-2; here the flat-pack form in part (c) is used mainly by the manufacturers; the other two forms are used on the circuit boards (either as soldered-in or plug-in units), the *transistor-can type T0-5*, shown in part (a), and the *dual-in-line package* (*DIP*), having usually 14 (and sometimes 16) leads, shown in part (b) of the figure.

[1]Discussed further, in one use as an *operational amplifier*, in Chapter 18.

Figure 13-2. Typical integrated-circuit packaging: (a) TO-5 can type; (b) dual-in-line package (DIP); (c) flat-pack. (Courtesy of Radio Corporation of America.)

13-2. CIRCUIT FUNCTIONS

Branching out from the original circuit function of *micro-digital-logic elements* (whence comes the original Fairchild designation of μL for micrologic), the rapid growth of these versatile midgets encompasses also a collection of *linear-integrated circuits*, which are designed for various circuit functions, such as integrated audio, wide-band, or operational amplifiers. With such diverse capabilities, therefore, it is easy to see why the integrated circuits are finding their way into a host of electronic-instrument applications. In addition to a very complete selection of *switching* (*or digital*) applications, the ICs are well suited for taking over many *amplification and oscillation applications*, in circuits that formerly required much interconnection among a great number of discrete components.

Possible exceptions to this trend toward microelectronics include instances in which high-voltage, high-power, or extreme-precision characteristics might be required. Moreover, there are naturally many components whose natures make it impractical for inclusion in the integrated-circuit package, such as large capacitors, coils, switches, and the like. Consequently, circuit configurations are quite likely to be *hybrid*, that is, a combination of ICs and discrete components.

Widespread acceptance of this healthy development goes far to make future circuits more compact, more reliable, and also much easier to maintain (often by the simple means of replacing a plug-in IC). However, it does not relieve the technical worker of the need for a basic understanding of the principles by which the IC device operates. A good understanding still requires careful study of the basic circuit actions on which the IC is based. Such a study can be met in the simplest way by close examination of the original circuits, using discrete components. Hence a balanced combination that studies both theoretical discrete elements and practical integrated circuits still remains the best method for building a broad base necessary for understanding the electronics involved in the resulting instrumentation.

13-3. EXAMPLES OF DIGITAL INTEGRATED CIRCUITS

The *digital type* of integrated circuit will be examined first, followed in the next section by the linear IC type. These digital ICs perform logic functions by means of *switching action* and are generally designated as *gates*. There are four main basic digital circuits: the *AND* gate, the *OR* gate, the inverting amplifier or *NOT* circuit, and the bistable or *FLIP-FLOP* circuit. [Combinations, such as NOT OR (called NOR) and also NOT AND (called NAND), are so widely used that they are usually considered as single digital elements.]

These digital elements are interconnected in various ways to accomplish the required digital logic; such arrangements are discussed later as they appear in counting and other digital-display instruments, in Chapter 17. In the present chapter we will examine one of these arrangements in its integrated-circuit form, to serve both as an IC example and also as a simple introduction to the more complicated digital circuits.

The NOR Gate

A very simple form of digital-switching action is the *two-input NOT–OR* (*usually called NOR*) *gate*, shown in Fig. 13-3(b), where an inverted output is obtained, if *either one or both* inputs are present. This is built up from the simple single-input gate of Fig. 13-3(a), where it is seen that the elementary gate circuit is simply a transistor having a 640-Ω load resistor (R_C) in its collector circuit and a base resistor (R_B) to limit the amount of base current. Using a +3-V supply for the n–p–n transistor, we distinguish between the "0 and the 1" of the binary system, in this case, by arbitrarily designating a voltage of +3 V above ground as "1" and ground (or practically 0) V as a "0", (called positive logic). We then note that with no input to the base, the transistor is in its off condition, drawing practically no collector current; therefore, the full +3 V appears at the output terminal (in logic short-hand, a "0" input gives a "1" output).

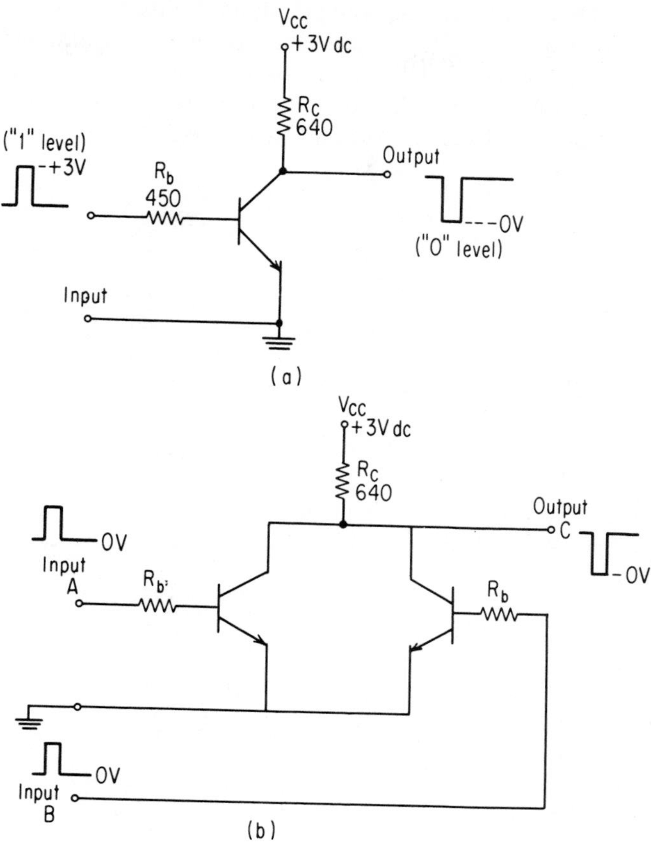

Figure 13-3. Gating circuits: (a) simple gate, an inverting transistor switch; (b) two-input gate, forming a NOR gate [a "1" level at either *A* or *B* input (or both) gives a "0" level output at *C*].

In a similar way, when the base input is at +3 V, the transistor conducts heavily in the on (or saturation) condition, so that almost all of the 3-V supply drops across load resistor (R_C), thus causing the output to drop to practically ground potential, or 0 V (expressed in logic, a "1" input gives a "0" output). *The single gate, therefore, is simply an inverting switch.*

When we add a second similar transistor, connected as in Fig. 13-3(b), we have a two-input gate. Now, if either input A or B receives a +3-V signal, collector current flows through the 640-Ω load resistor, and the output goes to ground potential. Since the same is true if both inputs are at the +3-V level, we have a *two-input NOR circuit*, where an input of 1 at either A or B (or both) inputs, produces a 0 at the output *C*.

The IC we are discussing contains another two-input gate on the same integrated-circuit chip, resulting in a dual, two-input gate (as found in the *Fairchild μL 914 IC*). The basing terminals and internal circuit are shown in Fig. 13-4, where it can be seen that the chip comprises a total of four transistors and six resistors (four for each base and two collector resistors).

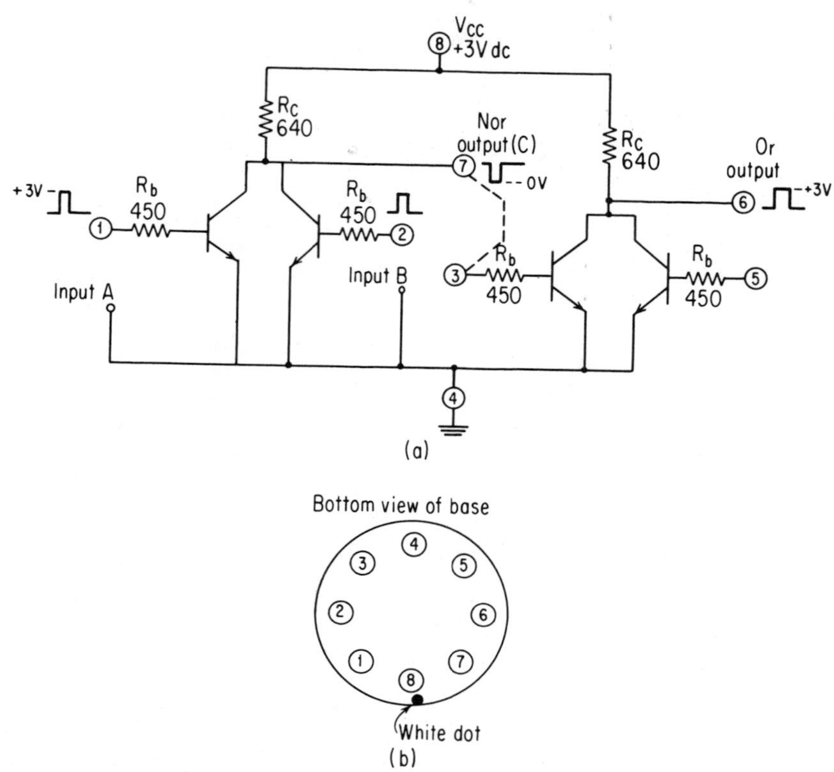

Figure 13-4. Circuit and basing terminals of dual two-input NOR gate integrated circuit; output (C) at ⑦ provides NOR gate for two inputs, A at ① and B at ②; dashed connection between ⑦ and ③ provides OR gate output at ⑥. (*Fairchild Semiconductor, μL 914 IC.*)

The addition of the second two-input NOR circuit to the first one allows us to build up a great many functional circuits from the basic NOR gate. For example, to produce an OR circuit, we may add an inverting gate to the first NOR circuit, as indicated by the dotted external connection between terminal ⑦ (output of first NOR gate) to terminal ③ (input to inverting gate of the second NOR circuit). Consequently, a "1" at either input terminal (or both) will now produce a "1" at ouptut terminal ⑥.

An *AND* circuit, as another example, can be constructed from two of these ICs, connected so that *both* inputs ① *and* ② must receive a logical "1" in order for the output to be a logical "1". Other examples that can be constructed from this relatively simple IC include a set-reset flip-flop, a Schmitt trigger, bistable and monostable multivibrators, and an elementary difference amplifier.

The types of logic circuits discussed so far are called *resistor–transistor logic* or RTL. Another (and much more widespread) class of logic circuit employing bipolar-junction transistors is TTL (transistor–transistor logic).

A four-input TTL NAND gate is shown in Fig. 13-5. The unique feature of

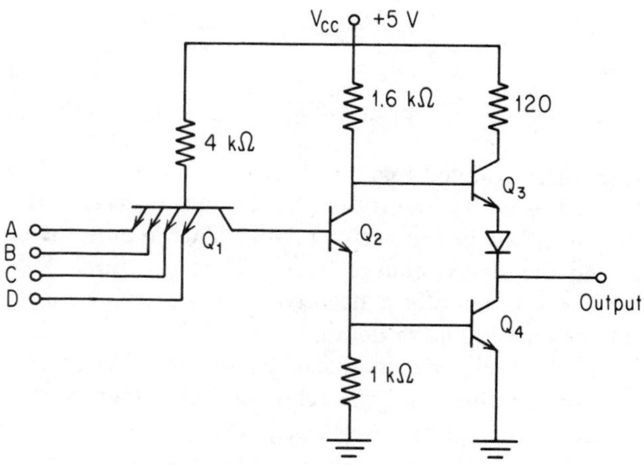

Figure 13-5. Four-input TTL NAND gate.

this circuit is the multiple-emitter input transistor. This device is formed by diffusing several (in this case, four) n-type emitter regions into a single p-type base region. If any emitter is at a logic low or ground state, normal transistor forward-bias conditions prevail, and the transistor is on. The collector of transistor Q_1 will drop to a low voltage in this case, pulling the base of Q_2 to nearly ground potential, and turning Q_2 off. Transistor Q_4 will now be off and Q_3 will be on making the output a high or logic "1." This condition will exist if any or all of the inputs are at a logic "0" level.

If, on the other hand, all the inputs are high, then all emitter–base junctions will be reverse-biased. Under these conditions the base–collector junction of Q_1 will become forward-biased, providing base current to transistor Q_2. This will cause Q_2 to turn on, turning off Q_3 and turning on Q_4. The output will then go to a low or logic "0" state.

The output stage shown here is called a *totem-pole* output stage. It has the advantage of an "active pull-up" as well as an "active pull-down," resulting in higher operating speed than is normally achieved with RTL.

An even-faster type of logic circuit is shown in Fig. 13-6. This is a two-input

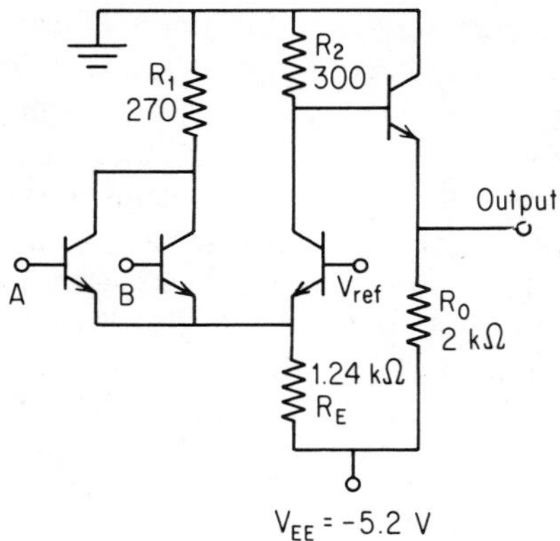

Figure 13-6. Two-input ECL OR gate.

ECL (emitter-coupled logic) OR gate. The reason ECL gates are faster than RTL or TTL gates is that the transistors are never driven into saturation. They operate either cut off or in the active region. This avoids the delay time associated with removing the stored charge from the base region of a saturated transistor. The ECL gate is inherently a negative-logic device. To avoid confusion, we will not discuss negative logic in detail.

All the logic circuits discussed so far suffer from the disadvantages of high power consumption and the relatively large chip areas required for each device. These problems can be overcome, with a sacrifice in speed, by using MOS technology. Metal-oxide field-effect transistors have such a high input resistance (typically thousands of megohms) that the input to a gate made with MOS devices produces only negligible loading of the previous gate's output stage. This allows the circuit to operate at extremely low currents. Digital watches and hand-held calculators would not be possible without these devices. In addition to low power drain, MOS devices require very little space on the chip, allowing more gates for a given amount of silicon "real estate." This is also very important for the applications mentioned above as well as for microprocessors in which the entire central processing unit of a computer is constructed on a single chip.

Although MOS integrated circuits take many forms, CMOS (complementary metal-oxide semiconductor) logic gates are perhaps the most popular. A typical form of CMOS NOR gate is shown in Fig. 13-7. Notice that there are no resistors in this circuit. This is important in that an integrated-circuit resistor takes up several times as much "real estate" as an MOS device. Thus, designers have resorted to using one MOS device as a load for another one. In the case of the CMOS gate shown, the transistors are connected as series pairs.

In the circuit of Fig. 13-7, all MOS transistors are enhancement-mode devices, as indicated by the broken line representing the channel. To analyze this circuit, let us start with both inputs in the low or logic "0" state. In this case, the n-channel devices will be on, resulting in a high or logic "1" output.

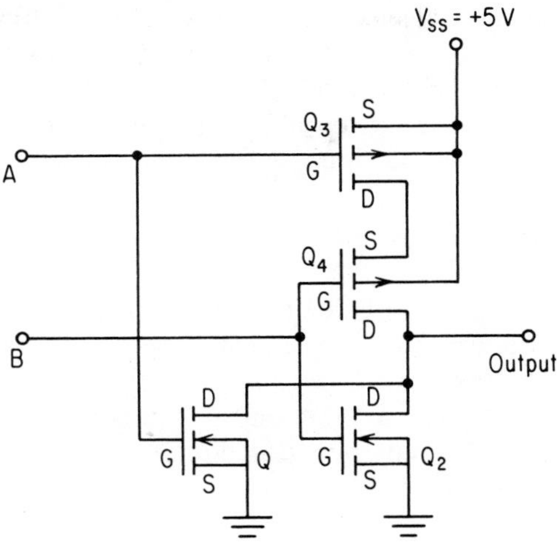

Figure 13-7. Two-input CMOS NOR gate.

Next, assume that one of the inputs, say input A, goes high while the other one remains low. Transistor Q_3 will turn off and transistor Q_1 will turn on. The output will go to a low state because of Q_1 turning on. Even though Q_4 is still on, no current will flow because its source is in series with the drain of Q_3, which is now off. A similar situation occurs when input B is high and input A is low.

If both inputs are high, the n-channel transistors, Q_1 and Q_2, will be on while the p-channel transistors, Q_3 and Q_4, will be off. Thus, the output will be in the low state. It is apparent, then, that this circuit is a NOR gate.

The one disadvantage of MOS logic circuits is that they are not as fast as bipolar circuits. The reason for this is that the impedances in the circuit are very high, and even very small stray capacitances can result in relatively long time constants which limit switching speed. For this reason MOS technology is used where small size and low power are the primary consideration, while bipolar technology is preferred where high speed is required as in large computers.

13-4. EXAMPLES OF LINEAR-INTEGRATED CIRCUITS

Just as the digital ICs are firmly established in their use in digital computers for switching applications, so, in a similar development, the *linear ICs* have naturally fitted into general instrumentation circuitry for *oscillation* and *amplification purposes*. Although a great variety of specialized ICs are available for specific instrumentation, the emerging trend is to regard the linear IC as a generalized building block, which can be substituted for a large number of discrete components (thus improving reliability), and around which specific circuit applications can be tailored for a particular instrument. It is therefore profitable to examine examples of two main building-block ICs, the simple *dc amplifier* and then the *operational amplifier*, both of which are sufficiently general to be useful in a wide variety of both dc and ac amplification applications. The *audio amplifier* may also be considered a building block, serving as a preamplifier or a driver for audio-power applications (an example of this application is given in a later section).

Direct-Current Amplifier IC

The first example, a simple dc amplifier, is shown in schematic form in Fig. 13-8. This circuit is particularly pertinent as a building block, since it clearly shows the basic *balanced form of dc amplifier circuitry* (or *emitter-coupled differential amplifier*); it is *especially* suitable and is extensively used in ICs as a basic amplifier circuit. This balanced transistor circuitry—suitable for both dc and ac amplifi-

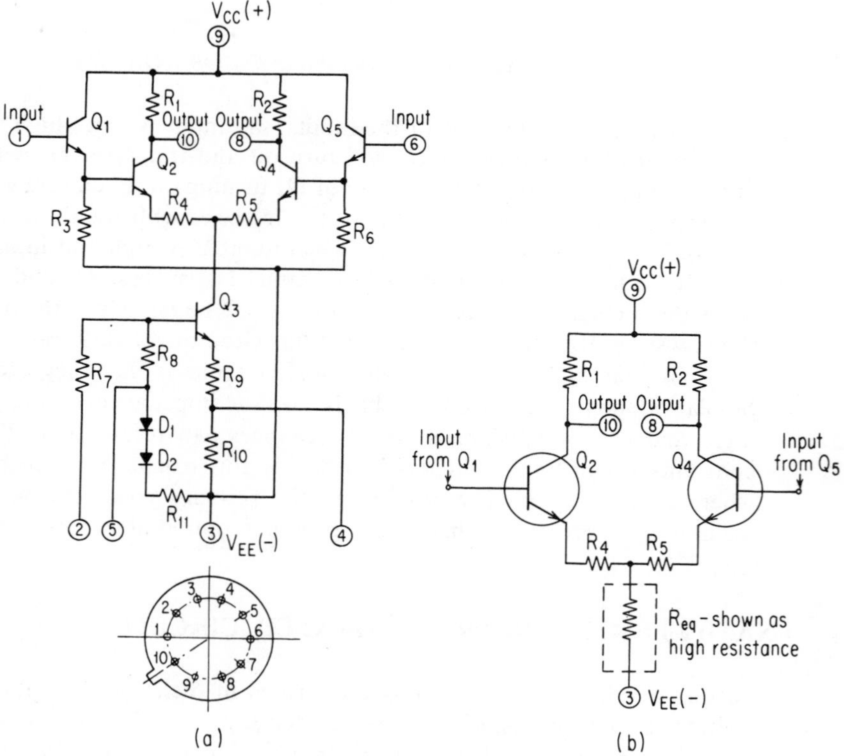

Figure 13-8. Integrated-circuit dc amplifier: (a) schematic diagram and 10-lead (TO-5 can) terminal connections; (b) functional circuit of emitter-coupled (differential) stage, showing equivalent current-source [Q_3 in part (a)] as R_{eq}. (*RCA Linear Integrated Circuit, CA3000.*)

cation—combines many advantageous features; it provides fairly high input impedance, and minimizes the drift problem, along with offering good wide-band performance and common-mode rejection of extraneous noise pick-up. These advantages, it will be noted, stem primarily from the emitter-coupled configuration, separately shown in Fig. 13-8(b). Here, in simplified functional form, the two transistors forming the balanced circuit (Q_2 and Q_4) are shown coupled by an equivalent high resistance in the common-emitter return (R_{eq}); that is, in Fig. 13-8(a) the equivalent resistor R_{eq} is shown functioning as a very high resistance obtained from the constant-current operation of Q_3 (and its associated options, which provide gain control). This basic form of circuit (as discussed in Chapter 7) does sacrifice some gain (since it has the gain of only a single transistor) but, at the same time, it provides improved protection against unwanted changes, which have equal effects on the two matched collector currents, and it also provides improved stability from the inherent feedback obtained from the large equivalent resistance in its emitter-coupled arrangement. The fact that the output is obtained as the difference between each collector voltage (at terminals ⑧ and ⑩) contributes to its common-mode rejection ability, since equal pick-up signals on both bases are cancelled out by producing a negligible difference.

The circuit is completed, as shown in the entire schematic [Fig. 13-8(a)], by the addition of common-collector (or emitter-follower) transistors Q_1 and Q_5 at the input to the corresponding bases of Q_2 and Q_4, in order to provide a fairly high (100-kΩ) input impedance. The basing terminals of the 10-lead T0-5 package of the dc amplifier are shown at the bottom of the figure.[2] Much higher input impedances (10^8 up to 10^{14} ohms) are available in ICs having FET input.

13-5. EXAMPLE OF IC OPERATIONAL AMPLIFIER (IC "op amp")

The designation "operational amplifier" was originally adopted for a series of high-performance dc amplifiers that were used in analog computers. These amplifiers were used to perform mathematical operations applicable to analog computation, such as summation, subtraction, integration, magnitude scaling, etc. (these are discussed further in Chapter 18). However, the application of operational amplifiers is now so widely diversified that this original terminology has been extended. In many electronic systems, the operational amplifier is used as a basic building block for many kinds of signal conditioning, including phase shifting, filtering, and detecting, and also in many nonlinear applications, such as comparators, oscillators, and multivibrators. In short, the operational amplifier is fast becoming an increasingly useful "*jack of all trades*," with the added advantage

[2] *RCA Application Note ICAN-5030* gives full information on this RCA-CA3000 integrated circuit, which has four operating modes that provide versatile gain options for various supply voltages from 6 to 12 V dc.

that it can easily be tailored for specific purposes by the addition of just a few discrete components.

The circuit of a generalized operational amplifier is shown in Fig. 13-9(a), while the complete schematic circuit of a representative high-performance operational amplifier in the form of a monolithic integrated circuit is shown in Fig. 13-9(b). It can conveniently be considered as made up of two sections: the first section consisting of an *input section* in the form of a two-stage differential amplifier (of the form previously discussed in Sec. 13.4 on a dc amplifier), followed by an *output section*. The differential input signal, starting at terminals ① and ②, is

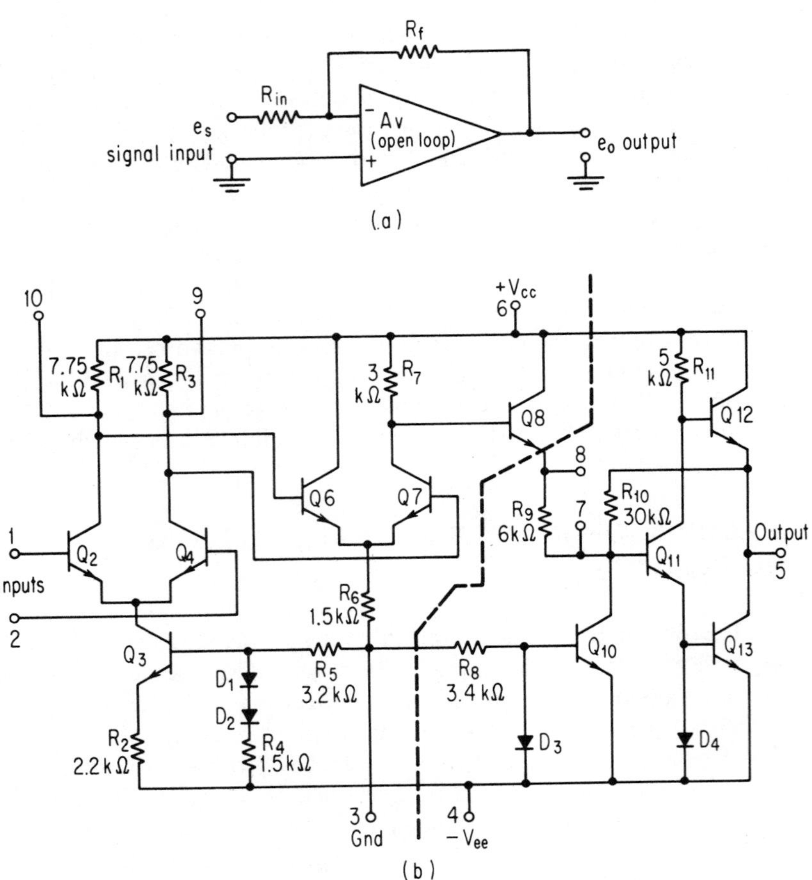

Figure 13-9. Operational amplifier: (a) symbolic diagram, showing relationship $e_0 = -e_s(R_f/R_{in})$; (b) schematic diagram of integrated circuit (IC) version, with input section of two differential stages to the left of the dashed line, and output section to the right. (*Motorola Linear Integrated Circuit, MC-1530.*)

amplified in one differential stage (transistors Q_2 and Q_4, together with the temperature-compensated current source Q_3); the result is then fed differentially to the second difference-amplifier stage (transistors Q_6 and Q_7) having a single-ended output, which is taken from Q_7 through emitter follower Q_8 to the output section. (*Note:* the Q numbers of the some of transistors are omitted, as in cases when fabricated transistors are internally connected to function as diodes, which are numbered D_1, D_2, etc.)

The circuit of the output stage is shown to the right of the dotted line in the figure. The output section of an operational amplifier, in general, must serve not only to produce some *power gain* but has another important function, namely, that of *translating the dc level of the output to zero*, for a zero-input signal. This level translation problem arises in practically all direct-coupled circuits and cannot be solved by the simple expedient of capacity coupling, if the capability of dc amplification is to be retained. The function of reducing the dc offset voltage (V_{OS}) to practically zero is accomplished by the action of transistor Q_{10}, arranged as a constant-current source, combined with Q_{11}, which feeds the series combination of Q_{12} and Q_{13}. The final output, it will be noted, is taken at the junction of Q_{12} and Q_{13}, where the dc output voltage to ground is practically zero, even though the voltage at the emitter of the dc-coupled preceding transistor (Q_8 to ground) is in the order of around +4 V.

Operational Amplifier Gain Specifications

The open-loop voltage gain (A_{VOL}) of the two-stage input section is approximately 60×20 or 1200, and this is then multiplied by 5 for the output section, to provide an open-loop gain of approximately 6000. Under normal closed-loop conditions [with external feedback resistor R_f connected, as in Fig. 13-9(a)], the output will drive a load resistor as low as 1000 Ω to a maximum peak-to-peak swing of about 10 V (±5 V). The closed-loop voltage gain (A_{VCL}) depends, directly on the ratio of R_f/R_{in}:

$$\text{closed-loop output} = -E_{\text{in}}(R_f/R_{\text{in}})$$

This value for closed-loop gain, while sufficiently close for most cases, depends for its accuracy upon a double assumption: (1) that A_{VOL} is very high, and (2) that there is negligible forward transfer of the signal through a very high value of feedback resistor R_f. A more pertinent specification for determining accuracy is called *the loop gain*, (not to be confused with either A_{VOL} or A_{VCL}). The loop gain, which is used in finding the overall performance of the actual system, in terms of stability and accuracy, is given by:

$$\text{loop gain} = A_{VOL}[R_{\text{in}}/(R_{\text{in}} + R_f)], \text{ or } A_{VOL}(\beta)$$

As a brief but general rule, it can be said that the accuracy limits of operational amplifiers depend upon a high value of β, which, of course, can drastically reduce the closed-loop gain.[3]

While the general-purpose operational amplifier is a basic building block, a great variety of off-the-shelf *op amps* are available from many manufacturers. Many of them are refined developments of the basic 709 monolithic approach, pioneered by *Fairchild Semiconductor* (1966), but special requirements (plus a great deal of competition for this promising development) have resulted in many various models having much higher open-loop gains (typically of the order of 10^5 or 100 dB), and various *external frequency-compensation* techniques for wide-band operation (including some unconditionally stable varieties, where the frequency compensation is included within the IC package) and also other varieties that include FET inputs, to obtain much higher values of *input impedance*.

Versatile Functions of the Operational Amplifier[4]

The *op amp* possesses the great versatility of performing many varied functions by the simple process of changing its external feedback Z_f. It can thus provide any reasonable value of desired gain by manipulating the R_f/R_{in} ratio; additionally it can readily change from a summer to an integrator by the simple substitution of a capacitor for the feedback resistor R_f.

Because of this great flexibility, the *op amp* is available in many "off-the-shelf" versions to be used as a versatile building-block element. Examples of such functions are discussed more fully in Sec. 18-3.

The next sections will concentrate on some practical applications of various forms of linear-integrated circuits.

13-6. TYPES OF LINEAR-INTEGRATED CIRCUITS

If we consider the main applications of electronic circuits as encompassing three broad functions, namely, amplification, oscillation, and switching, a very general grouping of applications may be made; the *switching function is the province of the digital ICs*, while the *amplification functions are performed by the use of linear ICs*. The third function of *oscillation may be handled by either type*, where the digital ICs are generally employed in relaxation oscillators (such as multivibrators for generating pulses), while the linear ICs usually provide sine-wave oscillators (such as audio oscillators of the Wien-bridge type).

Since most forms of instrumentation employ amplifiers of one form or another, it follows that the major emphasis on applications for instrumentation will be on the linear ICs. The selection of a particular linear IC for some specific

[3]The significance of the loop gain and other specifications of operational amplifiers (such as slew rate = $\Delta e/\Delta t$), are explained in great detail in Motorola Application Notes AN204—*High Perfomance Integrated Operational Amplifiers*, and AN273—*Interpreting IC Amplifier Data Sheets.*

[4]See M. Kahn, *The Versatile Op Amp* (New York: Holt, Rinehart and Winston, 1970).

application may be clarified by grouping the linear IC amplifiers into some five general application classes, as follows:

1. Audio-frequency amplifiers (both preamp and power-amp types)
2. High-frequency (or wide-band) amplifiers
3. Differential amplifiers
4. Operational amplifiers
5. Sense (or comparator) amplifiers

The matter of selecting an individual IC amplifier for a specific use is not a particularly simple one, as might be expected from the fact that each IC type is made up of a multiplicity of transistors, diodes, and resistors. (Of course, this is nothing unusual to instrumentation designers; even when faced with the task of selecting just one particular transistor for a discrete circuit, they are confronted with over 5000 different 2N type numbers—to say nothing of the thousands of special semiconductor 1N and 3N types.) However, the selection task has been greatly aided by the *application notes* available from each of the manufacturers of linear ICs.

13-7. FREQUENCY COMPENSATION FOR INTEGRATED-CIRCUIT AMPLIFIERS

It would seem, at first glance, that the use of progressively larger amounts of negative feedback would help to make the IC amplifier more stable, albeit at the expense of gain. This is not true, however, when carried beyond the point where we attempt to use increasingly larger amounts of feedback in an amplifier having a very high initial open-loop gain.

This can be shown in a simplified way, by considering an operational amplifier having three stages of amplification (as is true of a great many op amps). For the range of middle frequencies (around 1 kHz), the external feedback resistor supplies feedback from the output with relatively small phase shift, thus ensuring that the feedback remains negative over all of this middle range. As the frequency increases, however, the output falls off as expected, but also, the phase shift becomes progressively larger. Thus at some higher frequency, the phase shift of each stage would reach $-60°$, and the progressively smaller overall feedback voltage would be reversed a full 180° and thus become positive feedback. If, at that time, there is a sufficient amount of feedback, the positive sense of this feedback would cause the amplifier to oscillate. Hence, a large feedback fraction (β) of the output of a high-gain amplifier can generally cause instability at some high frequency.

The criterion for unconditional stability requires that *the loop gain at this critical frequency must be less than unity;* or, expressed in frequency-response terms, *the feedback factor* (open-loop gain A times feedback fraction β) *must be less than 1, at the time that the overall phase shift reaches 180°*. This stability criterion is satisfied by the addition of frequency-compensation capacitors, in order to cause a greater roll-off at the higher frequencies. The roll-off must be such that, at these frequencies, the gain with feedback becomes less than unity (0 dB), whenever the overall phase shift exceeds 180°.

Individual IC amplifiers prescribe various amounts of frequency-compensation capacity, as given in their data sheets, and depending on the particular design of the high-gain amplifier. Frequently, terminals are provided for connecting these capacitors externally, to provide compensation of both the input and output sections of the circuit. In the case of some popular operational amplifiers, the circuit is designed to be unconditionally stable without any external frequency compensation (for example, Fairchild μA 741, and 748 types). Most IC amplifiers include temperature-compensation arrangements, making them particularly useful for amplifying sensitive transducers having a very-low-level signal output.

13-8. EXAMPLES OF TYPICAL IC APPLICATIONS

Some examples are given in this section to illustrate particular values of voltage gain, input impedance, and frequency response, as applied to a specific system. A first example is a common system requirement in instrumentation, calling for amplifying the signal from a magnetic transducer (in this case, the output of a magnetic-recorder tape head) to produce an amplified output with sufficient power to operate a loudspeaker.

The signal obtained from the tape head of a magnetic-tape playback system is typically in the order of 1 mV or less and at a fairly low input resistance (say, around 1 kΩ). In order to satisfy the system requirement for operating a loudspeaker from this signal, it is necessary *to raise both the voltage level and the current capability* of the signal. In round numbers, we may assume that this requirement calls for an ac output of 2 W in 10 Ω. Expressed in another way, this requires an ac voltage of about 4.5 V rms across the 10-Ω load. We are therefore dealing with the need for a *preamplifier* (for voltage amplification) followed by a *power amplifier*. In addition, the bandwidth requirement, though relatively modest for the loudspeaker operation (to 20 kHz), should preferably be somewhat beyond audio frequencies for the preamp (say, to 100 kHz), since we might wish to use the preamp with other transducers, whose output does not necessarily operate a loudspeaker.

Integrated Circuit Preamplifier

An integrated-circuit preamplifier for a magnetic transducer (*Mallory, preamp, MIC 0101*) is illustrated in Fig. 13-10. The internal circuit, shown in Fig. 13-10(a), shows transistors Q_1 and Q_2 connected in a common-emitter Darlington configuration, producing some voltage and current gain at a relatively high input impedance. The next stage, Q_3, is again a common-emitter configuration for further

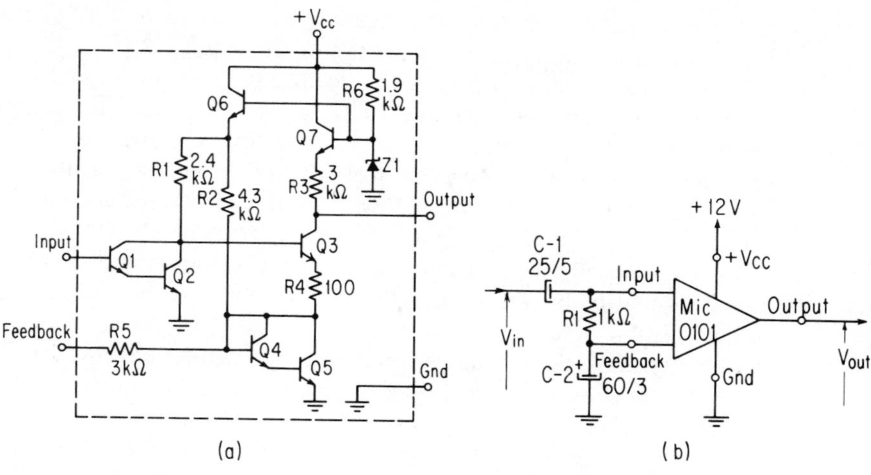

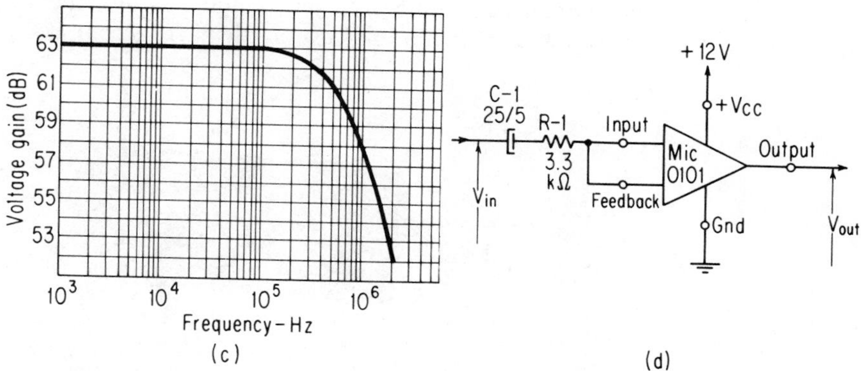

Figure 13-10. Integrated-circuit preamplifier: (a) circuit; (b) external connections; (c) open-loop frequency response; (d) feedback connection. (*P. R. Mallory type MIC 0101.*)

signal amplification. Transistors Q_4 and Q_5 are diode-connected to supply a constant self-bias, equivalent to two silicon diodes in series, or about 1.4 V dc. (This dc bias is obtained by the dc path offered between the input and feedback terminals.) The dual emitter-follower transistors Q_6 and Q_7, together with Zener diode Z_1 and resistor R_6, achieve power-supply regulation over the power-supply range of 9–24 V dc. The overall *voltage* gain achieved with this single-power-supply arrangement is typically around 60 dB (or 1000 times).

It will be noted that the input stage of this amplifier does not employ the basic difference-amplifier configuration that is so commonly used in general-purpose operational amplifiers, because, in the case of this audio application, there is no need for dc coupling for the input signal.

The *RC*-coupling method for injecting the ac input signal is shown in Fig. 13-10(b). The coupling capacitor C_1 (25 μF at 5 V) feeds the signal to the 1-kΩ input resistor, which completes the dc bias circuit. Capacitor C_2 (60 μF at 3 V) provides an ac ground to prevent loss of gain through ac feedback.

The frequency response of the *RC*-coupled circuit, shown in Fig. 13-10(c), shows the 3-dB bandwidth extending to approximately 700 kHz, which is well beyond the strictly audio-frequency range. The low-frequency response is determined primarily by the value chosen for the coupling capacitor ($C_1 = 25\ \mu$F in this case).

Extension of the bandwidth can easily be obtained by the use of negative feedback, as shown in Fig. 13-10(d). The feedback resistor of 3.3 kΩ in this case reduces the overall voltage gain to 100 (or 40 dB). In this configuration, the 3-dB bandwidth now extends from below 10 Hz to almost 3 MHz and is exceptionally flat from the low end to well beyond 1 MHz.

Power Amplifier

An integrated-circuit (monolithic) power amplifier is illustrated in the photo of Fig. 13-11. It is in the form of a low-cost plastic package, suitable for soldering on

Figure 13-11. Monolithic 2-W integrated-circuit (IC) amplifier; the tab at the left solders to the copper of a printed-circuit board for a heat sink. (*General Electric model PA237.*)

a printed-circuit board. It includes a copper tab, which allows internally generated heat to flow through this tab to any copper area of the printed-circuit board, thus serving as a self-contained heat sink. It may be fed by the preamplifier discussed above by utilizing the grounded-output connection, shown in the circuit diagram of Fig. 13-12(a). Here the *PA-237* circuit is shown with a gray background, along with the external components. Various power-output options are available, depending on the value of the power supply. Using a 24-V supply, 2 W into a 16-Ω load may be obtained, or with a lower supply of 14 V, the power output will be 1 W

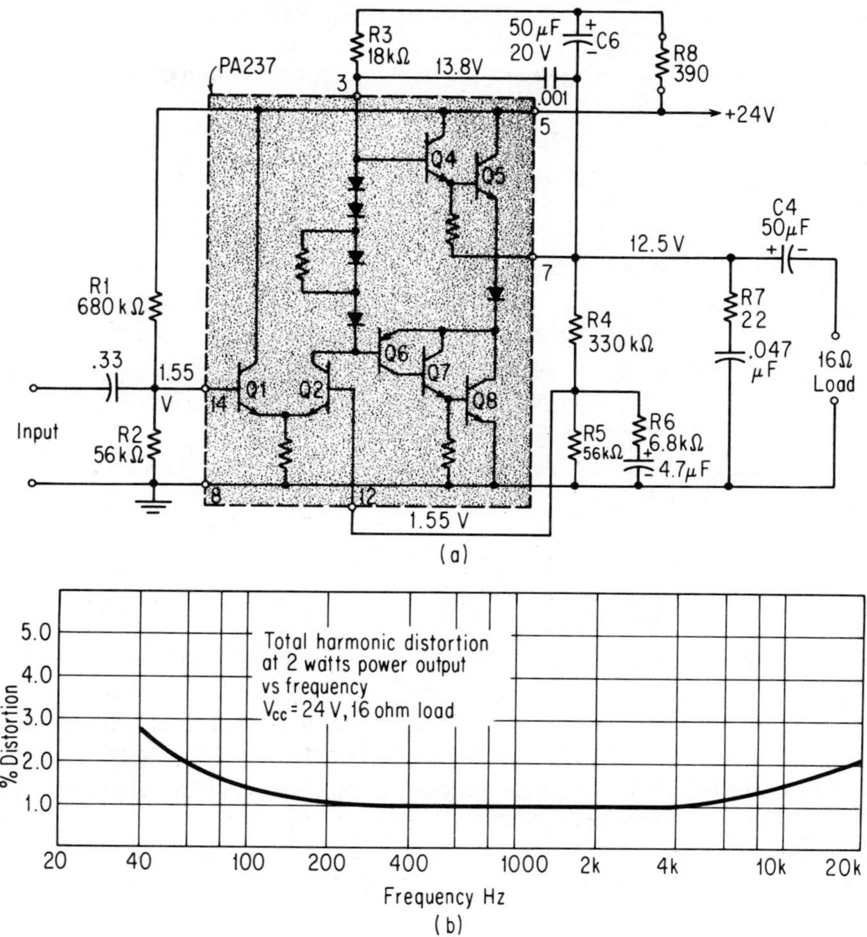

Figure 13-12. Integrated-circuit audio power amplifier, 2-W: (a) circuit of IC in gray rectangle; (b) graph of distortion vs. frequency. (*General Electric type PA237.*)

delivered to an 8-Ω load. Since full power output is obtained by an input of only 120 mV (into an input resistance of 40 kΩ), the output of the previously discussed preamplifier is more than ample to drive the power amplifier. Hence, a volume control of 50 kΩ can be used in place of the input resistor R_2, to reduce the input level as desired. The graph of total harmonic distortion vs. frequency, shown in Fig. 13-12(b), indicates less than 2% distortion over the entire frequency range, when feeding a 16-Ω loudspeaker at rated power.

Additional IC applications (op amps and dc instrumentation amplifiers) are given in Chapter 18.

QUESTIONS

Q13-1. Distinguish among monolithic, hybrid, and multichip types of integrated circuits.

Q13-2. For what application(s) do the functions of digital and linear ICs overlap?

Q13-3. State the primary advantages that ICs have over discrete components in the case of:
(a) Digital ICs.
(b) Linear ICs.

Q13-4. Justify the statement that an integrated-circuit chip generally uses more individual components than a discrete circuit would, to accomplish the same purpose.

Q13-5. Explain the statement that the difficulty in fabricating *exact resistor values* on a chip can generally be overcome by the scheme of holding *the ratio of resistor values* to a good tolerance.

Q13-6. State the chief advantage of each of the three types of IC packages:
(a) T0-5 can.
(b) Flat-pack.
(c) DIP.

Q13-7. Show by diagram how a simple differential amplifier (emitter-coupled stage) resembles a bridge circuit.

Q13-8. State two important advantages that make the differential amplifier an important building block for linear ICs.

Q13-9. How does the use of a transistor to replace the emitter resistor of an emitter-coupled pair improve the differential amplifier circuit?

Q13-10. Why is external frequency compensation often necessary in using a high-gain operational amplifier?

14

Untuned-Amplifier Test Methods

14-1. UNTUNED VOLTAGE-AMPLIFIER TESTS

Because of its importance as an element common to practically every major electronic instrument, the amplifier merits first consideration in the area of test methods. In terms of amplifying function, we may initially distinguish between two main groups, the untuned and the tuned (or RF) amplifiers. Each of these types will be discussed in a separate chapter; the *untuned type* (also commonly designated as an *audio-frequency or video amplifier*) will be the subject of this chapter, and the higher-frequency (or RF) tuned type will be discussed in Chapter 15.

In the large group of instruments that function as conditioners of transducer signals, one frequently finds a *voltage amplifier of the untuned type* which is used to raise the level of incoming signals without discriminating sharply in favor of any one frequency. Since gain is the primary function of the amplifier, a straightforward test for this function consists of checking the undistorted voltage gain at its output, over the frequency range of operation. While the familiar example of *audio-frequency amplifiers* in loud-speaker systems (both high-fidelity and otherwise) comes first to mind, this category also includes the group of *wide-band amplifiers* (often called video amplifiers), whose range of operation extends far beyond the audio frequencies into the megahertz region. Thus the video amplifier in television is called upon to provide practically uniform voltage amplification of signals from practically dc to beyond 4 MHz. The same is true for the compensated wide-band voltage amplifiers found in the initial stages of laboratory oscilloscopes and electronic voltmeters.

Since similar methods are used in testing both audio-frequency amplifiers and the untuned wide-band amplifiers that go beyond the strictly audible range, both types of untuned amplifiers are generally lumped together, for testing purposes; they will therefore be treated together in this chapter under the general heading of untuned amplifiers. The other large category, *tuned amplifiers*, is treated in Chapter 15.

14-2. TESTING AMPLIFIER GAIN AND FREQUENCY RESPONSE

Overall Gain Tests

An audio-frequency amplifier that is part of a loudspeaker system is generally tested for its voltage gain as a complete unit, including both the *voltage and power-amplification stages*. The instrument setup is shown in Fig. 14-1, where the test signal is impressed on the input of the amplifier by the variable-frequency audio oscillator. On the output side, the voice coil of the loud-speaker is usually replaced

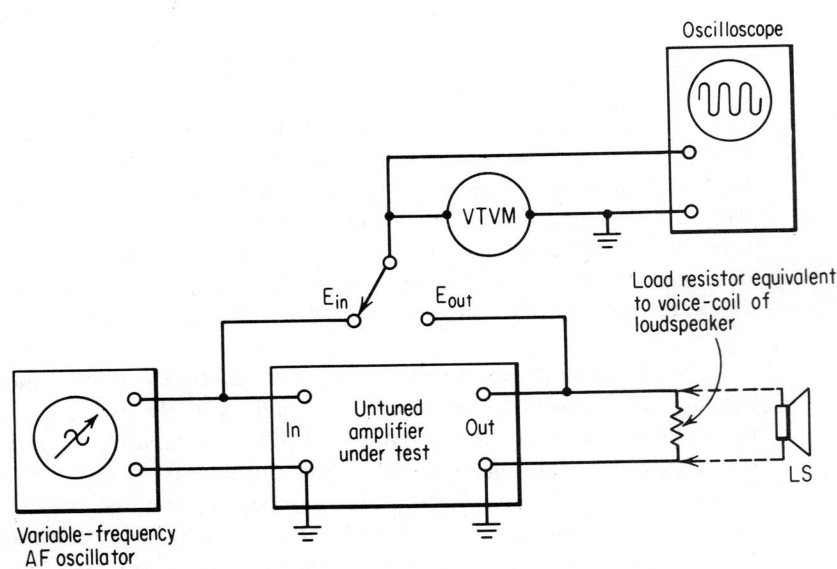

Figure 14-1. Setup for testing untuned (or AF) amplifiers for gain and frequency response.

by an equivalent resistor (generally around 4, 8, or 16 Ω of the proper wattage dissipation). The voltage output E_{out} across this resistor is then compared, by the switching arrangement shown, with the voltage input E_{in}, to arrive at the voltage gain from the following relation:

$$\text{voltage gain} = \frac{E_{out}}{E_{in}}$$

A rough idea of both the undistorted power level and the purity of the wave-form at the output can be obtained by observing the output wave-form on the oscilloscope for a small-signal input and then raising the input level until distortion becomes noticeable. Thus, when an output of 8 V across a 4-Ω load resistor is obtained under these conditions without noticeable distortion, for an input voltage of 10 mV, an undistorted voltage gain of 8000/10 mV or 800 is obtained, at the time when the amplifier is delivering a power output P_{out}. Since

$$P_{out} = \frac{E^2}{R}$$

the power output is $8 \times \frac{8}{4}$ or 16 W.

While this test method provides a simple means for obtaining comparative performance values, further tests would be required to give more definite absolute values. Voltage-gain tests on individual stages are frequently desired, since the final output stage contributes more to power than to voltage amplification. In the output stage, particularly, further tests on distortion and additional tests with the voice coil made operative would be performed for more complete evaluation.

Converting Voltage Gain to Decibel Values

Amplifier response is usually given in terms of decibel gain or loss compared to a reference output. The relationship of the *voltage-gain ratio* to decibel gain is

$$\text{(voltage gain) number of decibels} = 20 \log \frac{E_{out}}{E_{in}}$$

In our example of an 8-V output for an input of 10 mV, the voltage-gain ratio of 800: 1 would yield a decibel value of $20 \log \frac{800}{1}$, or about 58 dB for the voltage gain. If this gain were obtained at a midfrequency of 1000 Hz, the response of the amplifier at some other frequency, such as 20 kHz, might be expressed as within 0.5 dB of the reference value of 8 V. To convert this small decibel loss into a percentage figure, it is convenient to remember that for the small decibel values (up to about 1 dB), each 0.1-dB loss corresponds closely to a 1% loss; hence the response at 20 kHz in this example would be about 5% down from the reference output of 8 V.

Frequency-Response Tests

The overall frequency response of an audio-frequency amplifier is generally expressed on a graph (usually on semilog paper), which plots *output voltage* (expressed in decibels on the linear vertical scale) against the *frequency* plotted along the logarithmic horizontal scale; an example of such a frequency-response graph is shown in Fig. 14-2. The arrangement of the instruments for this plot is

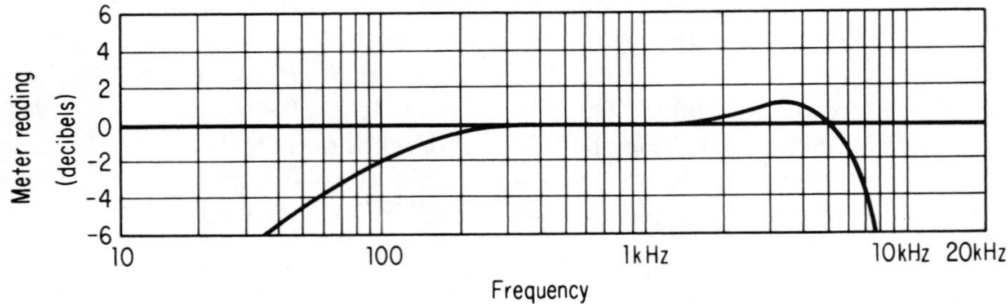

Figure 14-2. Typical graph of untuned-amplifier frequency response. [From J. F. Rider and S. D. Prensky, *How to Use Meters*, 2nd ed. (New York: Hayden–Rider, 1960).]

the same as that given for gain measurements in Fig. 14-1. Instead of the switching arrangement shown, a separate ac voltmeter is sometimes used to monitor the input voltage, which is kept constant, as the frequency is varied, at the value that provides rated power output at a standard frequency of 400 or 1000 Hz. In such a case, the output is monitored by another voltmeter, generally called an output meter, which contains a capacitor to block any dc, and which might carry decibel calibration, in addition to the regular voltage scales, on the meter face.

In testing high-fidelity amplifiers, the effect of the various controls, such as bass and treble boost and equalizing controls, can be checked at the same time that the frequency run is being made.

14-3. HARMONIC-DISTORTION TESTING

The extent to which the output wave-form of an amplifier differs from the wave-form of the input signal is a measure of the distortion introduced by an amplifier. This measurement is an important consideration in evaluating high-fidelity performance and also in detecting the maximum signal level that can be handled by an amplifier without exceeding the specified accuracy tolerance in the sensitive untuned amplifiers that are a part of so many electronic instruments.

When an amplifier is not operating in a linear fashion, either because of overloading or from some other cause, the output will contain *harmonics in addition to the fundamental frequency* of the input. When these harmonics are present in a considerable amount, their presence can be noticed in the oscilloscope display of a sine-wave output, either as an unbalance between the positive and negative portions of the sine wave, or as a change in the sine-wave form. Thus, as has been previously noted, the *oscilloscope can be used as a qualitative check of visible harmonic distortion*, generally when the distortion reaches values above 5–10%, depending on the experience of the observer.

In most testing situations, however, a better quantitative measure of harmonic distortion is required. The principle employed in *harmonic distortion meters*, shown in functional form in Fig. 14-3, is to measure the total voltage output, including the harmonic and fundamental components, and then to compare this output-voltage value to the voltage obtained when the fundamental frequency is filtered out by a rejection filter tuned to this fundamental frequency. The filter

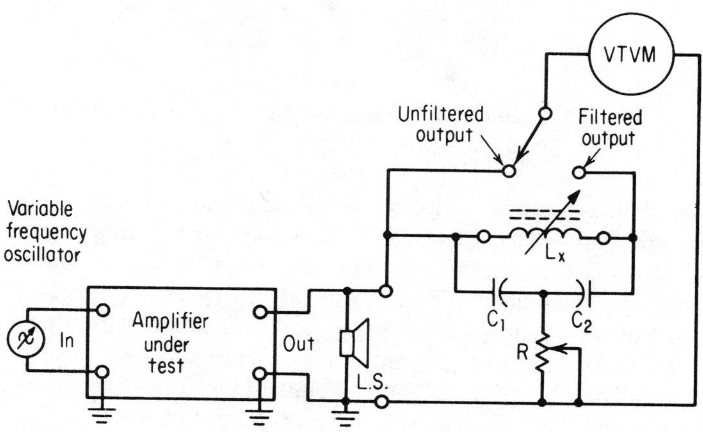

Figure 14-3. Principle of harmonic-distortion meter, showing bridged-*T* rejection filter composed of L, C_1, C_2, and R. [From R. P. Turner, *Basic Electronic Test Instruments* (New York: Holt, Rinehart and Winston, 1964).]

shown is a bridged-T network, which can be switched in and out of the circuit. The oscillator supplying the test signal must obviously be of a low-distortion type. With the oscillator set to give an output reading (E_{out}) of, say, 6 V with the filter out of the circuit, the filter is switched in and the oscillator is carefully tuned to obtain a minimum reading, representing the voltage output without the fundamental component. If this reading of minimum voltage E_{min} turns out to be, say, 120 mV (or 0.12 V), the per cent harmonic distortion is obtained from the relation

$$\text{per cent harmonic distortion} = \frac{E_{min}}{E_{out}} \times 100\%$$

In this case, the harmonic distortion would be (120/6000) × 100% or 2%.

Commercial Harmonic Distortion Analyzer

A typical distortion analyzer instrument for measuring total harmonic distortion is illustrated in Fig. 14-4. At any frequency in its range of 5 Hz to 600 kHz, accurate measurements of harmonic distortion as low as 0.1% may be made.

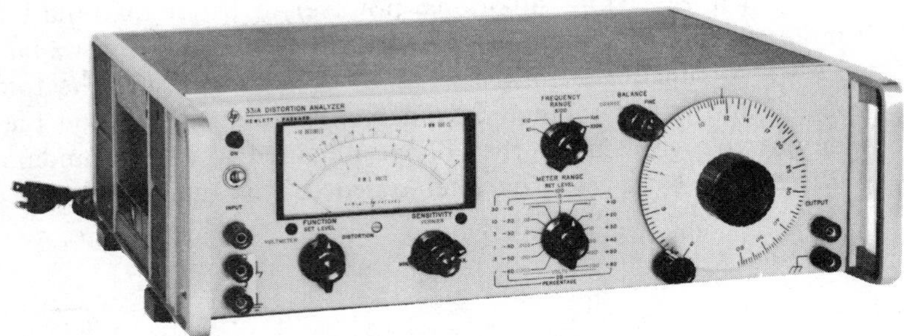

Figure 14-4. Harmonic (total) distortion analyzer. (*Hewlett-Packard model 331A.*)

Since the frequency-selective amplifier of the instrument has the task of providing a great amount of rejection to the fundamental frequency (greater than 100 dB), while still offering the least attenuation of the higher harmonics, its rejection and high-pass characteristics are of primary interest. A rejection characteristic is given in Fig. 14-5, showing the sharp notch or attenuation that can be produced by a bridged-T filter in the audio-frequency range.

The reduction of the fundamental in Fig. 14-4 is typically at least 99.9%, while the second-harmonic attenuation, for example, is less than 1 dB in the AF

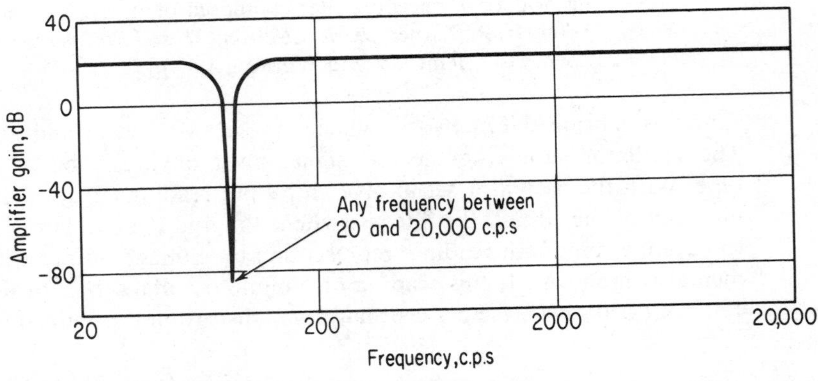

Figure 14-5. Rejection characteristic of a variable-tuned audio filter.

range, and less than 3 dB (about 30%) out to the 600 kHz limit. In order to pass the higher harmonics (at least up to the fifth harmonic), the response of the highly stable, solid-state amplifier extends to 3 MHz. Sensitivity (0–300 μV full-scale) is sufficient for distortion measurements of fundamental signals as low as 0.3 V.

Noise and Hum Measurements

In addition to the utility of the distortion-analyzing function of the instrument, advantage may be taken of the stable wide-band and high-gain amplifier characteristics to perform direct-reading measurements of *noise and hum levels.* Noise as small as 100 mV and hum levels down to -75 dBm may be satisfactorily measured. (The dBm value is the decibel figure obtained when 1 mW in 600 Ω is the 0-dB reference.)

14-4. INTERMODULATION-DISTORTION TESTING

When a test method using a single frequency at a time is employed, the test results, in general, are perfectly valid for quantitative measurement of amplifier gain. These would also give a true picture of the output when complex music is being amplified, provided that the amplification could be assumed to be *entirely linear* in its operation. In an absolutely linear amplifier, we can rely on the principle of superposition, which states that each component frequency of a complex wave will produce its own effect, regardless of the presence of other frequencies that are applied simultaneously. In such an ideal situation, the signals from a violin and piano (and from all other instruments of a symphony orchestra), all striking the microphone at the same moment, would produce an output containing the same instrument sounds in their true proportions, without interaction between the various simultaneous signals. In an actual amplifier, however, this situation remains true only to the extent that the amplifier is strictly linear; *any nonlinearity of operation would cause the modulation of one signal by another* (*called cross modulation or intermodulation*), to the degree that the amplifier operation was nonlinear. While most general-purpose amplifiers can approach the ideal linearity to a quite satisfactory degree, there will be some amount of nonlinearity even in well-designed amplifiers, and it is therefore desirable to determine the amount of the resulting *intermodulation distortion*, especially when dealing with high-fidelity amplifiers.

The measurement of intermodulation distortion takes on special added importance in the case of amplifiers used for reproducing music, because of the specific manner in which our sense of hearing operates. The ear generally accepts *harmonic* distortion without too much complaint, since the interaction between the harmonic components produces components also of a harmonic nature, and these need not be displeasing to the ear. But in the case of modulation of one frequency by another, the generation of the resulting *sum-and-difference frequencies* presents a variety of possibilities of nonharmonious combinations.

Oscilloscope Method for Intermodulation Testing

An oscilloscope can be used to view the result of any intermodulation produced when two signals of different frequencies are applied as a combined input to an amplifier. The block diagram of such an arrangement is shown in Fig. 14-6(a) and the resultant scope display in part (b). Since the intermodulation produced by the action of one signal frequency on another is essentially a *modulation process*, the same pattern and calculations used for the modulation process in general also apply to the intermodulation output.

The standard S.M.P.T.E.[1] method for intermodulation testing employs a low-frequency signal, generally 60 Hz, that will partially modulate a high-frequency signal, on the order of 50 times the frequency of the low one (3 kHz in this case). Because of the small degree of nonlinearity expected, the low-frequency (60 Hz) signal is applied at an amplitude four times as great as the high-frequency one, and the two are combined in a resistive circuit, arranged in bridge form, as shown. The combined input to the amplifier from the linear-resistive circuit does not contain any modulation products as yet, until it is passed through the amplifier. The amplifier output will then contain modulation products having a strength proportional to the degree of nonlinearity present in the amplifier. These modulation products contain harmonics together with sum-and-difference frequencies. The high-pass filter is designed to reject all the low-frequency components (its cutoff frequency rejects components up to twice the low-frequency value).

The resulting pattern of the high-frequency components of the modulated wave is displayed on the scope in the form shown in Fig. 14-6(b). The percent modulation that will be obtained from this pattern is also the per cent intermodulation distortion. As in the case for deliberate modulation, the percentage modulation is given by

$$\frac{\text{amplitude of modulation } E_m}{\text{amplitude of unmodulated carrier } E_c} \times 100\%$$

In the pattern shown, peak-to-peak values rather than single-peak amplitudes are easier to measure; their ratio, which gives the percentage intermodulation (or modulation), is obtained by finding $[(b - a)/a] \times 100\%$.

As an example, if the unmodulated p–p value a measures 20 divisions vertically, and the modulated p–p value b measures 23 divisions, the percentage intermodulation distortion is $\frac{3}{20} = 15\%$.

[1]Society of Motion-Picture and Television Engineers.

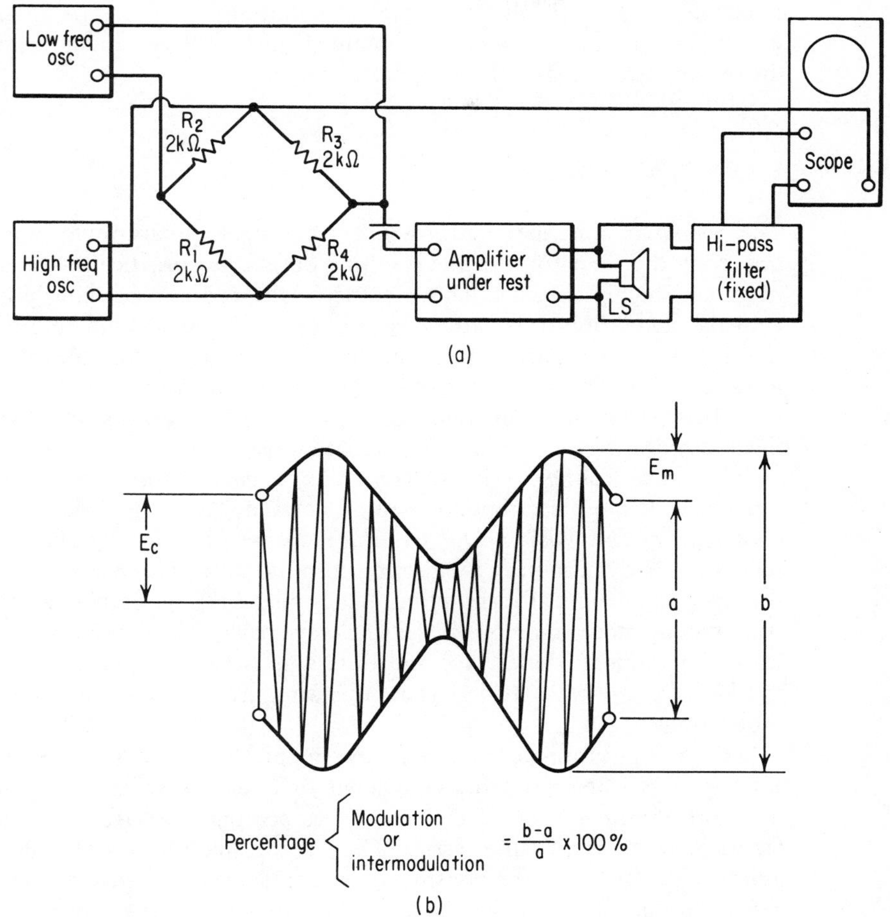

Figure 14-6. (a) Block diagram of scope method for IM testing; (b) determining

$$\text{percentage}\left\{\begin{array}{c}\text{modulation}\\ \text{or}\\ \text{intermodulation}\end{array}\right. = \frac{b-a}{a} \times 100\%$$

Intermodulation Meter

The intermodulation meter is quite similar to the instrument measuring modulation in a transmitter, with the exception that the high-frequency signal, corresponding to the carrier in transmitter measurements, is at an audio frequency for IM measurements. In the meter instrument, the signal containing the high frequency, with its modulation sidebands, is envelope-detected and then measured to determine the dc and the rms values of this detected output. The ratio of the rms modulation to the dc value so obtained gives the percentage intermodulation distortion, as defined in the S.M.P.T.E. method.

14-5. SQUARE-WAVE TESTING

The testing of amplifier performance by using a square-wave input signal and observing the wave-form of the resulting output is a practical method that is both rapid and highly effective. This method was introduced previously in connection with the generation of square waves in Sec. 11-4 and was illustrated in Fig. 11-8. In this figure, the departure from the ideal square wave could quickly be interpreted in terms of the frequency-response and rise-time characteristics of the amplifier.

It is pertinent at this point to inquire into the reasons underlying the effectiveness of this testing method. The square wave can be shown by the mathematical analysis of a Fourier series to consist of a *fundamental sine wave* (having the frequency of the pulse repetition rate), combined with an *infinite number of odd harmonics of this fundamental frequency, also sinusoidal in form*, and with each successive odd harmonic diminishing in amplitude. This view, of course, is not simply a mathematical assumption, but is physically realizable, as can be seen by recombining more and more of the sine-wave harmonics in their proper proportion with a fundamental sine wave and observing the closer and closer approach to the ideal square-wave form, as more harmonics are included in their proper proportions.

The application of a square-wave input to an amplifying system may, accordingly, be considered the equivalent of applying an infinite number of sine waves at one time, each of the sine waves bearing the harmonic relation to the others given by the Fourier series. Thus, if the amplifying system had a perfectly uniform response, all the constituent sinusoidal components of the square wave would be reproduced in the same relation that they were present in the original wave; as a result, then, in accordance with the super-position principle, a perfect reproduction of the square wave would appear in the output. By the same token, the *extent to which the form of the output departs from the ideal reproduction of the square-wave input also signifies the departure of the amplifier response from the perfect flatness of a uniform frequency response.*

Briefly summarizing the interpretation of the wave-forms previously illustrated in Fig. 11-8, on the one hand, we find evidence of *poor low-frequency response* in the tilting of the horizontal portions of the square wave (indicative of the tendency to differentiate rather than to couple faithfully at these low frequencies). At the high frequencies, on the other hand, the rounding of both corners (after the steep rise and fall portions of the square wave) is indicative of *poor high-frequency response*. This occurs, of course, because of the absence of full reproduction of the higher (odd) harmonics that are necessary to reproduce the sharp corners of the original square wave.

14-6. INTEGRATED-CIRCUIT TESTING

The testing of amplifiers in the linear integrated-circuit (IC) form (either monolithic or hybrid) is of equal importance to the testing of amplifiers made up of separate (or discrete) components. Because of the mass-production methods of fabricating these building-block ICs, and also because of the large number of active and passive elements that are present in these highly compact units, it is to be expected that the methods employed for testing the linear ICs will differ considerably from those used in testing discrete amplifiers. The primary requisite for an IC tester is the ability to measure quickly the numerous characteristics (both static and dynamic) that make up the complete specification of each type of linear IC amplifier. (Additionally, in the case of the manufacturer, it becomes essential that the multiplicity of tests be performed automatically.

For the sake of clarity, the example shown in Fig. 14-7 is of the *semi-automatic type of IC tester*, in order to avoid the extra complexity of the fully automatic models. This unit will perform both the static and dynamic characteristics of *IC discrete or hybrid operational amplifiers;* the static tests are designed to measure the dc or quiescent characteristics, while the dynamic tests are designed to evaluate the operating parameters.

Figure 14-7. Integrated-circuit operational amplifier tester provides semiautomatic tests of both static and dynamic characteristics of IC discrete and hybrid op amps. (*Philbrick/Nexus Research model 5102.*)

Static (DC) Tests

The dc tests check for the following characteristics (with no signal applied):

Oscillation (or Instability): Is checked by the indication of a red light in case of oscillation of the IC that exceeds 10 mV peak, 100 Hz–10 MHz, with the operational amplifier connected for 100% feedback.

Offset Voltage E_{os} *and Offset Current* (I_{os}): Are measured as equivalent input voltage and corresponding current.

Bias Current: Is measured at the plus (noninverting) and minus (inverting) input terminals, separately.

Supply Current: Is measured under the no-signal condition from the positive or negative supply, separately.

Dynamic (Signal) Tests

The application of low-frequency square waves, coupled with feedback and synchronous detection, allows the following operating readings:

Gain: Is measured in a stable, closed-loop mode of operating at full output under specified load. The readout is in decibels and volts per volt.

Common-Mode Rejection Ratio: Measures the ratio of amplified output of the desired signal to the output of the undesired pickup that is common to both plus and minus terminals.

Out-Voltage Swing ($\pm E_{out}$) *of Either Polarity:* Is read directly on the meter. Additionally, readings are provided for maximum common-mode voltage and power-supply rejection ratio of the op amp.

All the above dynamic tests are completely automatic, once the selection has been made on the function and range controls. The readout for amplification is displayed on a logarithmic scale (linear decibels) to accommodate the wide dynamic range encountered in the broad field of operational amplifiers that can be tested with various test-socket accessories.

QUESTIONS

Q14-1. In testing an audio-frequency amplifier, state what is meant by:
(a) Half-power (3-dB) frequency points.
(b) Bandwidth.

Q14-2. Explain the situation when there is no actual cutoff at the frequency f_2, commonly called the "high-frequency cutoff" point of an amplifier.

Q14-3. (a) In what respects is a comparison of amplifier gain in terms of decibels superior to a comparison of relative gain by percentage?

(b) In what respect is the decibel comparison inferior to a comparison in terms of voltage gain?

Q14-4. What is the primary cause of harmonic distortion, and how can it generally be minimized?

Q14-5. State how the presence of excessive harmonic distortion can be recognized by means of:
(a) An oscilloscope.
(b) Reading a dc meter in the plate circuit.

Q14-6. Distinguish between harmonic distortion and intermodulation distortion as to:
(a) Cause.
(b) Effect.

Q14-7. How does a wave-analyzer instrument differ from the more usual commercial instrument called a distortion analyzer (harmonic distortion meter)?

Q14-8. Describe the evidence provided by using square-wave testing on an amplifier to indicate:
(a) Poor high-frequency response.
(b) Poor low-frequency response.
(c) Ringing (oscillation).

Q14-9. Explain the fact that the same amplifier that produces a distorted output when fed by a 10-kHz square wave may provide almost perfectly flat amplification of a sine wave of the same fundamental frequency (10 kHz).

Q14-10. Explain why much importance is often attached to the voltage and current offset parameters E_{os} and I_{os} of *operational amplifiers*, despite perfectly satisfactory high-gain, low-distortion, and good frequency-response characteristics. (*Hint:* See Chapter 18 on analog applications.)

PROBLEMS

P14-1. (a) An amplifier produces an output of 84 V rms, with a sinusoidal input signal of 4 V rms. Find the voltage gain.
(b) What would the voltage amplification (or gain) be if the output were 84 V peak to peak for the same input signal as in part (a)?

P14-2. If the amplification of an amplifier, expressed as a voltage ratio, is 20, express the gain in decibels.

P14-3. If the reference level (0 dB) of an amplifier output is given as 10 V, find the voltage output at:
(a) +3 dB.
(b) +6 dB.
(c) +20 dB.
(d) +32 dB.
(e) +46 dB.

P14-4. An amplifier under test is connected to an oscilloscope in such a manner that the sine-wave input signal is connected to the vertical posts of the scope, and the amplifier output signal is connected to the horizontal posts. The patterns shown in Fig. P14-4 are obtained at various times during the test. Interpret the conditions producing each of the test patterns shown.

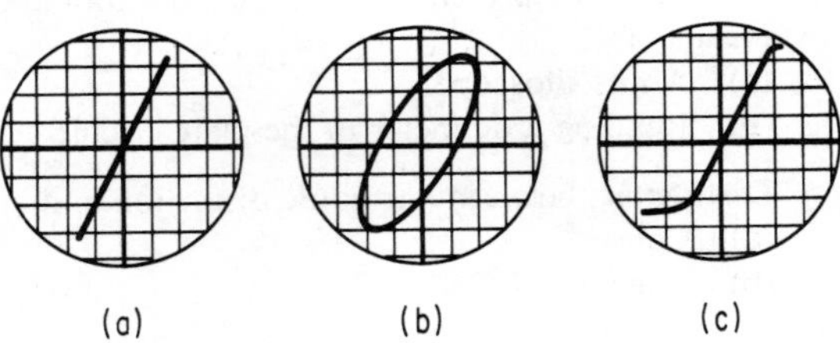

Figure P14-4

P14-5. Each of two amplifiers A and B has the same voltage output at the 0-dB mid-frequency reference level, and the amplifiers have the following individual specifications:

Specifications

Amplifier A is down 1 dB at 500 kHz.
Amplifier B is down 3 dB at 800 kHz.

State which one of the two has the wider frequency response by comparing the frequency at which each amplifier is down 1 dB (10%) from the reference level.

15

Radio-Frequency Test Methods

15-1. TUNED RADIO-FREQUENCY TESTING AREAS

The methods used for testing radio-frequency circuits differ in a number of important ways from those used in the untuned audio-frequency work. Although the general approach to the testing of both kinds of circuits is the same, in that we observe the effect produced by a known input test signal, the specific radio-frequency tests involve various kinds of *modulated and unmodulated input signals* and also include the marked effects produced by *tuned circuits* on these signals.

The uses of modulated RF signals are well known in the fields of radio and TV broadcasting (often designated as the *entertainment area*) and are also extensive in the general field of communication (which can be distinguished from broadcasting by calling it the *commercial/industrial area*).

In order to stay within reasonable limits, the major emphasis in this text, as before, will be placed on the commercial/industrial area; the large field of maintenance/service test methods will be pointed out and references will be cited to the extensive literature on such radio/TV servicing methods.[1]

[1]See the very practical treatment of such methods in Shunaman, cited in Appendix B.

Sources of Modulated Radio-Frequency Signal

The modulated RF signal input for testing most communication equipment may be obtained in two different ways. In one method, a signal having the required frequency (and per cent modulation) is obtained from an RF signal generator (such as the "standard" signal generators of Chapter 11). In the other method, advantage is taken of on-the-air broadcast signals (AM/FM/TV), in which case the received signal is monitored by means of a service-type signal-tracer instrument, used primarily for troubleshooting in receiving circuits.

15-2. MEASURING CONTINUOUS-WAVE RADIO-FREQUENCY SIGNALS: DIRECT AND INDIRECT METHODS

Test methods for RF signals will differ considerably, according to whether the signals are modulated or unmodulated (continuous-wave, or CW). For example, we are primarily concerned with the audio modulation of an RF carrier in the case of the program material forming the envelope of the modulated wave.

The monitoring of a broadcast RF signal by means of the simple detector probe of a signal tracer depends entirely on the fact that the RF signal is modulated. The problem of monitoring a continuous-wave (CW) or unmodulated RF signal *directly* is much more difficult, since the very act of connecting a meter will usually result in disturbing the original RF circuit to some degree, depending upon the inherent reactance of the measuring instrument. This troublesome loading effect is especially obvious in tuned RF circuits, where the measurement produces a detuning effect in addition to its loading action.

For these reasons, only a small minority of RF measurements are made directly by thermal instruments, such as hot-wire ammeters or thermocouple-type meters. A fairly common example is the use of an RF ammeter to measure the current flowing in the antenna circuit of a transmitter. Here it is significant that the direct-measuring instrument is generally incorporated in series with the other components as a permanent fixture, and as such, its reactance becomes a fixed part of the resonant system.

In the great majority of cases, the measurement of RF signals is accomplished *indirectly* by rectifying the RF energy and measuring the resulting dc value. The measuring instrument, such as a *wavemeter* or *field-strength meter*, is not connected directly to the RF circuit but is generally inductively coupled to it through loose coupling, so that both loading and detuning effects are minimized. By rectifying the induced RF energy, it becomes possible to use a dc meter for an indirect indication, corresponding to the level of the RF signal being measured.

15-3. IMPEDANCE MEASUREMENTS: RADIO-FREQUENCY BRIDGE

The testing of RF signals is greatly affected by the *radiation characteristics* of the device under test, and in particular, the antenna and transmission-line impedances are important governing factors (the testing of microwave signals is beyond the scope of this text). Figure 15-1 illustrates an example of an RF bridge for measuring resistive and reactive values at frequencies from 400 kHz to 60 MHz, direct-reading in ohms. It is adaptable to a variety of coaxial connectors including 50, 100, and 200-Ω standard terminations.

Figure 15-1. Radio-frequency bridge for determining antenna and transmission-line impedances. (*General Radio model 1606B.*)

As a bridge only, it requires an external signal source and an external detector, covering the desired frequency range. Since the reactance is direct-reading at a frequency of 1 MHz (up to $\pm 5000\ \Omega$), a generator providing a signal at this frequency would generally be satisfactory. For other frequencies, the dial reading must be divided by the frequency in megahertz. As an external detector, a well-shielded radio receiver would be satisfactory.

The testing of RF impedance in the polar form (magnitude of Z and phase angle ϕ) can be accomplished (both in circuit and out of circuit) by the vector-impedance meter (*Hewlett-Packard model 4815A*), which was previously illustrated in Fig. 12-11. As stated in Sec. 12-9, the frequency coverage of this model extends to 108 MHz (starting at 500 kHz) and provides the convenience of "probe and read," with provision for a self-calibration check of the direct reading.

15-4. RADIO-FREQUENCY VOLTMETERS

A simple-circuit arrangement for an RF voltmeter is shown in Fig. 15-2(a). It consists primarily of a half-wave rectifier feeding a dc meter through series resistance R_1, which sets the voltage scale and also improves the linearity of the rectifier

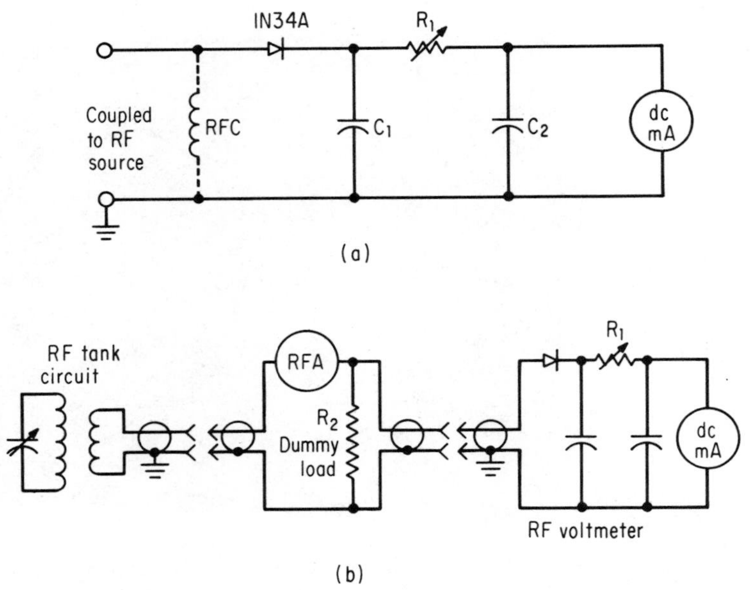

Figure 15-2. Simple RF voltmeter: (a) circuit showing peak rectification of RF input; (b) calibration by absorbing RF power in a known resistance (dummy load).

action. Capacitor C_1 provides peak rectification, making the indication on the dc meter correspond to the peak value of the input RF signal. In order for C_1 to remain charged near the peak value, the time constant RC_1 should be large compared with the period of the lowest radio frequency to be measured. This is easily obtained for a frequency as low as 200 kHz, for example (at which the period is 5 μsec), by using a resistance R_1 around 125 kΩ and a capacitor around 1000 pF for C_1, giving a time constant of 125 μsec or 25 times as great as the longest period expected in this example. The other capacitor C_2 provides for additional filtering around the dc meter. The radio-frequency choke (RFC), shown as a dashed connection, is not needed if the RF source provides a continuous dc return path of reasonable resistance.

An instrument as simple as a diode-tuned voltmeter is generally intended only for comparative readings, rather than for measuring precise values. It can be calibrated approximately by a method frequently used for a quick check on the power output of a small transmitter, shown in Fig. 15-2(b). In this illustration, the RF power is being dissipated in a dummy load of known resistance R_2, while the RF current is monitored by an RF ammeter. In a typical example, where the dummy load R_2 of 50 Ω would absorb in the neighborhood of 4 W of RF power from a small transmitter, the rms reading of the RF ammeter would be observed to be, say, 0.28 A. Since the IR voltage drop across R_2 would then be 0.28(50) or 14 V, the adjustable resistor R_1 would be set to indicate 14 V rms at this point of the scale. It should be noted that the actual rectified dc voltage is close to 20 V at this time, since the meter responds to the peak value of the RF input voltage, even though the calibrated scale reads in rms volts.

The RF voltmeter can also serve as a rough measure of power in this procedure. When the voltmeter is calibrated in rms volts, the power is obtained from E^2/R, or $(14 \times 14)/50 = 4$ W. If the voltmeter is calibrated in peak volts, the same value of power would be obtained from $(E \text{ peak})^2/2R$, or $(20 \times 20)/100$.

15-5. FIELD-STRENGTH AND FREQUENCY TESTS

Absorption Wavemeter

The *absorption wavemeter* offers an extremely simple means of monitoring an RF signal (modulated or unmodulated) from an oscillator or transmitter. All that is needed is a resonant LC circuit [Fig. 15-3(a)] and an indicator such as a pilot lamp. The coil L is held near an oscillating circuit to provide loose coupling, and the capacitor C is tuned to resonance, as indicated by the lighting of the indicator lamp. When the dial of the variable capacitor is calibrated in terms of frequency, the unit becomes very useful as a *rough frequency meter*. A rough calibration is advantageous for quickly identifying the fundamental frequency of oscillation, since the circuit is not sensitive enough to respond to harmonics. In fact, when the transmitter contains an indicating meter in its circuit, we can even dispense with the pilot lamp (or other indicator) in the wavemeter circuit. In that case, as the wavemeter is tuned to resonance it will absorb the greatest amount of energy and will cause a detectable change in an indicating meter of the transmitter. When used in this way, without its own indicator lamp, the lowered resistance of the circuit results in a higher Q and provides a sharper indication of the resonant frequency.

When the lighting of a pilot lamp in the wavemeter is used as a rough indicator of resonance, it should be of the low-current, low-resistance type, such as

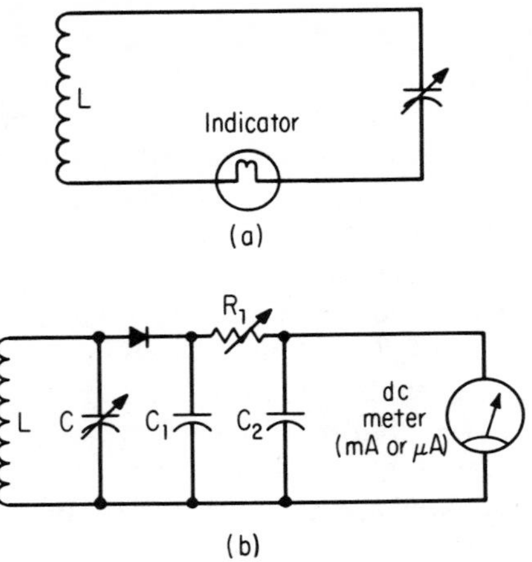

Figure 15-3. Absorption wavemeter: (a) with lamp indicator; (b) with meter indication.

the 49 pilot lamp that requires only 60 mA at 2 V. Even such a lamp, however, is much less sensitive than the RF voltmeter circuit using a meter. The wavemeter combined with the indicating meter [Fig. 15-3(b)] is much more sensitive and can be used with looser coupling to provide more definite readings at resonance.

Field-Strength Meters

The field-strength meter is used to measure the radiation intensity from a transmitting antenna at a given location. Using its own small antenna, it is essentially a simple receiver with an indicator. The wavemeter circuit with rectifier-meter indication of Fig. 15-3(b) is often equipped with a small whip antenna and designated as a field-strength meter. Although it is possible to obtain an indication with this setup when the whip antenna of the meter is positioned fairly close to the transmitting antenna, the sensitivity is generally not great enough for use with ordinary low-powered transmitters or test radiations.

The field-strength measurement should be made at a distance of several wavelengths from the transmitting antenna, to avoid misleading readings when the pickup is obtained from a combination of the radiation field with the induction field close to the transmitter. Thus, when monitoring the transmission from a 5-W transmitter such as used on the citizens' band (at 27 MHz), for example, a wavelength is roughly about 35 ft. The simple wavemeter combination of Fig. 15-3(b) even at such a distance would be too far away to respond, and when brought closer would be satisfactory only for a very rough check of the fact that the transmitter is on the air.

To enable the wavemeter combination to act as a field-strength meter, greater sensitivity can easily be obtained by the addition of a transistor dc amplifier, as shown in Fig. 15-4. The transistor provides ample current gain, so that satis-

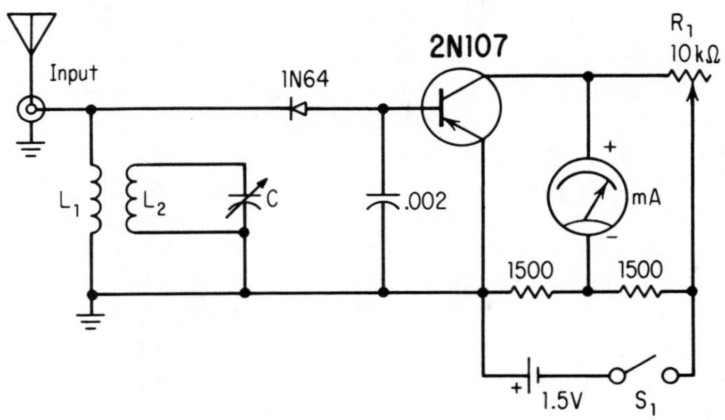

Figure 15-4. Elementary field-strength meter using a transistor to amplify the dc output of the wavemeter. (From *Radio Amateurs Handbook*.)

factory sensitivity is obtained at the 35-ft desired distance of one wavelength in the example, even with the use of a 0–1-mA meter in the circuit. Of course, greater sensitivity can be obtained by employing a microammeter as the indicating meter.

With no signal being received, the residual or quiescent current is balanced out by the variable resistor R_1. This zero balance should be checked at intervals, since the quiescent current is sensitive to temperature changes.

The collector current through the meter provides an indication of the strength of the RF wave being picked up. This current is not strictly proportional to the field strength, because of the combined nonlinearities of the semiconductor diode and transistor. However, the response is entirely satisfactory for *relative comparison* of field strengths, and for this purpose the scale is generally marked 0–100 in arbitrary units.

Specialized instruments for measuring field strength in units of microvolts-per-meter are much more complicated than the simple field-strength meter shown. They generally include a sensitive and stable receiver circuit and a linear calibration of the scale on the indicating meter. Because of the sensitive receiver circuits used, they are usable at great distances from the transmitter for field-strength mapping surveys.

15-6. TESTING WITH FREQUENCY METERS

Heterodyne-Frequency Meter

Accurate measurements of the frequency of an RF signal source are most often made by the heterodyne or "zero-beat" method. This method is based upon the fact that when two signals of different frequencies are mixed (in a nonlinear circuit), the output contains a signal at a heterodyne (or beat) frequency equal to the difference between the two original signals.

As an example, assume that we wish to obtain an accurate measurement of an RF signal having a frequency in the neighborhood of 13.5 MHz. The heterodyne frequency meter is shown in the block diagram of Fig. 15-5(a) in the operate position. In this position, the unknown signal is picked up by the antenna and mixed with the output of the calibrated *variable-frequency oscillator* (VFO) in the meter. As the frequency-meter dial tunes the oscillator output close to a frequency of, say, 13,490 kHz, the difference frequency of 10 kHz will be heard as a high-pitched heterodyne (or beat note) in the phones. As the dial is progressively turned closer to an output of 13,500 kHz, the pitch of the beat note will progressively get lower as the difference decreases, until, at exactly 13,500 kHz, the beat note becomes inaudible at zero beat. Then as the dial continues on progressively to a frequency of, say, 13,510 kHz, the beat note again reappears and rises in pitch to produce an audible 10 kHz squeal again. Further turning of the dial, of course, results in the pitch becoming too shrill to be audible. Thus, by turning the dial back to the zero-beat point at which the beat note became inaudible (between the two high-pitched notes on either side of it), we obtain an exact point; the frequency on the calibrated dial is then equal to the unknown frequency. In the military versions of this familiar type of frequency meter the tuning dial is geared down to obtain a very precise slow motion of the dial, which can be read to five figures. These five figures (in terms of the fundamental frequency of the VFO) are then interpreted on a calibrated chart in terms of the frequency of the unknown.

The range of frequencies that can be handled by this method is enormous because of the possibility of using various harmonics of the variable-frequency oscillator instead of its fundamental for producing the zero-beat condition. As a matter of fact, the example previously cited for identifying a 13,500-kHz signal was actually employed to calibrate a signal of 27,000 kHz (27 MHz) on the citizens' band, and the identical procedure was followed. The only difference in the method when employing harmonics is that the calibration chart is read in the column relating to the second harmonic, where, of course, the frequencies are exactly double those of the fundamental.

Since harmonics of the calibrated oscillator can also beat with the unknown signal, it is often necessary to have some method of identifying which harmonic is being employed. This is done, when necessary, by first obtaining a rough measurement of the unknown frequency by an absorption-frequency method, using a wavemeter circuit that is only sensitive enough to respond to the fundamental frequency. When the approximate value of the unknown frequency is known, there is usually no trouble in identifying which harmonic of the calibrated oscillator is producing the heterodyne note, since the harmonics are so widely spaced.

In order to ensure the validity of the calibration of the variable-frequency oscillator during any given measurement, a *check against a known crystal frequency* is provided by setting the switch to the CRYSTAL CALIBRATE position. In this position the accurately known crystal output is allowed to beat against the signal

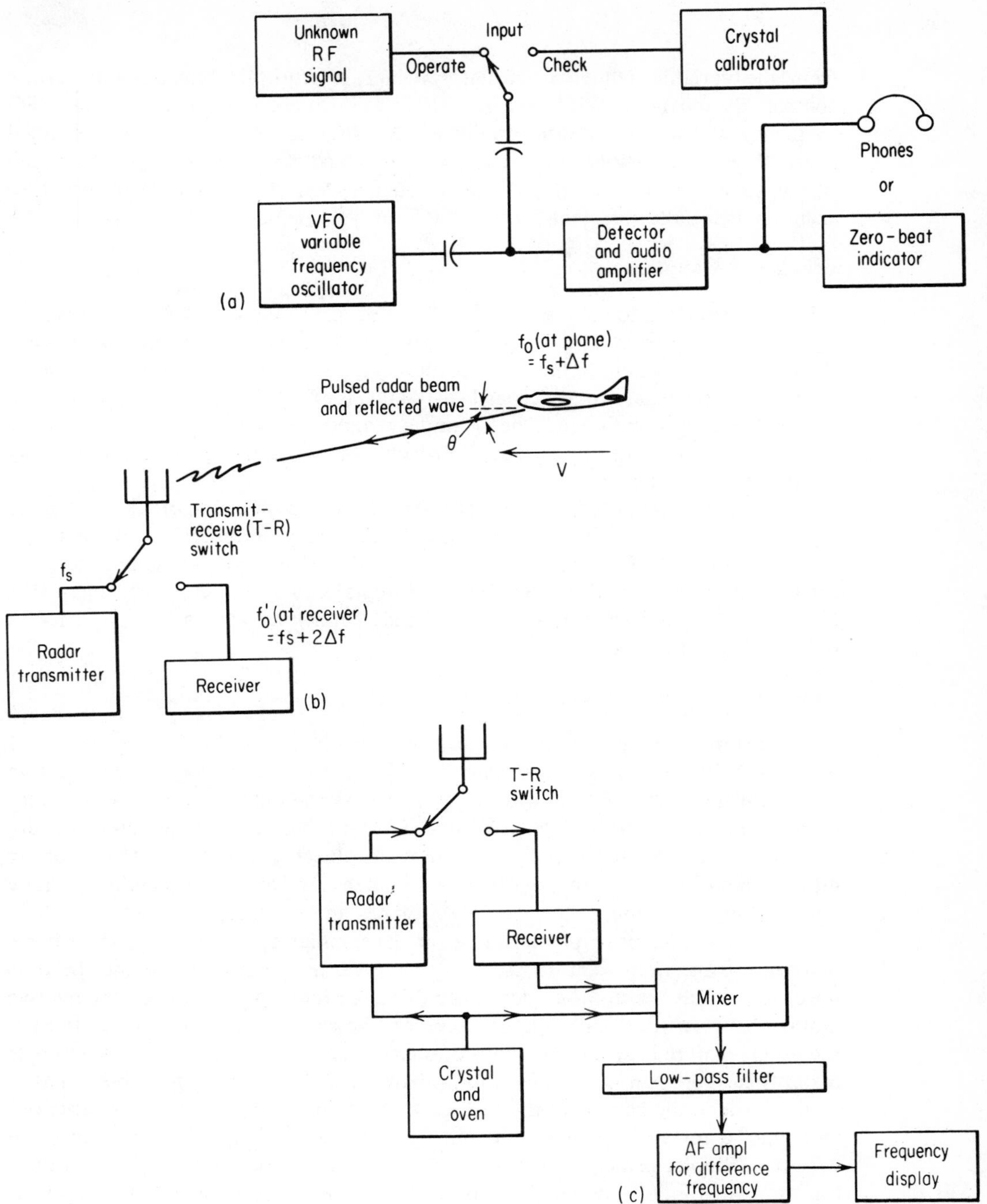

Figure 15-5. Heterodyne-frequency measurements: (a) block diagram of heterodyne-frequency (or beat-frequency) meter; (b) difference of frequency principle used in Doppler method for measuring velocity; (c) block diagram of Doppler velocity measurement.

from the variable-frequency oscillator, which is adjustable by a small vernier control. By means of this control, a slight adjustment is made to set the VFO frequency at the exact reading indicated for that checkpoint on the calibrated chart. When this is done, the readings of the chart can be relied upon to be accurate to better than 0.01% of the frequency being measured. This order of accuracy is usually sufficient for many RF measurements.

Frequency Counters

When it is required to have a higher order of accuracy than the 0.01% that can be provided by the heterodyne-frequency meter, it is necessary to employ secondary standards of frequency. In these standards, crystal oscillators are held to accuracies much better than the average 5 parts in 100,000 (0.005%) required for the citizens' band previously mentioned. The secondary standards must, in turn, depend for their calibration on the more accurate primary standards, in the manner previously discussed in Chapter 11.

The problem of quickly checking a frequency to such a high order of accuracy has been greatly simplified by the use of frequency and period counters that display the count either on a *rate meter* or in *digital form*. These counters are discussed separately in Sec. 17-1. With the digital-counter type of instrument that provides a 7-digit display (correct to a single count), an accuracy of 1 part in a million (1 ppm) can easily be obtained.

Doppler Velocity Measurement

An interesting and highly useful application of the heterodyne-frequency technique is the *Doppler method of determining the velocity of a moving target*. This method for measuring the velocity of a moving object has the outstanding advantage that it *does not require a determination of time* as most other velocity measurements do. In fact, the greater the speed of the moving object, the better is the resulting measurement—a factor, among others, that makes the Doppler principle especially effective for determining the speed of jet planes and satellites.

The Doppler principle accounts for the familiar situation where the whistle of a fast-moving train seems to rise in pitch as the train approaches and then become lower in pitch as the train recedes. Since this situation depends only on the relative motion between the train and the observer, the same effect would be noticed if the observer were moving toward a stationary source of the sound. In the opposite case, when there is any relative motion between the source and the observer that tends to widen the gap, as in the case of the receding train, the sound is perceived as having a lower frequency or pitch. This change in frequency, in accordance with the Doppler principle, is directly proportional to the relative speed, so that, for example, if an observer in an automobile at 30 mi/hr were closing the gap with a train approaching at 60 mi/hr, the resulting frequency change in the sound of the train whistle (a rise in this case) would be $1\frac{1}{2}$ times as great as it would be if the automobile were stationary.

In interpreting the Doppler effect, it is important to note that the observed change is a *frequency effect* and that the rise in frequency resulting from relative motion closing the gap is not only an apparent rise but an actual one. The frequency change Δf, therefore, can be measured on a frequency meter and so provide a measure of the relative velocity.

In applying the Doppler principle to measuring the velocity of an airplane, an RF wave of known frequency is transmitted from a ground station, as shown in Fig. 15-5(b). Either a passive or an active system might be used for returning this wave to the ground station from the airplane. In a passive system, the ground station would detect the wave reflected from the plane; in an active system, the signal received by the plane would automatically be retransmitted from the plane by its responder transmitter. In the latter case, some means (such as frequency doubling) would be employed to distinguish between the two transmitted signals. A simpler and clearer illustration will be obtained by considering the passive system, as it might be used by a Doppler radar transmitter.

The Doppler (or observed) frequency f_0, resulting from the original signal frequency f_s, is given by the expression

$$f_0 = \frac{(1 + v/c)\cos\theta}{[1 - (v/c)^2]^{1/2}} f_s$$

where v = velocity of the plane
c = velocity of radio waves (300×10^6 m/sec)

The presence of the $\cos\theta$ term in the expression supplies the horizontal component of the plane's velocity, which is effective in closing the gap and therefore in changing the observed frequency. The expression can be considerably simplified by two considerations:[2] first, we can consider θ to be small enough so that $\cos\theta$ is unity, if the plane is sufficiently far away; secondly, for planes, satellites, or any objects larger than atomic particles, we can consider $(v/c)^2$ as negligible. The simplified expression thus becomes

$$f_0 = f_s\left(1 + \frac{v}{c}\right)$$

Subtracting f_s from each side of the equation to obtain the change in frequency $f_0 - f_s = \Delta f$:

$$\text{frequency shift } \Delta f = \frac{v}{c} f_s$$

Since the frequency shift depends on the ratio of the plane's velocity v to the velocity of the radio wave c, it can be seen that the shift will be quite small

[2]Following the treatment in G. R. Partridge, *Principles of Electronic Instruments* (Englewood Cliffs, N.J., Prentice-Hall, 1958).

for ordinary communication frequencies. If, for example, the transmitted frequency f_s is 100 MHz and the speed of the plane v is 400 mi/hr, the shift Δf is only about 60 Hz. If the ground station receives the signal reflected from the plane, the frequency is shifted again by 60 Hz, making a total shift of 120 Hz in the signal received back by the ground station. This represents a change of 120 Hz in 100 MHz, or about 1 part in 10^6 (1 ppm). Small as this percentage change is, the techniques of heterodyne-frequency measurement make it possible to measure the velocity fairly easily. Moreover, the frequency transmitted by radar in Doppler measurements is measured in thousands of megahertz, rather than hundreds. Thus, the use of a 10-cm radar wave, at a frequency of 3000 MHz = 3 GHz would result in a frequency shift (Δf) 30 times as great as in our original example, or equal to 3600 Hz.

By means of available techniques that employ crystals operated in constant-temperature ovens, the transmitted signal can be maintained to within 1 part in 10^8, which is 30 Hz in the 3 GHz of this example. We must also take into account the fact that any change in transmitted frequency that does occur is not greatly effective in changing the Doppler shift, because a simultaneous change in the frequency of both transmitted and reflected waves still leaves essentially the same difference between them, thus providing the same Doppler shift.

A consideration of all these factors leads to an appreciation of the fact that the heterodyne method makes it possible to detect comparatively minute differences in very large quantities by accurate detection of the differences between them. With reasonable care in the associated circuitry for generating this difference (3600 Hz in this case), a digital-frequency counter is able to provide the velocity measurement to a highly satisfactory degree of precision.

The circuit blocks for measuring the Doppler-frequency shift are shown in Fig. 15-5(c). The output of the mixer contains the low-frequency difference component along with the original and reflected superhigh frequencies. Hence the difference frequency can easily be separated by a low-pass filter, then amplified and applied to a frequency meter or digital-frequency counter for display.

15-7. GRID-DIP METER

The discussion of RF signal measurement applications thus far has emphasized methods for handling RF signals from transmitters or other signal-generator sources. Another form of signal source providing a very handy method for investigating RF circuits is the *grid-dip meter*. Though not strictly classed as a signal generator, it conveniently combines the function of a variable-frequency oscillator with that of an absorption wavemeter, with a meter indication for both functions. In its oscillator function, the dc meter is switched in series with the grid circuit of the oscillator, making is possible to recognize quickly the resonant frequency of any *LC* circuit held in the field of the oscillating circuit. This is evidenced by a pronounced dip in the oscillator dc grid current at the frequency where the oscillatory energy is absorbed by the external resonant circuit. The resulting decrease in grid excitation causes the drop in dc grid current. In this fashion, the grid-dip meter is able to identify the resonant frequencies of any LC combination in an unenergized condition, whether wired into a circuit or even when entirely separate from any other circuitry.

The circuit of the grid-dip meter is shown in Fig. 15-6. With the function

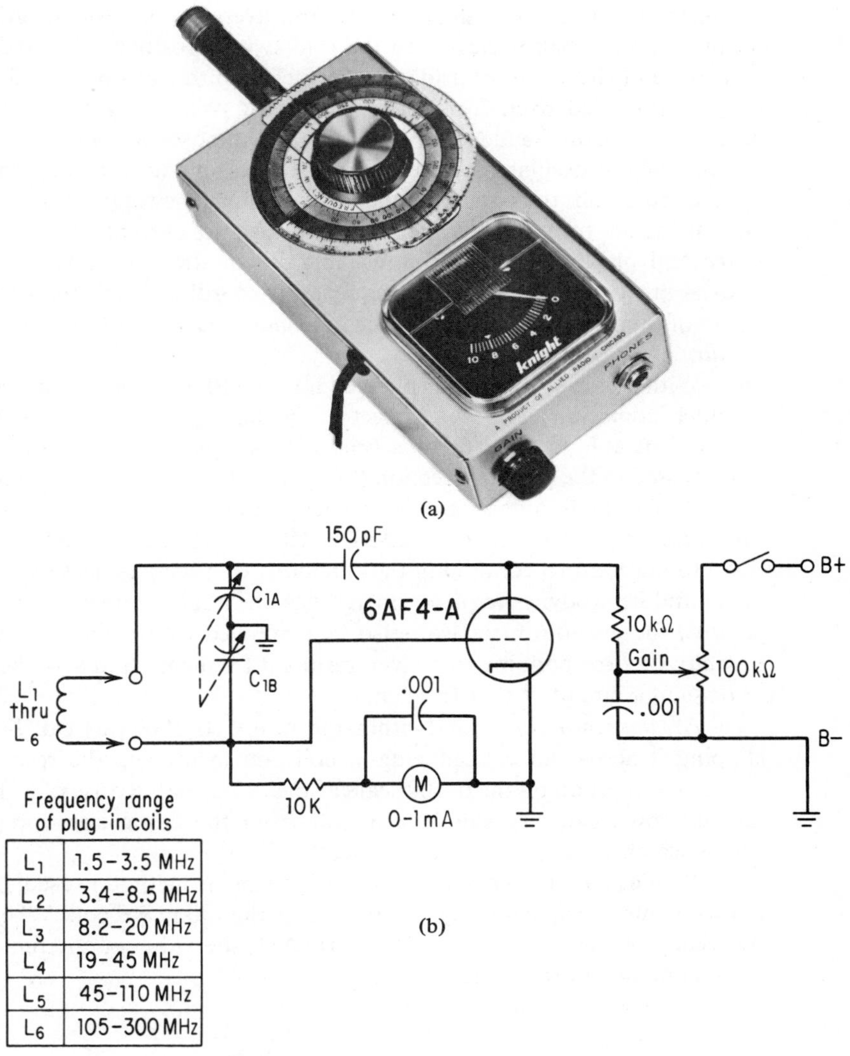

L_1	1.5–3.5 MHz
L_2	3.4–8.5 MHz
L_3	8.2–20 MHz
L_4	19–45 MHz
L_5	45–110 MHz
L_6	105–300 MHz

Figure 15-6. (a) Grid-dip meter and (b) its functional schematic diagram. (*Knight kit, Allied Radio model G-30.*)

switch in its closed (OSC) position, the instrument functions as a variable-frequency oscillator with plug-in coils, providing continuous coverage with a range of 400 kHz–250 MHz. When the function switch is in the diode position, the plate voltage is removed from the oscillator tube, which then functions as a diode detector, with the grid and cathode acting as the diode elements. The dc meter then reads the resulting diode current.

When the grid-dip meter functions in OSC position, where the plate voltage is applied to the tube, its circuit acts as a Colpitts-type oscillator. While oscillating, the meter in the grid circuit monitors and averages the flow of grid current that results from the half cycles when the grid swings positive. This average current is a measure of the extent of grid swing, which in turn is proportional to the strength of oscillation fed back from the tank circuit. When any external tuned circuit, having a resonant frequency close to that of the oscillating circuit, is placed near the coil of the oscillator, the external LC combination absorbs energy from the tank circuit and thus reduces the extent of grid swing fed back. The smaller grid swing in turn results in a decrease of grid current. This decrease is very sharp and produces an easily noticeable dip in the meter reading when the two circuits are tuned to the same frequency. Accordingly, the frequency read at the point of maximum grid dip is the *resonant frequency of the external absorbing circuit*.

With a pair of earphones plugged into the PHONE jack, the versatile grid-dip circuit becomes an oscillating detector. In this condition, when held near an RF source, it is able to function as a *heterodyne-frequency meter*, in the same manner as discussed in the previous section (but without the exact crystal-frequency check).

When functioning as a diode detector, the grid-dip meter acts as an *absorption wavemeter* over the same 400-kHz–250-MHz frequency range.

In addition to combining the functions of a rough-check type of absorption meter and heterodyne-frequency meter with the ability to measure passive resonant circuits, the grid-dip meter can also be employed to measure the *inductance L* or *capacity C* corresponding to a given resonant frequency and also the *storage factor Q* of such a circuit at that frequency.

An *unknown capacitor* (within the range 50–7000 pF) can be measured by clipping it across an unused plug-in coil and measuring the resonant frequency of the combination on the 0–100 scale of the meter dial. By use of a chart furnished, the unknown capacity can be read off from the curve appropriate to the coil across which the capacitor was connected.

To measure an *unknown inductance* a similar method is used, by employing a known value of capacity, say, 75 pF, across the unknown coil. When the resonant frequency of this combination is determined, the value of the inductance L may be obtained from the basic relation

$$f_r = \frac{1}{2\pi\sqrt{LC}}$$

which can be converted into a more convenient form:

$$L = \frac{1}{39.5f^2C}$$

An approximate value for the Q of an unknown resonant circuit may be determined by using a VTVM connected across the tuned circuit and obtaining the 3-dB bandwidth of the circuit. Couple the grid-dip meter close enough to give a convenient reading on the sensitive ac range of the VTVM and tune the grid-dip meter to give a maximum reading on the VTVM. Record this resonant dial reading as f_r (see Fig. 15-7). Keeping the coupling constant, retune the grid-dip meter to

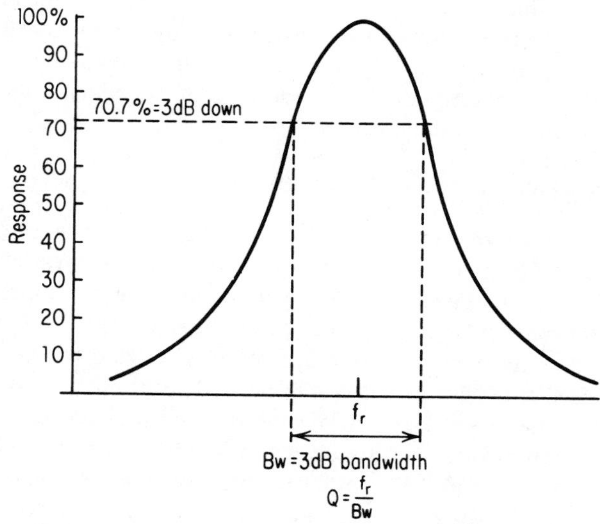

Figure 15-7. Finding Q of a coil by locating both 3-dB points on resonance curve.

obtain the dial reading for the lower 3-dB frequency f_1, at which the VTVM reads 0.707 of its previous value and repeat to obtain the dial reading for the higher 3-dB frequency f_2, at which the VTVM again reads the 0.707 value. The Q of the resonant circuit is then obtained from the relation

$$Q = \frac{f_r}{\text{3-dB bandwidth}} = \frac{f_r}{f_2 - f_1}$$

From the variety of uses that have been mentioned, it can be seen that the grid-dip meter, although not a precision instrument, is an extremely handy instrument for obtaining a first approximation of the resonant frequency and operating characteristics of all sorts of tuned circuits, both active and passive. It is particularly effective in distinguishing a fundamental frequency from the many possible harmonics that may be encountered in a particular measurement.

15-8. SWEEP-GENERATOR TEST METHOD (VISUAL INDICATION OF FREQUENCY-RESPONSE CHARACTERISTICS)

There are many occasions when it is desired to find the *frequency response of a tuned circuit*, without resorting to the step-by-step method of finding each point of the curve. As was indicated in the discussion of sweep generators in Sec. 11-10, it is possible to display the complete circuit response over a given frequency band by using a generator to sweep over this desired band, while the overall response is displayed on an oscilloscope screen. This sweep-generator method is especially useful for investigating the response of tuned circuits in AM and FM radio, television, and radar circuits.

Sweep-Generator Characteristics

In order to obtain an oscilloscope display of the overall response of a tuned circuit, the injected signal must be continuously varied over the passband at a periodic rate. The width of the frequency sweep required will depend, of course, on the particular circuit being investigated. Thus, where a 20-kHz sweep width would be ample for an IF stage of an AM radio receiver, the greater sweep width required for the tuned stages of an FM receiver extends to at least 150 kHz, while in the video section of a television receiver it covers all of 4.5 MHz. By employing a generator where the sweep width is adjustable, a single sweep generator is able to handle all of these requirements. Such a sweeping generator for laboratory applications was discussed in a general way in Sec. 11-10.

Since sweeping across the full video response of a television receiver offers the most stringent requirement, this example will be employed to illustrate the test method for displaying a frequency-response curve. Over a full television channel width of up to 6 MHz, a sweep 10 MHz wide is employed. In general, the frequency will be swept at a 60 Hz line rate, producing a forward trace (from left to right on the oscilloscope) and a return trace (from right to left) at the rate of 60 times/s. Each of these traces is shown, before being superimposed, in Fig. 15-8(a). This 60-Hz rate is also used as the *SYNC FREQUENCY* for the oscilloscope, either by selecting the line position on the SYNC selector of the scope or, preferably, by feeding the 60-Hz *signal from the sweep generator* to the *EXT SYNC* terminals of the oscilloscope. The latter method will most easily allow the proper phasing of the *SYNC* signal. By means of an adjustment of the phasing control, the forward and return traces are made to coincide, as in Fig. 15-8(b). It should be understood that what we are looking at here is the relative output of a particular stage (in this case a 25-MHz video IF transformer) as the frequency is caused to vary between

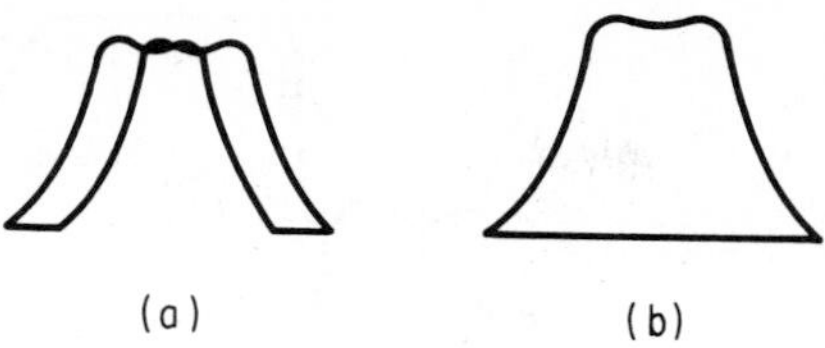

Figure 15-8. Appearance of double-peak response curve: (a) when phasing control is improperly set; (b) typical response curve with correct phasing.

20 and 30 MHz at a 60-Hz rate. Even though, at any one instant, the sweep-generator signal applied to the test circuit is an unmodulated RF signal, the detected response varies at a 60-Hz rate, in accordance with the rate of frequency variation. We are therefore able to display the instantaneous dc output of the detector stage, since this dc voltage varies at the 60-Hz rate of the frequency variation.

Blanking

Many sweep generators avoid the possible confusion of the double trace by suppressing the RF signal from the sweep generator during the return trace. The oscilloscope beam then returns without any vertical deflection and, as an added advantage, forms a base line for the response in the forward direction. This is accomplished by a blanking tube, which develops a high negative voltage only during the return trace time, and so cuts off the signal-generator output during that time.

Markers

When a sweep generator is employed, it becomes necessary to have some means of identifying exact frequency points on the response curve, since the output of the sweep generator is constantly changing in frequency. This identification is done by a *marker generator*, which produces an unmodulated RF signal as a marker, at a desired accurate frequency. This marker voltage is then superimposed on the sweep voltage to mark the point at which the exact frequency of the so-called marker "pip" appears, as shown in Fig. 15-9. This pip results from the interaction of the sweep-generator signal at the instant when its frequency coincides with that of the marker, producing a heterodyne note containing a zero beat at the point of frequency coincidence. The pip that is actually observed is the audible beat producing one peak when the sweep frequency is slightly below the marker, then a zero beat, then up to another peak when the sweep frequency is slightly higher than the marker. However, since the space occupied between these two peaks is insignificant when drawn to the proper scale of frequencies, the pip appears condensed as a single peak when viewed on the scope.

By changing the frequency of the marker generator, the pip can be moved along to any desired place on the response curve. It can thus be used to delineate the frequency limits of response or, if desired, the response amplitude at any

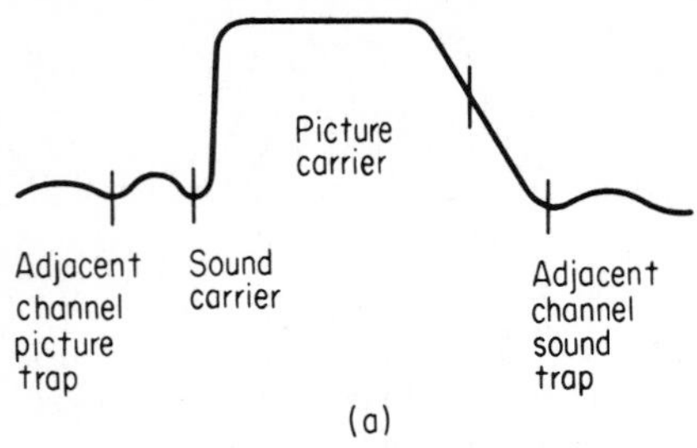

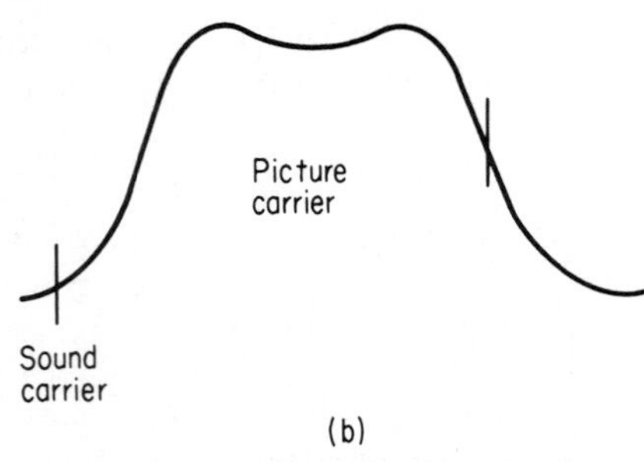

Figure 15-9. Typical video IF response curves showing markers: (a) dual-channel video IF response curve; (b) intercarrier video IF response curve. (*Hickok TV Sweep and Marker Alignment instructions, model 615.*)

given frequency within the sweep range. *Absorption markers* may also be used for identification purposes. Such markers require only the presence of a tuned circuit to absorb energy at a particular frequency, rather than an actual generation of a signal. The absorption marker produces a dip in the response curve rather than a peak. Because of the attenuation due to absorption, this dip indication is usually less pronounced than a peak indication.

The primary requirement for marker generation obviously is the production of an accurately known frequency. For this purpose, any unmodulated RF signal generator that covers the range is suitable, provided its frequency can be checked by crystal calibration or its equivalent. The marker amplitude should also be variable, to ensure the proper amount of marker injection.

Examples of TV Sweep Generators

Sweep generators for the VHF television channels (Nos. 2–13) also straddle the FM frequencies (88–108 MHz) and can therefore be used for both services. The *RCA type WR-69A*, illustrated in Fig. 15-10(a), covers the video IF range

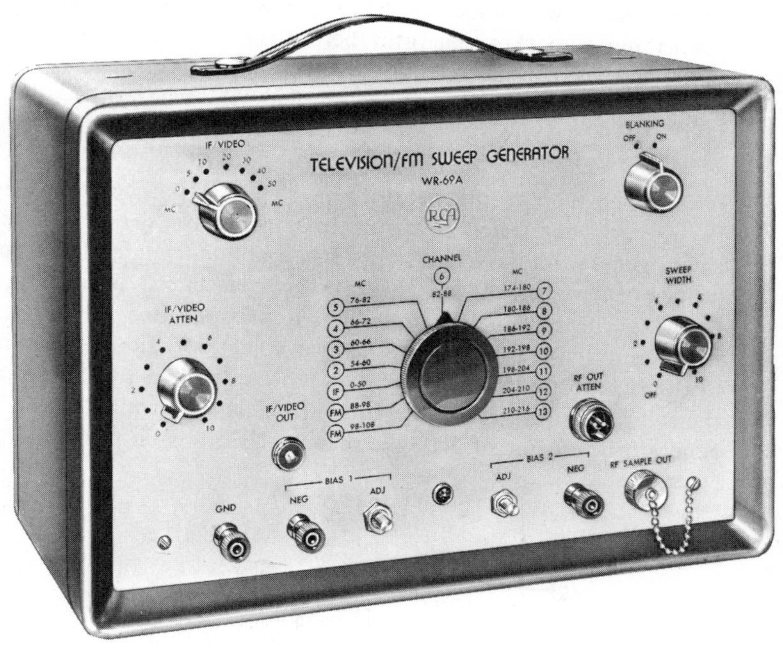

(a)

(b)

Figure 15-10. TV/FM sweep generators: (a) *RCA TV/FM sweep generator type WR-69A;* (b) *B&K sweep and marker generator model 415* (courtesy Dynascan Corporation).

(0–50 MHz), two ranges of FM, and the 12 TV channels with provisions for connecting a marker generator before the receiver or a marker-adder after the signal has passed through the receiver. The *B-K model 415*, in Fig. 15-10(b) is a combination TV sweep and marker-alignment generator.

Sweep-Generator Setup

When the sweep generator has no self-contained marker provisions, it must be properly connected to an external generator for marker points. Since most signal generators have low-impedance outputs, ranging between 30 and 300 Ω, to reduce loading effects on the generator circuit, it is important that the sweep generator be terminated in such a low impedance for reliable operation. The instruction manual for each generator will supply the required information. When in doubt, most undesirable effects can be avoided by the arrangement shown in the setup of Fig. 15-11. Here, a low-impedance termination of 30 Ω is seen by the generator, and one of 300 Ω is seen by the antenna terminals of the receiver.

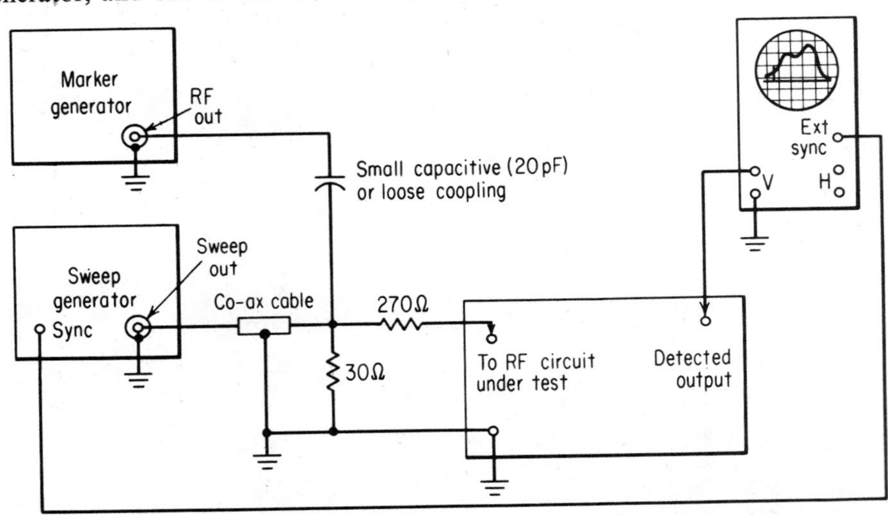

Figure 15-11. Setup for sweep and marker generators to produce RF response curve on oscilloscope.

With the SYNC output of the sweep generator connected to the EXT SYNC terminal of the scope and the center frequency and width controls of sweep properly set on the sweep generator, a clear pattern showing the response of the circuit under test should appear in stable form as a display of the detected output. The base line is produced only if the return trace is blanked out; otherwise a phasing control is generally required to make the forward and reverse traces coincide.

The connection of the marker generator may reduce the overall height of the response curve, but it should not alter its shape, provided that the coupling is sufficiently loose. The frequency and amplitude of the marker pip are then adjusted to identify the extremes and the middle of the response curve (or to mark any other desired portion of the curve), as shown for the video IF response curve in Fig. 15-9.

15-9. FREQUENCY SYNTHESIZERS

As a tunable frequency standard, the frequency synthesizer represents an advance over the crystal-controlled oscillator (which is a part of it), by the fact that it is tunable to a *precise frequency in discrete steps*, progressing digitally in steps of 0.01 Hz each (or better), up to the top frequency in megahertz.

Figure 15-12 shows a coherent decade frequency synthesizer (*General Radio type 1162A*), covering 0–1 MHz. The seven-decade boxlike dials display the

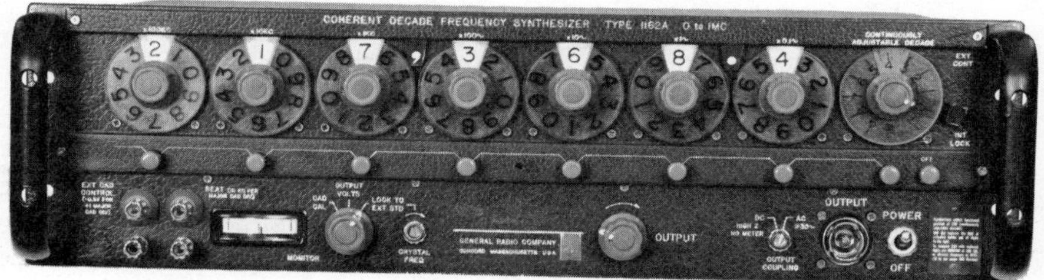

Figure 15-12. Coherent decade frequency synthesizer. (*General Radio type 1162-A.*)

selected frequency in digits (in this case: 217,368.4 Hz), without the ±-digit ambiguity of a digital type of frequency counter. Each of these seven dials operates a *digit insertion* (DI) unit [the eighth dial shown is for an optional, continuously adjustable decade (CAD), which can replace any of the seven DI units for sweeping purposes]. Thus, with this seven-step decade model, a desired frequency can be set by steps of 0.1 Hz up to 999.9 kHz, with the assurance that any selected frequency will have a rocklike stability, obtained through a phase lock to the internal crystal oscillator. For even greater precision, a similar model allows setability in steps of 0.01 Hz/step, with a top range of 100 kHz.

The process by which each digit insertion (DI) unit produces its part of the desired frequency (and then passes it on to the next DI) involves the *addition of two frequencies* accompanied by *frequency division*, as discussed below, in connection with Fig. 15-13, which shows the signals transversing a typical DI unit.

Synthesizing Principle

Starting with a fixed crystal-controlled "carrier" frequency of 5000 kHz, each input and output signal of this and subsequent DI units is made up of two signals, namely, this invariant 5000-kHz "carrier" frequency f_c plus a signal component

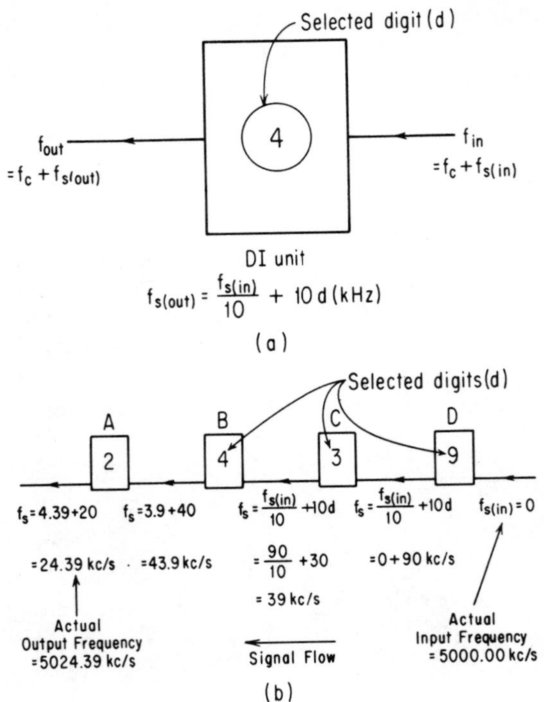

Figure 15-13. Numerical example of frequency synthesizer: (a) signal through a DI unit; (b) signal through four DI units. (From General Radio, "Experimenter," Sept. 1964.)

f_s that always lies between 0 and 100 kHz. Thus, when the first DI unit (progressing from the right) has a selected digit d, of 4 [as shown in Fig. 15-13(a)], the input signal represents the addition of 5000 kHz (f_c) plus the signal components, f_s (*in*), which are 10d or 40 kHz in this case. Because the invariant 5000 Hz passes through unchanged (since it is added at the initial input and subtracted from the final output), we can therefore, for convenience, disregard the carrier frequency f_c in our numerical example and concentrate on only the signal components f_s.

Numerical Example

We will use the general relation for each DI unit, with f_c disregarded, for four such units, proceeding from the right [as shown in Fig. 15-13(b)] to generate a frequency of 24.39 kHz. (Note that the smallest increment in this case is 0.01 Hz, and we will therefore have an output frequency range of only 0–100 kHz for a seven-step decade in this instance.)

The right-hand DI unit *block D* [as shown in part (b) of the figure] has a selected digit d, of 9. The frequency of the signal out of this unit f_s (*out*) will therefore be

Block D $d = 9$	f_s (out) $= f_s$ (in)/10 + 10d kHz $= 0/10 + 10(9) = 90$ kHz
Block C $d = 3$	f_s (out) $= 90/10 + 10(3) = 39$ kHz
Block B $d = 4$	f_s (out) $= 39/10 + 10(4) = 43.9$ kHz
Block A $d = 2$	f_s (out) $= 43.9/10 + 10(2)$ $= 24.39$ kHz, as desired

The actual output frequency of block A, as shown on the bottom line of the figure, is 5024.39 kHz, since it contains the invariant f_c of 5000 kHz. This invariant component, however, is subtracted by mixing and filtering to provide the desired frequency of 24.39 kHz at the final output.

The accuracy of the final output frequency will be seen to depend on both the crystal-controlled fixed "carrier" frequency of 5000 kHz and on the frequency generated by the signal oscillator (0–100 kHz) as commanded by the setting of digit d. In the above examples these signal frequencies ($d \times 10$) were successively 90, 30, 40, and 20 kHz. The accuracy of these signal frequencies is assured by having each one pass through a phase detector, whose capture range is ample to produce phase locking to the proper submultiple of the crystal-controlled "carrier" frequency (5000 kHz).

As a result of this arrangement, the stability (even when the crystal is not temperature-controlled), is within 2 parts in 10^8 (or 0.02 ppm), a generally adequate stability under reasonably constant ambient temperatures (such as the temperature cycling between a heated and unheated room). With more stable standards, the phase-locking circuitry of the instrument can be used for even tighter stability.

The frequency-synthesizing method (first disclosed in an Australian patent in 1951) is now used by many other manufacturers to provide a variety of signal outputs. In the exemplified model, the available ac output voltage is 2 V rms into 50 Ω and can also be dc-coupled to provide an open-circuit voltage of approximately 1 V.

QUESTIONS

Q15-1. State which of the following functions are necessary in a signal tracer instrument suitable for use on a broadcast radio set, and give reasons for your answer for each function (whether required or not):

(a) Audio-frequency amplifier and loudspeaker.
(b) Demodulator.
(c) High-voltage (500 V) power supply.
(d) VTVM.
(e) Tuned radio-frequency amplifier.

Q15-2. When a dc meter is connected to the output of a demodulator (or detector) probe, what indication on the meter may be expected when the probe is connected to:
(a) An unmodulated RF signal source?
(b) A modulated RF signal source?

Q15-3. When an oscilloscope is used to monitor the output of a detector probe connected to a 27-MHz RF signal source, modulated at 400 Hz, what should be the minimum bandwidth of the frequency response of the oscilloscope's
(a) Vertical amplifiers?
(b) Horizontal amplifiers?

Q15-4. Explain how the sensitivity of a radio receiver may be measured.

Q15-5. (a) What is meant by a signal-to-noise ratio measurement?
(b) How may such a measurement be made?

Q15-6. How can a radio receiver be tested for its ability to reject image frequencies?

Q15-7. Explain why the use of an ordinary voltmeter (moving-coil) is not generally satisfactory for testing AVC voltages in a radio receiver.

Q15-8. Explain the difference in the situations where a field-strength meter is used as contrasted with the situation where an absorption-type wave meter is employed.

Q15-9. Explain why a transistor amplifier is very helpful as a part of an RF voltmeter.

Q15-10. Distinguish between the principle employed in making a frequency measurement with a heterodyne-frequency meter, as contrasted with the principle used in an instrument employing a frequency-counting method.

Q15-11. Explain how the Doppler effect is used in making a velocity measurement.

Q15-12. Explain the setup used in a visual alignment of the IF strip in an AM radio receiver.

Q15-13. Explain how the setup for a visual alignment of the IF strip in an FM receiver differs from the setup in Question Q15-12 for an AM receiver.

Q15-14. Explain how a dc VTVM is used in checking an FM detector (or discriminator).

Q15-15. Explain the use of a marker generator in the visual alignment of a TV tuner.

16

Nuclear-Radiation Detection Instruments

16-1. NUCLEAR OR IONIZING RADIATIONS

The discovery of atomic energy and the subsequent intense developments in this new field have unleashed nuclear forces possessing tremendous potential benefits along with terrible dangers. Understandably, there is a vital interest in methods for the detection of these radiations of atomic or nuclear origin. Such radiations are technically termed *nuclear or ionizing radiations* and are also commonly called atomic radiation, radioactive rays, "fallout" radiation, or often simply radiation.

In referring to radioactive effects in general, it is well to include some adjective together with the term radiation, to eliminate confusion with other forms of electromagnetic radiation. By using the preferable designation of nuclear or ionizing radiation, we clearly distinguish this form of wave from the other forms that are also properly classed as electromagnetic radiations, i.e., light waves (including the ultraviolet and infrared), heat waves, and radio waves. (See Fig. 19-7 for the wavelengths of the complete electromagnetic spectrum.)

We may divide the nuclear radiations, into *two main groups, according to their source*—natural or man-made—as follows:

1. *Natural background nuclear radiation*, from *cosmic rays* and also from naturally *radioactive minerals* occurring in rocks, concrete, and water.

2. *Man-made nuclear radiation*, subdivided into peaceful and military uses:
   (a) Peaceful applications, including *generation of atomic power, radioisotopes* for tracer activity in research or medical therapeutic use, and *x-rays.*
   (b) Military atomic explosions of both the *fission (atomic) bomb* and *fusion (hydrogen or thermonuclear) bomb types.*

Since x-rays are essentially similar to the gamma rays produced from some of the radioisotopes, we might simplify the group of man-made nuclear radiations by lumping all the peaceful applications together under the term *radioisotope activity*, leaving the term *nuclear explosion* for the military applications. Similarly, the natural nuclear radiations may be combined in the single term *background radioactivity.*

Major Types of Nuclear Radiation

The property common to all of the nuclear radiations listed, from whatever source, is the ability of the radiation *to interact with the atoms that constitute all matter.* The nature of the interaction of a given radiation with any form of matter varies with the different components contained in the radiation. In this introductory approach we shall confine ourselves to the *three major types of radiation:* the *alpha* (α) *particles*, the *beta* (β) *particles* (or electrons), and the *gamma* (γ) *rays* (or x-rays).

Alpha particles are the relatively heavy nuclei of ionized helium atoms (or He^{++}), bearing a double positive charge owing to the loss of helium's normal two electrons per atom.

Beta particles or electrons are the extremely lightweight fundamental particles, each bearing a single negative charge.

Gamma rays (or x-rays) have no charge and are characterized as photons, i.e., wave packets bearing unit energies dependent on their wavelength (or frequency).

These three major components of nuclear radiation are responsible for the most commonly observed interactions with matter, generally producing *ionization* of the air, or other gas, through which they pass. This ionization is the main effect used in the detection of the presence of nuclear radiation in any of its forms. (In passing through certain liquids and solids, the *luminescent or scintillation* effect may also be used, as will be discussed later.)

Penetrating Power of Nuclear Radiations

Each type of radiation is able to penetrate matter to a quite different extent. The three types, arranged in the order of increasing penetrating power, are alphas (α), betas (β), and gammas (γ). The penetrating qualities of each of these radiations (and also neutrons) are illustrated in Fig. 16-1.

In concentrating on the α, β, and γ radiations as the three major types, it should be mentioned that we are deliberately omitting other types, associated with the *fundamental particles of the atom.* Thus rays of protons (H^+, or hydrogen nuclei bearing a single positive charge) and neutrons (uncharged particles) are also part of

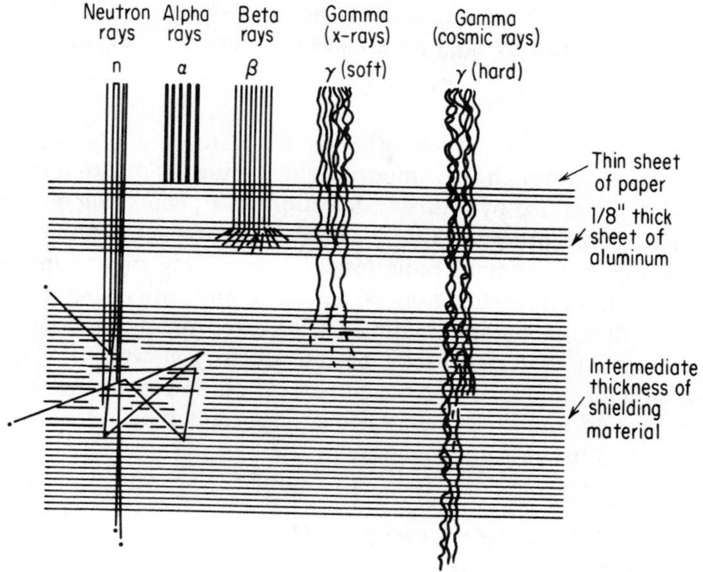

Figure 16-1. Penetration of α, β, and hard and soft γ rays; neutron rays (of importance only in the vicinity of reactors) are also penetrating radiation. [From *Principles of Radiation and Contamination Control*, Vol. II, Navy Dept., Navships 250-341-3.)

the complete picture, to say nothing of the positrons, deuterons, tritons, neutrinos, and mesons. For the purposes of this text, however, these more detailed aspects of nuclear physics will not be needed.[1]

Common Forms of Ionizing Radiations

The rays to which people are ordinarily exposed are the *medical x-rays* and the *gamma rays of background radiation*, which include the cosmic rays from space (augmented to some degree by fallout) as well as environmental radiations from rocks, water, and even the luminous wrist-watch dials that are painted with a diluted radium compound. (The distinction between x-rays and gamma rays is only in their origins; the x-rays are rays given off from an ordinarily stable nucleus of some target being bombarded by high-speed electrons in an x-ray tube.)

The beta or alpha rays are not often encountered, and since they are so easily stopped, are not usually taken into account when considering possible harmful effects of radiation, unless special conditions exist. A thickness of less than $\frac{1}{8}$ in.

[1]For a more in-depth treatment of nuclear radiation and nuclear physics, *Basic Nuclear Engineering*, 3rd ed., by A. R. Foster and R. L. Wright, Jr., published by Allyn and Bacon, 1977, is suggested.

of aluminum, for example, easily stops most beta rays or electrons, while alpha rays can be stopped by even a thin sheet of paper.

Survey Meters

Forms of radiation other than the three major types, such as protons or neutrons, although very important in reactors or research laboratory work, are not usually recorded by ordinary detection instruments. *Survey meters* most commonly encountered are generally constructed to measure gamma radiation, or less frequently, beta/gamma radiations. These survey meters in the main perform the vitally important function of detecting and measuring gamma radiation arising from fallout contamination or from exposure to medical or industrial uses of radioactive material. From the introductory viewpoint of this chapter, therefore, the major emphasis is directed toward an understanding of the electronic instrumentation underlying the various important types of nuclear-radiation detectors that find use in survey meters.

16-2. CHARACTERIZING STRENGTH OF NUCLEAR RADIATION

In measuring the so-called "strength" of a given nuclear-radiation situation, we must identify a number of factors before we can arrive at a value of *dose rate*, which is generally the figure of immediate interest and almost always given in some equivalent of *roentgens per hour* (R/hr), and *total exposure in roentgens* or equivalent.

Disintegration Activity (Curie)

The standard unit of radioactivity is the curie (derived from the original unit of measurement for radium), which is measured in terms of disintegration rate. *One curie* is defined as 37 billion disintegrations per second. The disintegration rate is characteristically unique for each radioisotope and is independent of the type radiation, holding equally well for alpha particles, beta particles, and gamma rays. More frequently, the activity is expressed in microcuries (10^{-6} Ci) or $\mu\mu$Ci (10^{-12} Ci).

In contrast to the curie measurement of the *activity* of the source, the *roentgen* measures the *effect of the radiation*. The roentgen is defined in terms of the *amount of ionization* produced within a unit volume of air under standard conditions and is dependent upon both the type of radiation and its energy.

Rate of Ionizing Effect (Roentgens per Hour, Dose Rate)

The *dose rate, in roentgens per hour* (or milliroentgens per hour, mR/hr), is the preferred indication on a survey meter; it is obtained as *product of activity rate* (*C* in curies) *times energy* (*E* in mega-electron volts). The energy figure is the sum of the emanations being measured. In the ordinary type of survey, which responds in all but special cases to gamma energies, the energy figure E is the sum of all the gamma energies. Thus cobalt 60 (^{60}Co) emits two gamma rays per disintegration, of energies 1.1 and 1.3 MeV, for an energy sum E of 2.5 MeV. Using a 500-microcurie (μCi) sample of ^{60}Co as an example, the ionizing effect would be proportional to the CE product of $500(C \times 10^{-6})$ times 2.4 MeV, or $1200(CE \times 10^{-6})$.

A rule-of-thumb expression for estimating the dose rate of a gamma point-source (reliable within 20% only) uses this CE product as follows:

$$\textit{dose rate}\ (1\ \text{ft from source}) = 6CE\ \text{R/hr}$$

where C = activity strength, *curies*
E = *energy* sum of gamma energies in mega-electron volts, MeV
R/hr = roentgens per hour

Thus, in the example above, the dose rate is

$$6(1200CE \times 10^{-6}) = 7200(10^{-6})\ \text{R/hr}$$

or, expressed as it usually is in milliroentgens per hour, 7.2 mR/hr at 1 ft from the source.

Reference Dose-Rate Values for Survey Instruments

In attempting a rough estimate for some value of dose rate (in milliroentgens per hour) that constitutes *unsafe or dangerous exposure*, it is necessary to keep clearly in mind the *time duration* of the exposure, if the figure is to have any valid meaning at all. We should naturally expect that, in the case of *momentary exposure* during an x-ray examination, we would willingly tolerate dose rates that would be entirely unacceptable for *continuous workaday exposures*.

The order of magnitude for typical dose-rate values for these two widely different exposure conditions is listed in Table 16-1 as roughly average (or "ball-park") figures for comparison purposes only. They can, however, serve to indicate the kind of full-scale ranges (in milliroentgens per hour, mR/hr) that might be required of survey instruments in each case.

TABLE 16-1. Typical Dose-Rate Values
(Order of Magnitude for Comparison Purposes)

For Momentary Exposure per X-Ray Examination		*Exposure Limits under Continuous Working Conditions*	
Chest	6–30 mR/hr	Working area in nuclear industry (revised from original 7.5 mR/hr)	2.5 mR/hr
Teeth (whole mouth)	4000 mR/hr	Home color-TV set (proposed limit at specified close distance from set)	0.5 mR/hr

Note: For the most commonly required sensitivity for surveying doubtful working areas, a full-scale range of at least 0–5 mR/hr would be indicated; greater sensitivity would obviously be required in radioisotope detection, while other survey needs might easily be four orders of magnitude higher (10,000–1), perhaps up to 50,000 mR/hr (0–50 R/hr).

Where relatively high external levels of radiation rate are necessary because of the nature of the particular activity, the Nuclear Regulatory Commission requires that all areas where the external radiation level exceeds 100 mR/hr be posted with a "*High Radiation Area*" *sign*, as illustrated in Fig. 16-2.

Figure 16-2. Caution sign required to be posted in areas exceeding a 100-mR/hr radiation level. [From *Peacetime Hazards in the Fire Service*, Office of Education, Circular 657 (OE-84019), U.S. Government Printing Office, Washington, D.C.]

16-3. RADIATION HAZARDS

Any meaningful discussion of the harmful effects of ionizing radiation on humans must necessarily include much highly specialized biological aspects for a proper interpretation of the readings obtained from the monitoring instruments and is therefore beyond the scope of this book. It is nevertheless instructive to emphasize the elements that contribute to possible radiation hazards.

The major factor to be considered in almost all practical situations is the *total accumulated exposure*, since the harmful effects of radiation are cumulative. The total radiation exposure is measured in *roentgens* and is usually expressed in its *equivalent REMS* (roentgen equivalent man); this measure, of course, is not obtained directly from a survey meter reading of counts per minute (cpm) or milliroentgens per hour (mR/hr); it must be obtained from either an addition calculation from a cumulative record or an integrating method in the detection instrument. The accumulated external dose provides a recommended maximum limit over a period of one or more years; this limit forms the basis of a reasonably safe "*bank*" on which a worker in the radiation field can draw, without experiencing any observable symptoms of illness.

Another vital distinction as to radiation hazard must be made between the effects of *external* and *internal radiation.* The matter of internal radiation is the much more complicated one, since the chemical source of each form of radioactive material taken into the body must be taken into account as well as the specific biological activity of such a source. The measure of the effects of *internal radiation is expressed by the "body-burden"* for the radioisotope, and is generally handled by tables in the study of biomedical radiation effects. This aspect of biomedical specialization will not be pursued here, and emphasis in this text will be directed only to the area of *external radiation hazard,* as it concerns the readings obtained on survey-type radiation-detection instruments.

In interpreting the degree of hazard due to external radiation, a distinction must first be made between detection methods used to count the rate of disintegrations in *counts per minute* (cpm), and those that measure dosage rate in *roentgens per hour* (R/hr). The count-rate measurement, while most useful for nuclear experimentation (as in half-life determination and the like), is not suitable for determining safe limits to humans unless it can be interpreted in terms of ionization effects, that is, in terms of *exposure rate in* roentgens per hour or *total exposure in roentgens.*

A valid simplification can be made here, however, if we restrict the discussion to an initial interpretation of the readings obtained by *general purpose gamma-survey instruments, measuring external radiation intensity in milliroentgens* (*or roentgens*) *per hour.* Some pertinent facts can be summarized as follows:

1. The unit of 1 *roentgen* (R) is generally equivalent to 1 *REM* (roentgen equivalent man) of body effect.
2. A "*High Radiation Area,*" which must be posted with the caution sign of Fig. 16-2, is defined[2] in the *Code of Federal Regulations, No. 10, Part 20,* as an area where the *external radiation level exceeds 100 mR/hr.* It is important to note that this does not include airborne or other *internal radiation hazards,* which require posting as an "Airborne Radiation Area," according to separate standards in the same Code.
3. For *industrial workers,* a figure of *3-REM*/quarter (or 12 REM/yr) is set by radiation safety personnel, primarily for safety administration purposes, and is not to be regarded as an absolute limit. For radiation "banking" purposes, an exposure rate of 5 REM/yr for each year after the age of 18 is set, but the National Committee on Radiation Protection (Bureau of Standards Handbook No. 59) also suggests that a single, once-in-a-lifetime whole-body 25-REM

[2] *Peacetime Radiation Hazards in the Fire Service,* Office of Education, Circular 657 (OE-84019), U.S. Government Printing Office, Washington, D.C.

emergency exposure may be taken, without effect on the 5-REM/yr radiation-banking exposure rate. Charts prepared on radiation sickness effects[3] indicate no expected radiation sickness from whole-body exposure[4] to external radiation in the 25–100-REM range.

4. The estimated average yearly radiation/dose to all people in the United States is estimated[5] as 250 mR ($\frac{1}{4}$ R), distributed as shown in the chart of Fig. 16-3.

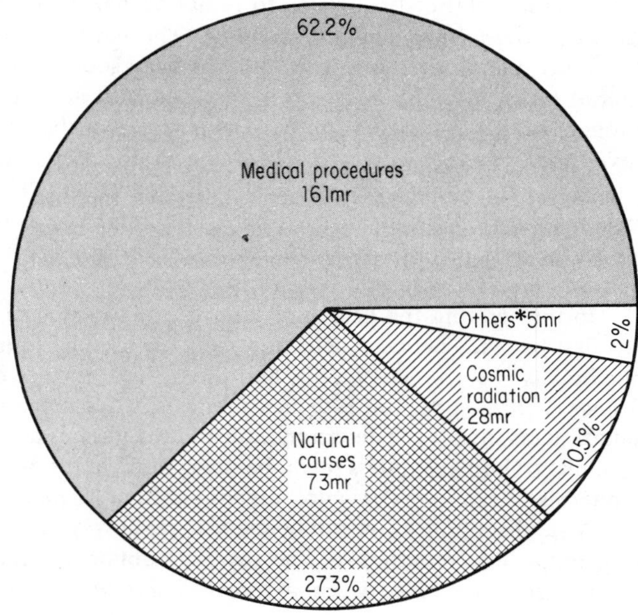

*Others
3.1mr Fallout from nuclear tests
1.4mr Luminous watch dials, shoe fluroscopes, TV tubes, radioactive industrial wastes, etc.
0.5mr Occupational exposures in atomic energy work
5.0mr

Figure 16-3. Estimated yearly average radiation doses to all people in the United States [based on total accumulated exposure of 250 mR ($\frac{1}{4}$ R) in 1 yr]. [From *Peacetime Radiation Hazards in the Fire Service*, Office of Education, Circular 657 (OE-84019), U.S. Government Printing Office, Washington, D.C.]

[3]Ibid.

[4]When the radiation is directed at the more sensitive parts of the body, such as the lens of the eye, the gonads, or the blood-forming organs, the lower 5-REM/yr figure applies.

[5]*Peacetime Radiation Hazards in the Fire Service.*

It is hoped that what has been said here in introducing the study of radiation survey meters is sufficient to indicate the need for care in intelligent interpretations of the meter indications and the equally compelling necessity for wide-spread public education in clarifying these vitally important considerations.

16-4. DETECTION METHODS FOR NUCLEAR RADIATION

The common methods used for detecting nuclear radiation are based on various means for monitoring the *ionizing effects* of these radiations. They divide into two large groups: the first group detects each ionizing event as a *pulse*; that is, the ionizing events produce discrete electrical signals and are of the types listed below; the second group is concerned with relatively steady-state *ionization current.*

Group I. Pulse-Type Detectors

1. *Geiger–Mueller Tube Detector* with
   (a) indication of *individual counts on a counter or scaler.*
   (b) indication of *count rate on a meter giving a count-per-minute* (*cpm*) *type of survey meter.*
   (c) indication of *radiation intensity in terms of roentgens per hour in dose-rate survey meters.*
   (d) *accumulated total exposure survey meters* (in roentgens).
2. *Proportional Counters,* where only pulses of a certain size are detected and counted in special tubes similar to GM tubes, to discriminate between the energy values of the radiation components.
3. *Scintillator Detector,* again with
   (a) individual-count indication, or
   (b) count-rate indication, capable of much greater sensitivity than the GM-tube types.

In contrast to the pulse type, the second group of detectors uses an *ionization chamber* to measure the *average effect* produced by the radiation in terms of an increase or decrease of current in the chamber, without any attempt to resolve the individual events. This method is obviously necessary in cases where the pulse-type method would not work, because of the very high rate at which the ionization events occur. It is also preferred in certain other cases, since the current output of the ionization chamber more closely simulates the biological action of radiation on the body. In spite of the fact that ion-chamber instruments require more careful construction and more precise calibration (and are correspondingly more expensive), the advantage of obtaining reliably significant readings in milliroentgens per hour, rather than in counts per minute, outweighs the drawbacks of greater cost, for important health-physics survey work. Types of commonly used ionization-current detectors are listed below as Group II.

Group II. Ionization-Current Detectors

1. *Ion-Chamber Detectors*, including the following:
   (a) *Ion-chamber survey meters*, indicating on a meter in milliroentgens per hour.
   (b) *Dosimeters* (total-dose meters) of the charged electroscope/electrometer type in a greatly simplified ion-chamber form, indicating total leakage as a measure of accumulated dose.

In the nonelectronic category, the *film badge* may be mentioned briefly, as a means for determining total exposure by the extent of blackening of special photographic film.

Relative Sensitivity

The four main radiation detectors of the electronic type may be arranged in a decreasing order of sensitivity, as follows:

Most sensitive (generally solid-state detectors)	Germanium-Lithium Drifted (GiLi detector) Scintillation detector
Gas-filled detectors	Geiger–Mueller counter Proportional counter Ion-chamber detector

Since the gas-filled detectors, though not the most sensitive, are the kind most widely used, they will be considered first, followed by the scintillation-detector type.

16-5. GAS-FILLED DETECTORS

The detection of nuclear radiation by means of a gas-filled tube is based upon the ionizing effect of the radiation on the gas under low pressure in the tube, as shown in Fig. 16-4. The tube is fitted with a central electrode, generally in the form of a thin rod or wire, to which a relatively high positive voltage is applied, forming a wire anode; the enclosing cylinder is connected to the negative side of the battery as the cathode. As the battery voltage is increased, a flow of current (or collection of charged gas particles) takes place in the tube; the character of the voltampere relationship of this collection current, however, becomes quite different as the voltage becomes greater. It is therefore convenient to consider the action of the current in a number of different regions, depending on the range of voltage applied to the gas-diode arrangement.

Five distinct regions may be distinguished as the degree of ionization or number of charged particles—electrons and gas ions—becomes progressively greater, as shown in Fig. 16-4. Of these five regions, three are particularly suited for radiation detection: region II for *ion-chamber* detectors, region III for *proportional*

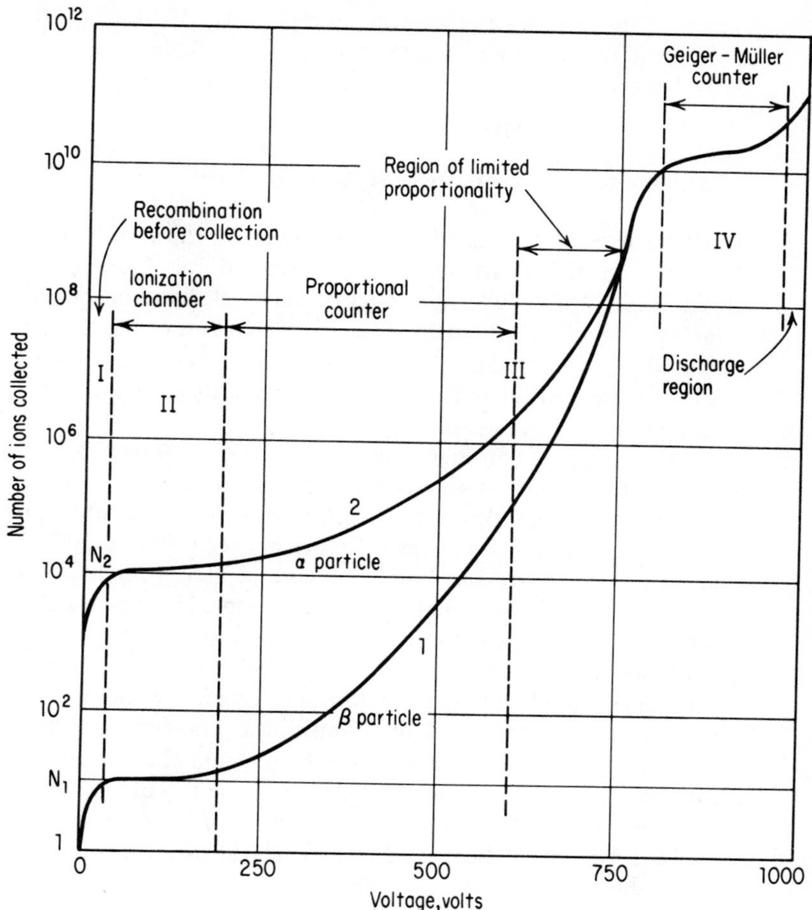

Figure 16-4. Ionization curves for ion collection in a gas-filled tube showing four successive regions: I, recombination region; II, saturation or ion-chamber region; III, proportional region; IV Geiger–Mueller-counter region; followed by arc discharge. [From W. J. Price, *Nuclear Radiation Detection*, (New York: McGraw-Hill, 1958).]

counters, and region IV for the so-called "*Geiger counters.*" In the short span of region I, most of the charged particles resulting from ionization recombine before they can be collected; as a result, the value of current is too small and unreliable to be useful for ordinary detection purposes. At the other extreme in region V (at about 1000 V of applied voltage), the current rises sharply as a glow or arc discharge is established, and, if unchecked, might result in excessively heavy currents capable of destroying the tube. In the regions between the two extremes of too small and excessively high currents—i.e., in regions II, III, and IV—the current is greatly dependent on the amount and kind of nuclear radiation to which the tube is exposed, as well as on the applied voltage, thus giving rise to the three different forms of detection in each of the three regions.

Characteristics of Low-Pressure Gas Discharge

It will be noted that separate curves are given for the alpha (α) and beta (β) particles. The curve for the β particles will be considered first. It should be kept in mind that this β curve also includes the effect of the much more penetrating gamma (γ) rays, which produce both β particles (or electrons) and charged atoms (ions), when the enclosed gas is ionized by the gamma ray. (The curve for α particles will be considered separately, in later discussions, particularly in connection with proportional counters.)

Returning to consideration of the curve for β particles (which we can interpret, for survey work, to include also the effect of gamma rays), a flattening of the curve, indicating a saturation effect, can be observed in *region II, where the ionization-chamber detector operates* (roughly between 50 and 200 V). In this region, it will be noted that an increase in voltage produces no significant rise in current. This situation is quite similar to the saturation effect observed in thermionic diodes operated at a point where there is a scarcity of thermally produced electrons. In both cases, the few electrons present are all swept to the collector anode, so that an increase in applied voltage (within this range) does not produce a corresponding rise in current. Under these conditions in the ionization chamber, the small current that flows is dependent primarily on the ionizing effect of the radiation, with practically no contribution from a limited increase in voltage. In this region also, the ion chamber most closely duplicates the effect of ionizing radiations on the body.

The situation changes markedly in the *proportional-counter sector* (*region III*, roughly between 200 and 600 V). Here, the increased range of voltage is sufficient to accelerate the few free electrons near the wire anode to a velocity sufficient to produce more electrons and gas ions by collision. The extra electrons, in turn, cause still more collision-produced ionization, resulting in a *local avalanche, or* "*gas amplification.*" The output current accordingly rises rapidly in this region and is very sensitive both to applied voltage and to the energy of the original ionizing event. Proportional-counter detectors can therefore be set to respond to certain minimum energies of the radiation and produce outputs proportional to the values of these energies. Since alpha particles in general are much more energetic than beta particles, the proportional counters can easily be set to distinguish between the two radiations, even when both are present simultaneously.

The avalanche effect (called the Townsend avalanche) in proportional counters is a local one, confined to the close proximity of the central anode wire where a high potential field exists. The gas amplification does not become a fullscale Townsend avalanche (one propagated through the entire tube) until the *Geiger–Mueller sector* (*region IV*) is reached, at voltages approaching about 850 V. (The portion between 600 and 850 V is labeled "region of limited proportionality" to indicate that it is a transition region, with contribution in indefinite amounts from both regions III and IV; it is therefore not as useful as the other regions.) Note, however, that the avalanche effect, common to regions III and IV, produces a pulse-type output, since the avalanche is not self-sustaining within these limits. (The sustained glow or arc discharge after region IV is again too unstable for radiation-detection purposes.)

Summary of Useful Gas-Tube Detection Regions

The effects in the various useful regions can be briefly summarized by recalling the previous classification, comprising one group, the *pulse-type detectors*, which include the proportional counter and the Geiger counter; the other classification group refers to the lower-voltage (and correspondingly less sensitive) detectors depending on *ionization current*, such as the ion-chamber type. Each of these gas-filled detectors will be discussed in sequence, followed by consideration of the scintillation detector, which appears in a liquid or solid-state form.

16-6. GEIGER–MUELLER TUBE-DETECTOR SYSTEMS

This is the type of detector most commonly seen and known under the name of "Geiger counter" or "GM-tube" detector. In its usual general-purpose form, this pulse-type detector indicates radiation in counts per minute (cpm). Although it is possible to operate a GM tube in the proportional region (III, Fig. 16-4), additional controls for selecting pulse heights would be needed. The usual Geiger counter operates in the level (or plateau) portion of region IV.

The Geiger-tube voltage is set at approximately 900 V so that the tube operates around the center of its "plateau range," and it is then exposed to the radioactive source. Beta particles will produce an electrical pulse for practically each particle entering the detector tube, when the movable shield in the probe is opened. Gamma rays (generally detected with relatively low efficiency in practically all survey-type meters) produce a pulse for approximately one out of every hundred gamma quanta entering the detector. (Alpha particles do not penetrate the wall of the detector tube.)

When ionization is initiated by a single ionizing event, the following sequence takes place: a gamma ray (or beta particle) enters the detector tube, it ionizes atoms of the gas enclosed in the GM tube, whose circuit is drawn in Fig. 16-5, to show the central anode wire and the enclosing thin-walled metallic cylinder. The gas mixture in the most usual "self-quenching" GM tube is made up of an inert gas, such as argon or helium, and a quenching gas of the organic or halogen type at a pressure of about $\frac{1}{4}$ atmosphere.

The primary ionization produces some free electrons and positively charged ions. The negative electrons are quickly accelerated toward the center wire, which

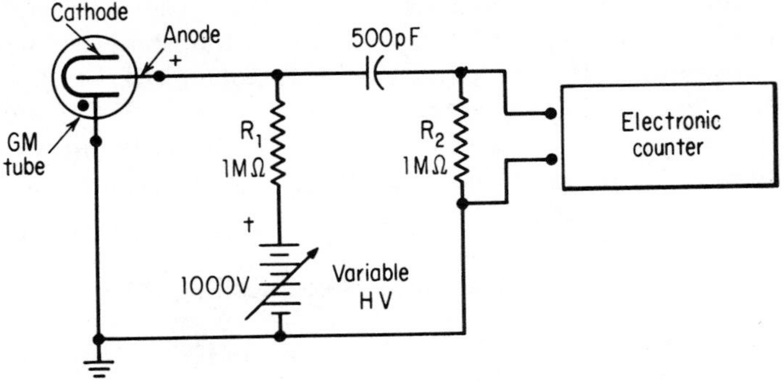

Figure 16-5. GM-tube circuit, feeding an electric counter.

bears a positive charge of around 900 V. As their speed increases, the liberated electrons progressively free more electrons from collision with other gas atoms. This greater number of free electrons in turn free still more electrons, producing a build-up of ionization, known as a Townsend "avalanche." A continuous discharge is prevented by the combined effect of a growth of a positive ion sheath and the action of the quenching gas, which absorbs excess released energy by the dissociation of its molecules. The tube is accordingly classified as a "self-quenching" type. The free electrons are collected very quickly by the positive wire anode (in a fraction of a microsecond), leaving the positive ions to migrate to the negatively charged cylindrical cathode. This ion current is also part of the discharge and contributes to the "dead time" of the tube, since additional particles will not trigger a pulse while the positive ion sheath remains. The dead time, which is variable, lies typically between 100 and 200 μsec, and thus limits the resolving power of the tube to around 250 μsec. This involves a correction for dead time when high count rates are being measured. The correction for counts below 10,000 cpm, however, is generally negligible.

An important consequence of the detecting action of the GM tube is the fact that the count is not influenced to any appreciable degree by the initial energy of the ionizing particle. With the GM tube operated in the region around 900 V, the requirements of the electronic counter are met rather easily (as discussed later), resulting in a fairly sensitive and generally useful type of detector.

16-7. ION-CHAMBER DETECTORS

The detection of radiation by the resulting current in an ionization chamber, as previously noted, provides an indication that is closer to the biological effect of radiation than the pulse type of detector. The ion chamber, therefore, even though it requires greater amplification and care in calibration, is extensively used, especially in health-physics applications in industrial areas subject to relatively high levels of radiation. One model, called a "cutie-pie" monitor, has long been popular for measuring high levels of radiation. It has also been successfully modified for survey work in the field, as illustrated by the model in Fig. 16-6.

Figure 16-6. "Cutie-pie," a popular type of portable ion-chamber survey instrument. This extended version of the previous *740A model* has direct calibration of four ranges of exposure rate (0–25/250/2500/25,000 mR/hr)—ZERO-SET and RANGE controls are on top. (*Victoreen model 740D.*)

"Cutie-Pie" Monitor

The principle of the ion-chamber detector is shown in the functional circuit of Fig. 16-7. As previously shown in the curves of Fig. 16-4, the very small ionization currents produced in the chamber by the radiation correspond to a source of very high resistance. The preamplification, therefore, must be done by electrometer tubes having very high input resistance and extremely small grid currents. As another consequence, the effects of noise and of leakage between the input terminals must be taken into account, as is true for all amplifiers of feeble currents so far discussed.

Ion-Chamber Circuit

On exposure to alpha, beta, or gamma radiation through its very thin window (0.0005-in. Mylar), the current I_e that flows through the ion chamber develops a corresponding positive voltage across the high-megohm resistor in the grid circuit

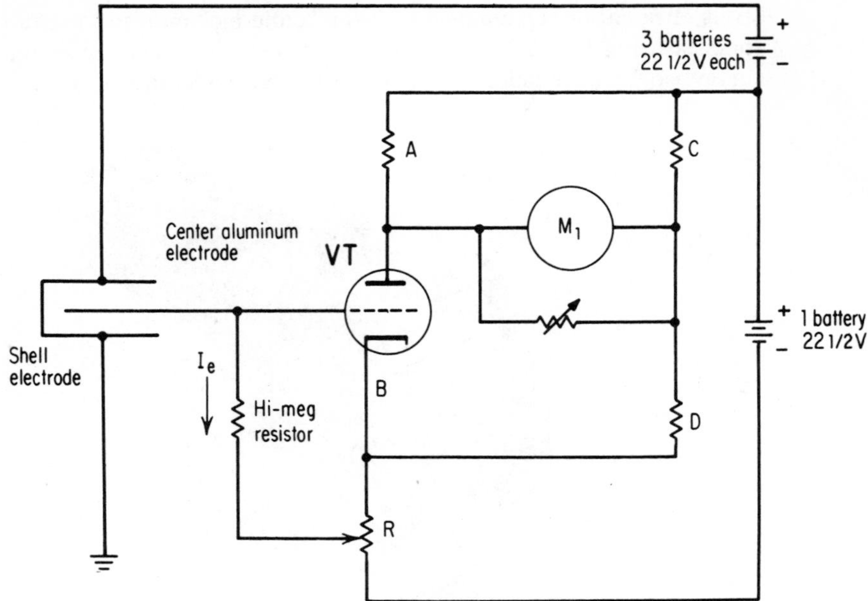

Figure 16-7. Functional circuit of the "cutie-pie" survey instrument. The electrometer tube (VT) connected to the ion chamber is made one arm of the resistor bridge. (*Victoreen model 740A*.)

of the electrometer tube. The magnitude of the positive signal required for full-scale deflection of the meter is approximately +0.5 V at the grid. The electrometer tube constitutes one arm (*B*) of bridge *ABCD*. With the bridge arms *A*, *B*, *C*, and *D* balanced under zero-set conditions, this positive grid voltage causes a decrease in the effective resistance of the electrometer tube (arm *B*), unbalancing the bridge. The metering circuit is properly adjusted to produce meter current proportional to the grid voltage, within each operating range (0–25/250/2500/25,000 mR/hr for the *740D model*).

The power supply for the ion chamber consists of four $22\frac{1}{2}$-V batteries to supply 90 V for the chamber (one of these batteries also supplies the electrometer plate circuit). A *Mallory* mercury cell (1.34 V) for the filament of the electrometer tube completes the supply.

Personnel Dosimeters

Devices to measure the *accumulated exposure* of personnel are worn in two forms: film badges and pocket dosimeters. *Film badges* using photographic film in appropriate holders are worn over periods such as a week and provide estimation of the absorbed dose by the relative blackening of the film after development. The laboratory developing the film keeps a record of each individual's exposure and reports its results periodically.

A *self-reading type pocket dosimeter* utilizes the principle of the *electroscope* in a form that can be worn clipped to the user's clothing, like a fountain pen. A well-insulated quartz-fiber electroscope (called a string electrometer) is mounted in the dosimeter, carefully positioned in an optical arrangement so that the position of the movable arm of the electroscope may be seen through the eyepiece against a reticule bearing an exposure scale (Fig. 16-8). The electrometer arrangement of course involves no electronic amplification but, by means of the carefully arranged optical system, is available in ranges of 200 mR, 1, 5, and 10 R and higher. The one illustrated has a 0–0.2 R range.

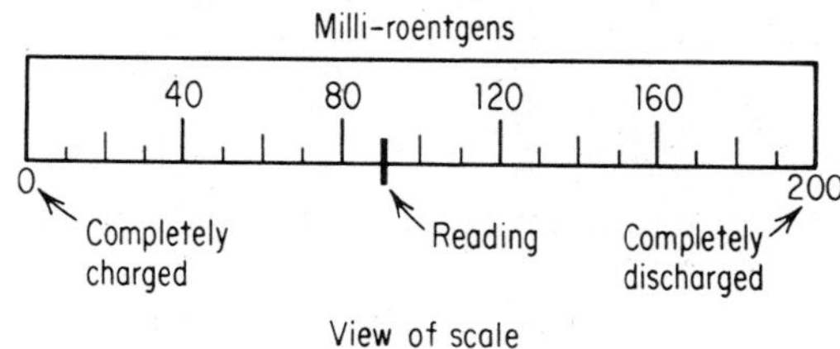

Figure 16-8. Personnel dosimeter scale. (*Victoreen models 541/A dosimeter* and *2000A charger*.)

A charging device is required for placing an initial charge on the electroscope, by the charger. In the absence of radiation this charge would be retained, with practically negligible leakage, throughout a working day. Upon any exposure to radiation, the resulting ionization causes a slow discharge of the electroscope, and the position of the quartz fiber at the end of the day can be read in terms of the total ionization encountered. After a reading for the day's exposure is taken, the electroscope is recharged. This is done generally at the start of another day's operation by pressing the end of the dosimeter down firmly on the transistorized charger (powered by a single flashlight cell) and adjusting the charger knob to set the dosimeter hairline to zero.

16-8. SCINTILLATION AND OTHER DETECTORS

Scintillations are flashes of visible light produced when ionizing radiation strikes certain liquids or solids. One of the oldest methods for showing the presence of x-rays and radioactivity, this detection method has been radically improved by the introduction of the photomultiplier tube and has become one of the *most sensitive methods* presently available for *counting individual radioactive disintegrations* and for other highly refined measurements of low-energy radiations.

Forms of Luminescence

The general name of *luminescence* is given to the phenomenon by which visible light is produced by a substance receiving photons whose wavelengths are too small to be visible. Certain rocks have long been known to *fluoresce* with a visible (sometimes eerie) glow, in the presence of ultraviolet waves (or so-called black light). The emission of the visible light may occur some time after it has been initiated by the transition of an orbital electron in the atom from one energy level to another (the transition time generally averaging about 10^{-8} s). In such cases of delayed production of the visible light, the phenomenon is known as *phosphorescence*. The terms phosphorescence and fluorescence are sometimes used interchangeably, but for the purpose of rapid counting by a scintillation detector only the fast-acting fluorescent materials are used.

Detection Process

The task of counting the individual flashes or scintillations initiated by ionizing events occurring in rapid succession is obviously not one for which the eye is suited, even when aided by a microscope. A photoelectric cathode, however, is well qualified for this work, producing secondary electrons (or photoelectrons) each time the sensitive surface receives a photon of sufficient energy. The invention of the *photomultiplier tube* made use of the sensitive photocathode in an arrangement capable of amplifying the effect of the resulting photoelectrons many hundreds of thousands of times; it brought about the intensive development of a great variety of forms of the scintillation detector, which is now the most versatile of the nuclear-radiation detectors.

The block diagram of a scintillation detector is shown in Fig. 16-9. The scintillator crystal, often a sodium-iodide crystal activated by thallium [NaI(Tl)], among other types, is placed in good optical contact with the photocathode of a photomultiplier tube. The nuclear particle, being detected as a result of the radiation, produces a scintillation (or flash of light) in the crystal. Most of this light is sensed by the photomultiplier tube, which produces a greatly amplified current pulse that is transmitted to the preamplifier as a voltage pulse. The pulse-shaping and counting circuits that follow are then the same as those used in a Geiger–Mueller pulse detector and produce similar indications, as previously discussed.

Photomultiplier-Tube Action

For the amplification of the relatively feeble flashes of light, highly sensitive *photomultiplier tubes* are used, as illustrated by the two commonly used types: the *RCA 6342* and the *DuMont 6292* tubes, in Fig. 16-10. The photocathode is generally either a semitransparent coating of an antimony–cesium (Sb–Cs) or silver–magnesium (Ag–Mg) combination, producing a high sensitivity to the main wavelengths of the emitted scintillation light. These wavelengths are in the visible light region

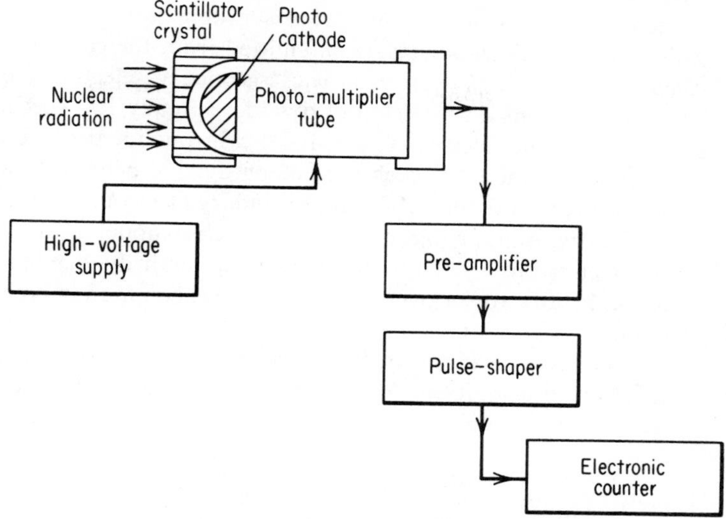

Figure 16-9. Block diagram of scintillation detector.

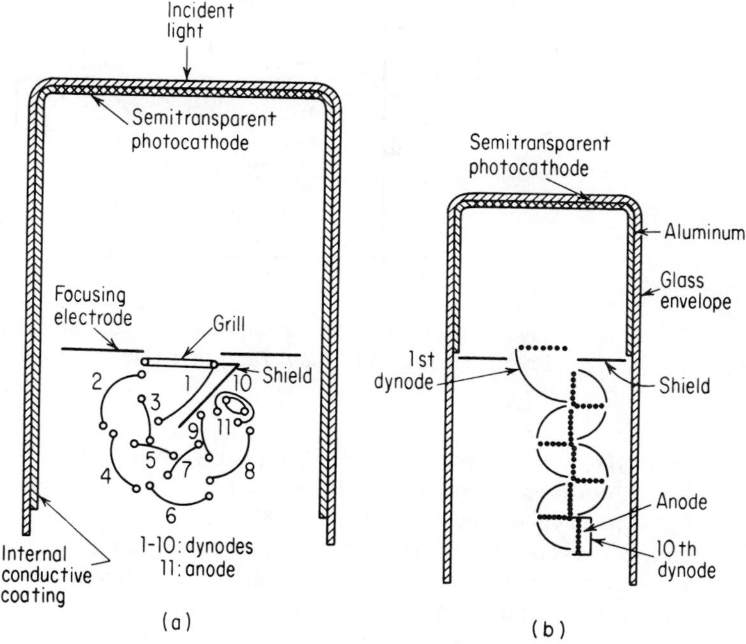

Figure 16-10. Schematic representation of photomultiplier tubes: (a) *RCA 6342*; (b) *DuMont 6292*.

(roughly from 0.4 to 0.6 micron or μm). The photomultiplier operates by cumulative scondary emission of electrons from a progression of specially shaped elements called *dynodes. Each photoelectron* emitted from the cathode is *accelerated by the electric field to knock more than one* "secondary" electron from the first dynode. The tube contains a succession of dynodes (usually 10), specially coated and each shaped to produce electric lines of force, which guide the secondary electrons from one dynode to the next, each succeeding dynode being held at a higher potential than the preceding one. Thus the secondary electrons continue to be accelerated progressively, multiplying in number at each dynode, until they are collected as a burst of electrons by the anode facing the tenth dynode. In this fashion, multiplication by a factor of over a million (10^6) is common in commercial photomultiplier tubes.

The circuit diagram in Fig. 16-11 shows the photomultiplier tube, followed by a 6AK5 preamplifier. A voltage divider across the high voltage is tapped at

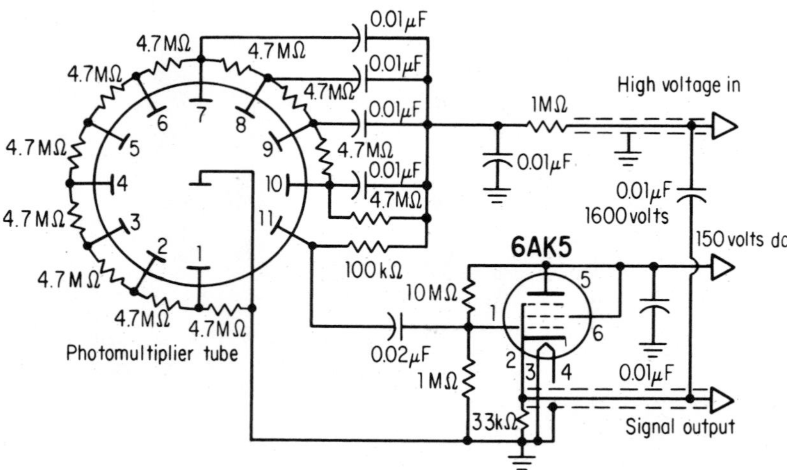

Figure 16-11. Circuit for scintillation counting. [From W. J. Price, *Nuclear Radiation Detection* (New York: McGraw-Hill, 1958).]

successive 4.7-MΩ points to provide progressively lower voltages for each of the lower-numbered dynodes. In the *DuMont 6292* tube, the difference in voltage between each of the 10 stages is 190 V per dynode stage, and the high-voltage supply is accordingly about 2000 V. The pulse output of the photomultiplier tube is capacity-coupled to the 6AK5 amplifier tube, operating as a cathode follower. The signal output from the amplifier cathode is a negative pulse, for connection to counters having an input sensitivity of around 0.25 V. Production of this value of output might require an electron multiplication of from 10^4 to 10^5 times. Greater multiplications for weaker signals may be obtained, requiring special care to ensure that the noise level (typically around 3 μV rms) remains sufficiently below the signal level.

Scintillation Counters

Scintillators are primarily used for gamma counting. Scintillators discriminate energy levels thus can be used to detect unknown gamma radiation. Figure 16-12 is the schematic of a scintillation crystal and a photomultiplier. As radiation enters the crystal, it causes luminescence. This light is reflected to the photomultiplier.

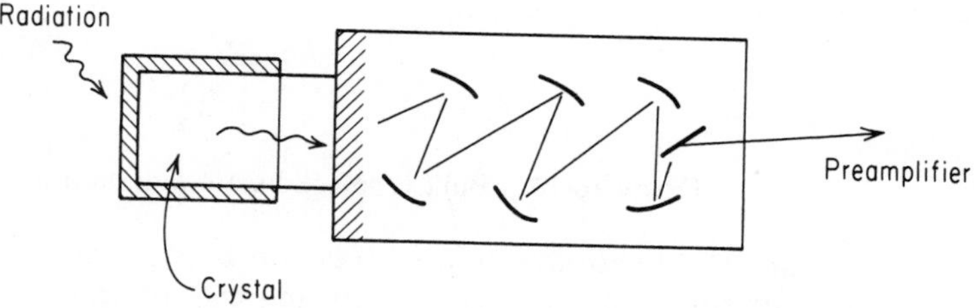

Figure 16-12. Scintillator schematic.

Good scintillators will emit light in less than 8–10 sec. These light flashes are emitted as a single pulse. It takes one time constant for the emersion of 63.2% of the photons. The time constant varies for the type of crystal used. Table 16-2 gives the time constant (τ) for several materials.

TABLE 16-2.

Use	*Material*	τ *(sec)*
Gamma ray	NaI (Tl)	0.25×10^{-16}
Beta neutron	Anthracene	0.27×10^{-7}
Alpha particles	ZnS	10^{-5}
Beta particles	Plastics	5×10^{-9}

Solid-State Detectors

There are two primary types of solid-state radiation detectors: bulk and barrier layer. Solid-state detectors offer some advantages over other methods of detection. They have high-energy resolution and offer a linear output with particle energy regardless of the nature of the particle. They are also stable, small, and fairly inexpensive. On the negative side, solid-state detectors are sensitive to high temperatures and have a limited life under strong radiation.

Figure 16-13 shows a bulk-type solid-state radiation detector. It uses a silicon crystal. Simply stated, the operating cycle is as follows. As radiation enters the detector, hole–electron pairs are created. These hole–electron pairs then produce an output pulse that is amplified and can be counted.

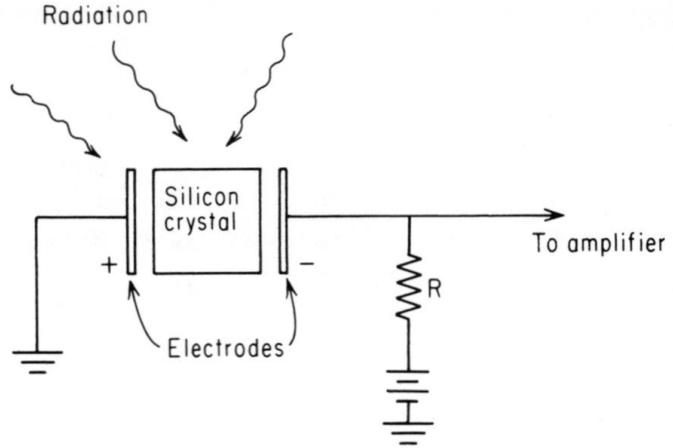

Figure 16-13. Bulk-type solid-state radiation detector.

Figure 16-14 shows a diffused junction barrier-type solid-state detector. As radiation strikes the device, the minority and majority carriers on either side of the junction receive energy (3.6 eV/pair for silicon and 3.0 eV/pair for germanium). This energy allows the holes and electrons to cross the junction. When they cross, they produce an output pulse that can be counted.

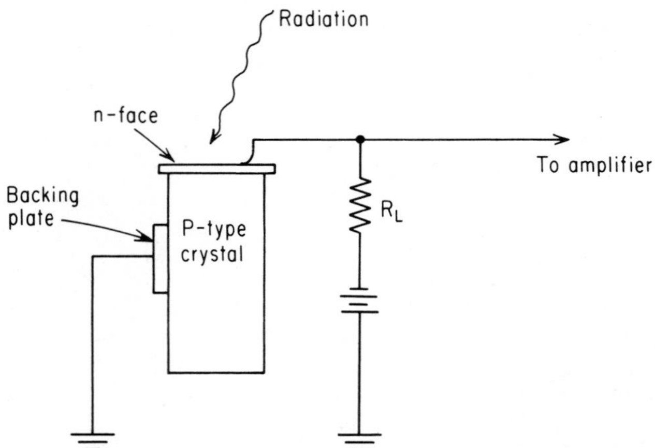

Figure 16-14. Diffused junction barrier-type solid-state radiation detector.

Thermoluminescent Dosimetry

A new development in the field of radiation measurement is the use of thermoluminescent materials. The technique is commonly known as thermoluminescent dosimetry (TLD). The action that takes place is as follows. Ionizing radiation may release an electron from the valence band to the conductance band in some insulating materials. Sometimes these electrons/holes do not recombine but become trapped between the bands. If the material is heated, the electron (or hole) may gain enough energy to recombine. When this happens, a thermoluminescent photon is emitted. The energy possessed by the electron will determine at what temperature the photon will be released.

Figure 16-15 and Table 16-3 show a curve of light emitted with relation to time heated and a table of TLD material characteristics.

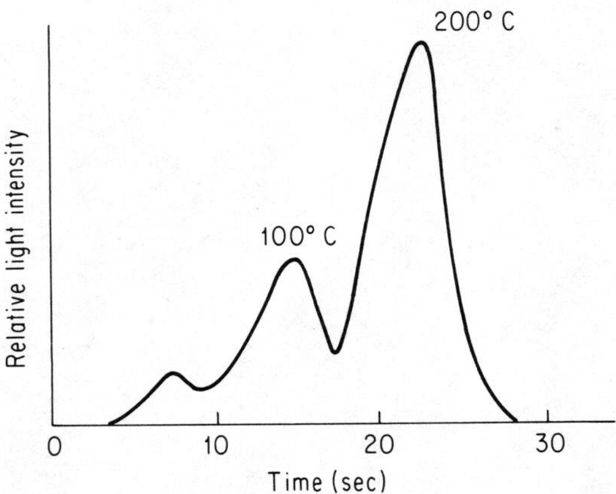

Figure 16-15. Typical thermoluminescent phosphor glow curve.

TABLE 16-3. Characteristics of TLD Materials

Characteristic	*LiF*	*CaF_2 : nat*	*CaF_2 : Mn*	*$Li_2B_4O_7$: Mn*	*BeO*
Temperature of useful glow peak	190°C	260°C	260°C	200°C	180°C 220°C
Relative response to ^{60}Co	1.0	23	3	0.3	
Useful range	1 mR–10^5 R	1 mR–10^4 R	1 mR–3 × 10^5 R	1 mR–10^6 R	1 mR–10^5 R
Sensitivity to light	None	High	None	None	None
Peak fading	$<$ 5%/wk	None	10% (1st month)	8% (1st month)	None
Physical forms	Powder, glass capillaries, embedded, extruded		Powder, glass capillaries, embedded chips	Powder, embedded	
Energy response	Essentially flat	Flat above 300 keV	Flat above 300 keV	Flat	

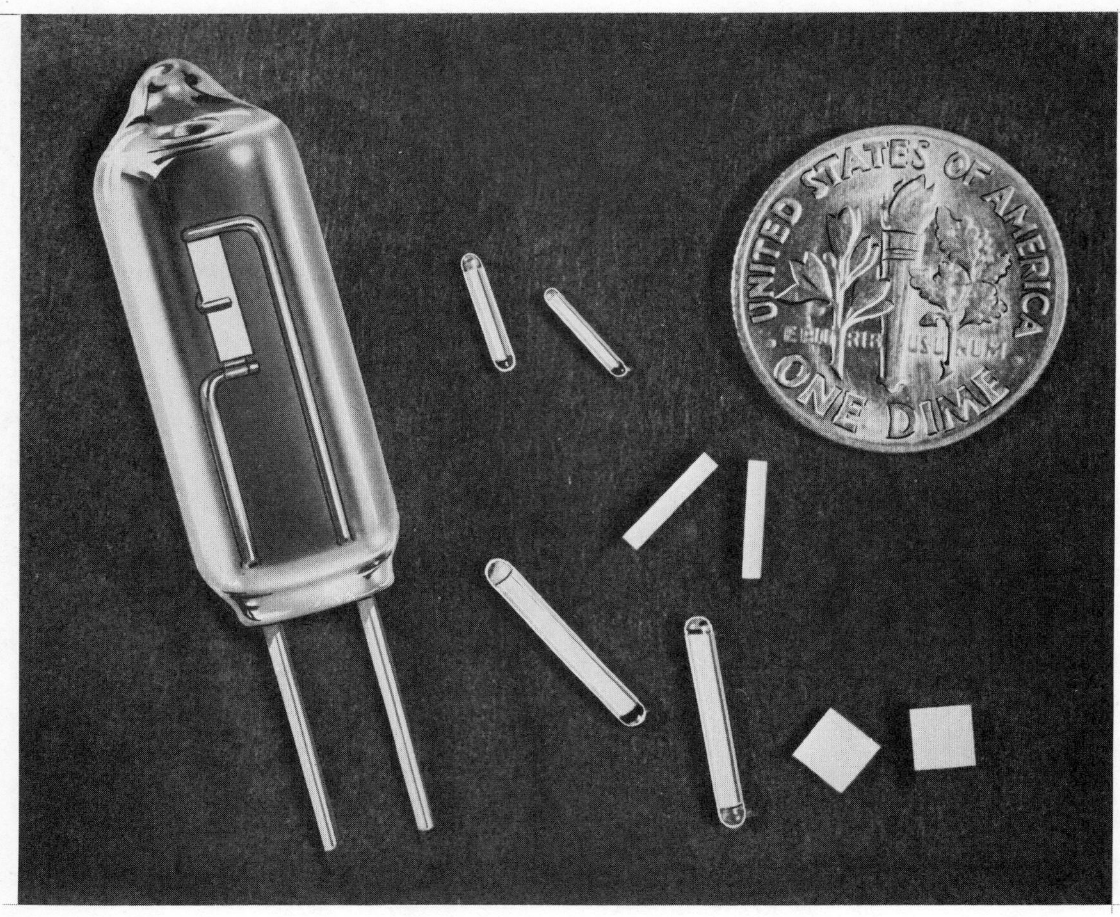

(a)

Figure 16-16. (a) Victoreen $C_aF_2 : M_n$ thermoluminescent dosimeter; (b) *Victoreen model 2810 ultra sensitive TLD reader*. (Courtesy Victoreen, Inc.)

Other Applications of Nuclear-Radiation Detection

Besides monitoring applications of survey meters, there is a broad area where the detection of nuclear radiation is employed in *nondestructive testing* (NDT). Measurement of thin-layer coatings and determination of flaws in metal castings are but two of many examples where x-ray (or gamma-ray), beta, and alpha rays (in addition to ultrasonic waves) are detected after absorption or reflection in the test unit. Another potent analysis method, utilizing characteristic radiation from chemical elements (*Neutron Activation Analysis, NAA*) is discussed in Chapter 19.

In this rapidly expanding field of radiation detection, important advances are being made in *solid-state detectors* of all types (including semiconductors alloyed with various rare earths) for highly selective and sensitive detection.

(b)

Figure 16-16. *Continued.*

Mention may also be made of other types of radiation detectors of specialized importance, even though they will not be discussed here in detail. These include detectors that trace the path of nuclear particles, in the form of *cloud chambers, bubble chambers,* and *nuclear-track plates;* also, in nuclear reactor work one encounters *Cerenkov counters* and various types of *neutron detectors.*

QUESTIONS

Q16-1. Distinguish between *ionizing radiation* and *electromagnetic radiation.*

Q16-2. Compare the penetrating power of the following three forms of nuclear radiation:

(a) Alpha (α) particles.
(b) Beta (β) particles.
(c) Gamma (γ) rays.

Q16-3. Compare the penetrating power of *x-rays* to that of each of the forms of nuclear radiation in Question Q16-2.

Q16-4. Distinguish between the radioactivity units of the *curie* and the *roentgen.*

Q16-5. Distinguish between the radioactivity units of *roentgens per hour* (R/hr) and *REMS.*

Q16-6. Explain how to correct for background count in determining the level of radioactivity at a given point.

Q16-7. A Geiger–Mueller (GM) tube is fitted with a beta shield. Explain how the activity due to *beta* (β) *particles alone* may be determined.

Q16-8. Discuss how the distance from a radioactive source will affect the count obtained on a GM tube, when source is primarily:
(a) A beta (β) emitter.
(b) A gamma (γ) emitter.
(c) An emitter of both β and γ rays.

Q16-9. Distinguish between the results obtained by two survey meters monitoring a single source, one calibrated in *counts per minute* (*cpm*) and the other calibrated in milliroentgens per hour (*mR/hr*).

Q16-10. Compare the outputs of two types of gas detectors as indicated by the curve for a gas discharge given in the text, one detector type operating in the *ionization region* of the curve and the other detector operating in the *Geiger–Mueller* (*GM*) *tube region.*

Q16-11. Give examples of *primary* and *secondary* ionizing events.

Q16-12. Explain the principle involved in the electronic action of a rate meter, comprising an *RC* circuit having a given time constant, connected to a dc meter.

Q16-13. (a) How does the detector action in a *scintillation counter* differ from the detector action in a *GM tube*?
(b) Compare the results obtained with each type of detector.

Q16-14. Explain the basis for the high order of amplification obtained in the photomultiplier tube of the scintillation detector.

Q16-15. State the difference between a scaler and:
(a) A GM survey meter.
(b) A proportional counter.

Q16-16. In what respect is the *ion-chamber* form of detector to be preferred to other detectors of nuclear radiation, such as the GM or scintillation detectors?

Q16-17. How can the error due to coincidence counting be estimated?

Q16-18. For measuring accumulated dose, compare the action of a *film-badge* type with that of a *self-reading pocket dosimeter*.

17

Digital Instruments

17-1. DIGITAL INSTRUMENTS

We are all familiar with the rapid progress in the development of digital equipment and have naturally been impressed with the great impact that digital computers have exerted in business, industrial, and scientific fields. In its penetration of the instrument field, equipment of the digital type has also become very important, by offering such highly desirable features as *convenience of reading and greatly improved accuracies.*

Historically, a great step toward digital display was taken by the replacement of electromechanical relays with the electronic switching of vacuum tubes; this step in turn, has been rapidly followed by equally great advances with the introduction of transistors and integrated circuits. These advances have given digital equipment a commanding role in a wide variety of instruments, such as digital counters, voltmeters, multimeters, and many other applications.

Basically, any digital device (including even the computer) may be considered to be an intricate arrangement of extremely fast-acting switches, able to change from one state to another in millionths of a second (i.e., *gating* in microseconds or even nanoseconds). If we then combine this switching speed with the ability to store either of the resulting (*binary*) *states in a flip-flop arrangement* (i.e., *in mem-*

ory), the combination achieves the highly useful result of displaying the output in a numerical (or digital) form.

Consideration of the whole field of digital equipment (including, as it does, the digital computer) is obviously beyond the scope of the primary emphasis on instruments in this book. It is, however, important for good understanding to have a fundamental grounding, as an introduction to the operation of the basic digital elements. In combination, these simple basic elements of gates and flip-flops produce a *digital readout*, as opposed to the more familiar *analog instruments*, which display the result in the form of meter readings or wave-form displays.

As one of the simplest approaches to the digital instruments, the *digital counter* offers a most direct example of the digital process.

The *frequency/period counter* may well be considered as one of the basic measuring tools in the laboratory, along with the voltmeter, oscilloscope, and signal generator. The photos of Fig. 17-1 illustrate the general-purpose type of

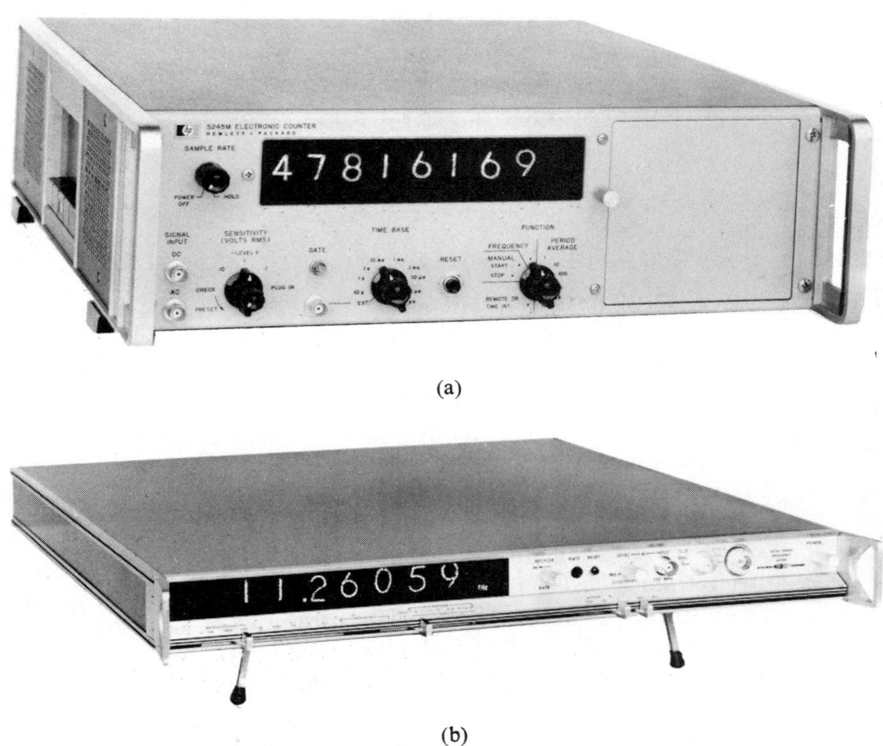

(a)

(b)

Figure 17-1. Solid-state digital counters: (a) an eight-digit frequency/period counter with plug-in arrangement to extend 50 MHz range to 18 GHz (*Hewlett-Packard model 5245M*); (b) "Thin-Line IC Series," IC to 12.4 GHz in panel height of $1\frac{3}{4}$ in. (*Systron-Donner model 6316A*).

frequency/period counter, which generally employs an in-line form of readout. The display also shows the unit of measurement (kHz, MHz for frequency or ms, μs for period) and thus provides a conveniently clear and unambiguous readout.

17-2. DIGITAL COUNTING: LOGIC ELEMENTS

Before discussing the details of digital-counting instruments, it is well to recall the elements on which the electronic-switching processes are based. The fast-switching ability of the counter and, even more to the point, the enormous capabilities of the digital computer (which obviously go far beyond the relatively simple functions of counting frequency or period) are all obtained by the massive use of *digital logic elements*. The many varieties that make up digital logic may be considered as derived from just *three basic units* (other than the flip-flop memory): the *NOR* gate, the *NAND* gate, and the *NOT* (or *INVERTER*) circuit.

In order to achieve the computing ability of a digital computer, it is clear that a very large number of logic elements must be combined, in highly intricate arrangements. In addition to the basic logic gates discussed above, mention should be made of other functions that are required in a digital computer: *comparators* (for so-called "decision making"), *clocks* (for synchronizing pulses), *adders*, *shift registers*, and *memory* elements, to mention just a few. The necessity for the many disparate elements and the large quantity of each element required in digital computers has been a great spur to the development of the *digital integrated circuits;* this development then expanded into the design of *linear-integrated circuits* to include many other electronic functions.

As a result of remarkable advances in the fabrication of the IC chips, digital and linear-integrated circuits have been widely adopted for the design of electronic counters (as has been true of many other digital instruments). The generous use of *medium-scale integration* (*MSI*), for example, accounts for such arrangements as the "thin-line series," illustrated in Fig. 17-1(b). In some cases, as will be seen in the multimeters discussed later, an even greater compactness has been achieved by the use of *large-scale integration* (*LSI*).

17-3. DIGITAL COUNTING VERSUS COMPUTING

In comparing the processes of counting with those of computing, it becomes fairly obvious that the capabilities of the digital computer go well beyond the relatively simple fast-switching ability of the counter. These superior capabilities of the digital computer must be achieved by a veritable maze of interconnected circuits that are required for more intricate functions, such as arithmetic, memory, and control, in addition to the gating function. The resulting product is the highly developed "electronic brain," with its untiring ability to take all kinds of commands and to execute them so much faster and more efficiently than the mere man at the controls.

The governing word here is "commands"; these orders must be fed into the computer by the human programmer in a long series of separate steps, each step being spelled out very deliberately and in a language so simple that no more than a simple "yes-or-no" response is required. This much is all that the computer can be required to do, and all that it is essentially capable of doing, by virtue of its counting ability or its ability to distinguish between a yes and no, which is equivalent in machine language to a 1 or 0. Let the programmer slip on just one step in the long sequence of necessary 1 and 0 recognitions, and the computer has not only reached the limit of its cleverness (or stupidity) but will most likely go on multiplying the error, many thousands of times each second! It is thus just as important to be aware of the limited "intelligence" of a digital computer (even when it is automatically programmed) as it is to appreciate its enormous ability to produce accurate computations at lightning speeds, when correctly programmed.

17-4. COUNTING DISPLAY: ANALOG VERSUS DIGITAL DISPLAY

Returning our attention now to the predominant high-speed-counting feature of the electronic counter, we distinguish between two methods of displaying the result of a count: in the *analog case*, as in the *rate meter* of the Geiger counter, a meter pointer displays the rate at which pulses arrive in terms of, say, 500 cpm; on the other hand, the read out of the *digital counter* would supply the count result as an *explicit number* for frequency, in this case, 30,000 Hz or 30 kHz. In either case, before being counted, the incoming signal must be shaped into pulses.

The wave-shaping process for producing sharp pulses from a sine-wave generator generally involves squaring and then differentiating the input signal to obtain peaked pulses of both polarities or, if desired, of a single polarity, through rectification. In many counting cases, the incoming signal is already in a pulsed form, as might be obtained from the phototransistor counting circuit Fig. 17-2, where output pulses are obtained for each interception of the light beam.

Starting then with the signal that has been shaped into sharp pulses, suitable as triggers, the analog method *employs a count-rate circuit*, repeated here for convenience in Fig. 17-3. Here, each positive pulse places a charge on C_2. In the interval between pulses, C_2 loses a portion of charge in accordance with the discharge time constant RC_2. Also, during this interval, C_1 (which is generally a small capacitor) loses practically all of its charge through the discharge path consisting of the other diode and the source resistance R_s. The circuit is then ready for the arrival of the second pulse in the train.

The result of the integrating action of this circuit is to produce an average value of voltage on capacitor C_2, which is read as a voltage proportional to the rate at which the pulses arrive.

The advantage of this type of display is that it presents a continuous value in the form of a voltage that represents the count rate, and can therefore easily

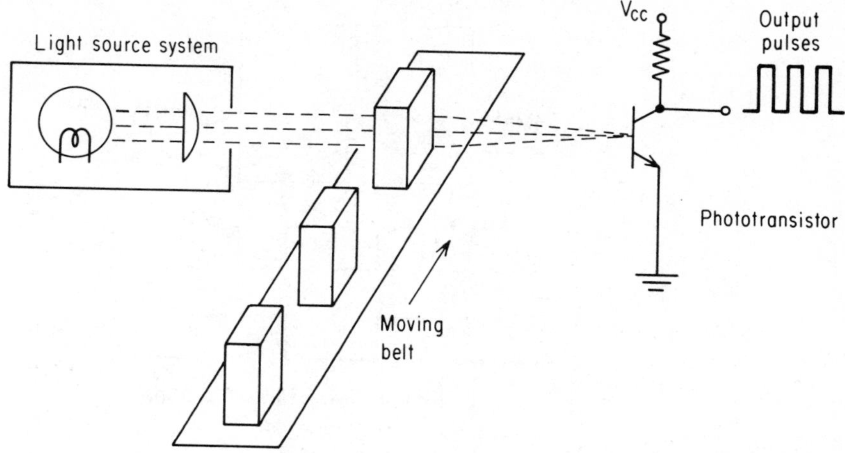

Figure 17-2. Counting arrangement in which a pulse is produced by a phototransistor.

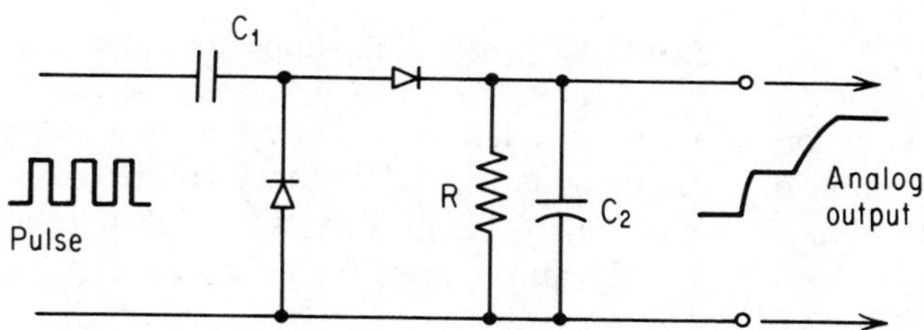

Figure 17-3. Count-rate meter presents the count in analog form.

be continuously recorded by any voltage recorder or retransmitted as desired, for remote indication. The count-rate meter analog output can be presented as a curve, or function, but cannot achieve the accuracy of a digital presentation, which can be made accurate to a single digit—with as many digits displayed as the accuracy requires.

17-5. BINARY COUNTER

Figure 17-4 shows an IC flip-flop. This unit will change states at its output terminal for each negative-going pulse that arrives at its input. The flip-flop or bistable multivibrator is also called a divide-by-two circuit. By placing several of these flip-flops in a series, we can achieve a binary counter. This is shown in Fig. 17-5.

Today, the IC manufacturers have made life easy for the designer. They

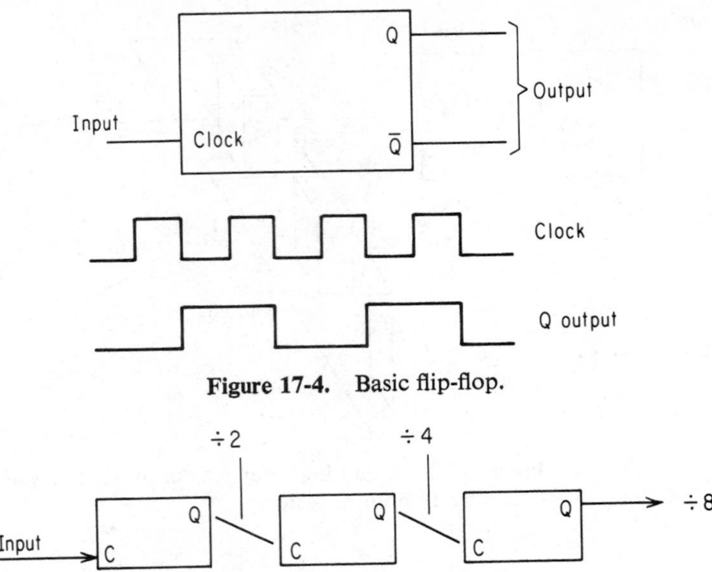

Figure 17-4. Basic flip-flop.

Figure 17-5. Series of flip-flops to produce a binary counter.

have incorporated a complete binary counter on a single chip. Probably the most common of these chips is the 7490 binary counter. Figure 17-6 shows this chip. The chip produces a binary number at its four output terminals which is the binary equivalent of the decimal count of the pulses at its input terminal.

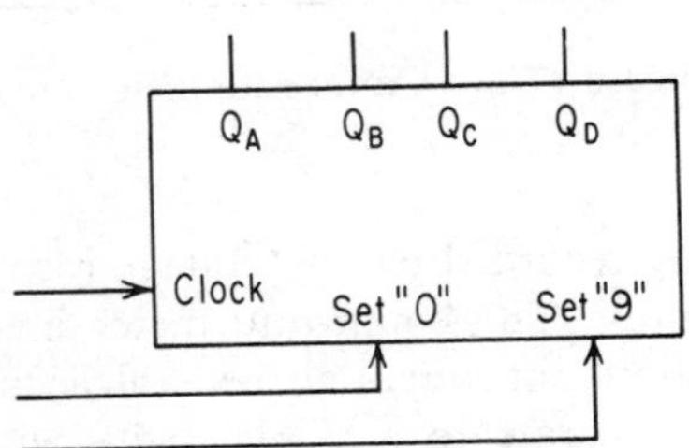

Figure 17-6. Block diagram for typical binary counter such as the 7490.

The input terminal is called the clock input. The set "0" and set "9" lines are used to set the output to a reading of zero (0000) or nine (1001) at the Q_A, Q_B, Q_C, and Q_D output lines. For more information on the use of this device, the reader is referred to any digital IC handbook.

The flip-flop is called a bistable multivibrator. This is because it has two stable states. As noted, the flip-flop will change states during a pulse transition at its input. It will take another, similar transition to change the output again. These are two other types of multivibrators: the astable and the monostable. Basically, the astable produces a series of output pulses as long as the circuit is energized. The frequency or pulse width of these pulses is usually determined by external *R* and *C* components. The monostable multivibrator is similar to the bistable in that it requires an input pulse to change its output state. The difference is that the output after going through this change will return to its original state after a predetermined time interval. This time is also determined by external *R* and *C* components. Figure 17-7 shows the three types of multivibrators.

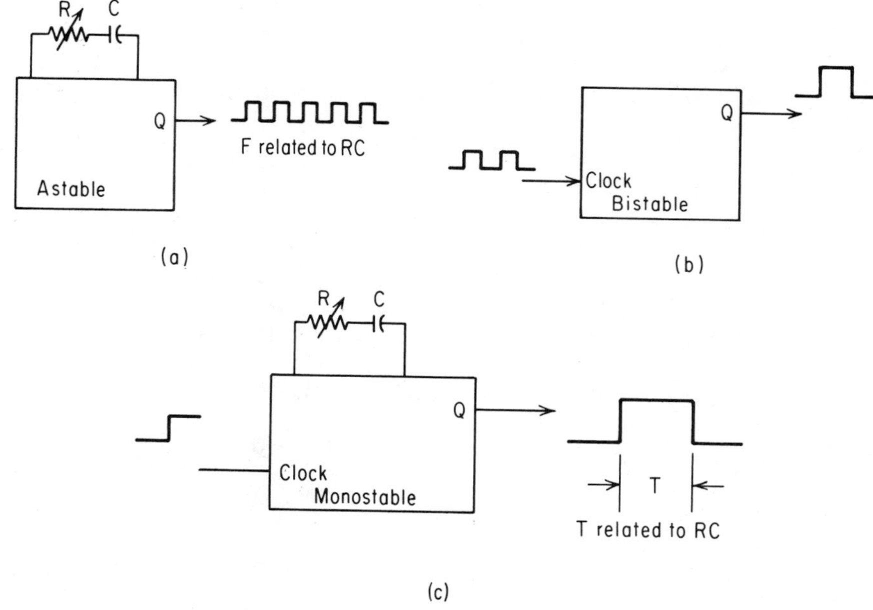

Figure 17-7. The three basic types of multivibrators: (a) the astable multivibrator; (b) the bistable multivibrator; (c) the monostable multivibrator.

17-6. BINARY AND DIGITAL COMPUTATION

Although the study of the important subject of digital computers, as such, is not part of our subject, it is well to point out that not only *counting*, but also *arithmetic computing*, can be done in the binary system, which is the basis of practically all digital computation. The processes of *addition* and *multiplication* (and their inverses) can be performed in the binary system as well as in the decimal system and, as a matter of fact, in a much simpler manner. Of course, in the new binary system, the rules for addition and multiplication may sound strange, but they are actually much simpler than the familiar rules of the decimal system, since only ones and zeros are involved. In the *addition process*, for example, only three rules need be learned:

Rules for Binary Addition:

$$0 + 0 = 0$$

$$\left.\begin{array}{c} 0 + 1 \\ \text{or} \\ 1 + 0 \end{array}\right\} = 1$$

$1 + 1 = 0$ and carry one, producing a binary indication of 10, read as "one-zero" (not "ten")

Example. Thus in the decimal example of $4 + 13$, we have the following binary addition:

$$\begin{array}{rlr} & \textit{Binary} & \textit{Decimal} \\ & 0100 = (2^2) & = \quad 4 \\ + & \underline{1101} = (2^3 + 2^2 + 0 + 2^0) & = +\underline{13} \\ = & 10001 = (2^4 + 0 + 0 + 0 + 2^0) = & 17 \end{array}$$

For *multiplication* also, there are only three simple rules:

Rules for Binary Multiplication:

$$0 \times 0 = 0$$

$$\left.\begin{array}{c} 0 \times 1 \\ \text{or} \\ 1 \times 0 \end{array}\right\} = 0$$

$$1 \times 1 = 1$$

Example. Using the decimal example 4×13, we have the following binary multiplication:

$$\begin{array}{rcr} \textit{Binary} & & \textit{Decimal} \\ 0100 & = & 4 \\ \times\ \underline{1101} & = & \times\underline{13} \\ 0100 & & 12 \\ 0000 & & \underline{4} \\ 0100 & & 52 \\ \underline{0100} & & \\ 110100 & = (2^5 + 2^4 + 0 + 2^2 + 0 + 0) & \\ & = (32 + 16 + 0 + 4 + 0 + 0) & = 52 \end{array}$$

It can be seen from these examples that computing differs from counting in both systems, in that the *addition process involves a carry*, and *multiplication involves both a shift* and *a carry*. These functions are performed in a digital computer by appropriate *logic* circuits.

Notwithstanding the fact that the simple examples given for the binary system seem to be a long way around compared with the shorter methods of the decimal systems, the overriding fact is that the binary computer, limited only to zeros and ones, can perform these steps at comparatively high speed, easily handling hundreds of thousands or even millions of binary digits (or bits) each second.

Binary Coding: Binary-Coded Decimal (BCD)

To overcome the awkward situation where a number with three decimal digits, such as 892, would require a very large number of bits to be expressed as a binary number, a variety of coding systems are employed. As an example of a simple *binary-coded decimal* (*BCD*) system, each decimal digit of 892 would be expressed by a group of four bits, and the position of each group would be interpreted in units, tens, etc., as in the decimal system. In this fashion decimal 892 is fed into the computer as 1000 1001 0010, involving no more than 12 bits for any number up to 999. The development of many other codes for such things as storage location positions (or addresses), commands for computer operation, and others, makes up the highly important technique of *computer programming*. The logic functions involved in the required switching actions are handled with great power by the applications of *Boolean algebra*.

17-7. DIGITAL FREQUENCY AND PERIOD COUNTER

The use of logic elements to provide a digital indication of frequency (or period) has greatly advanced the convenience and precision with which these measurements can be made. A wide variety of models are available, mostly solid-state and, in many cases, incorporating integrated circuits for compactness and improved maintainability. The frequency coverage ranges into the gigahertz region, and the accuracy specification, which is primarily dependent on the reference time base, can go well beyond 1 ppm ($\pm 0.0001\%$).

Commercial Examples

The degree of refinement (and consequently the cost) generally follows the number of digits in the readout. Thus, the lower-cost four-digit counters, covering a frequency range in the kilohertz region, generally depend upon the 60-Hz power-line time base (usually good to 0.1%), with gate times of 0.1 and 1 sec, and are suitable for many industrial measurements.

The generalized functional arrangement for frequency measurement is illustrated in the block diagram of Fig. 17-8. Each cycle of the input signal causes the input trigger channel to generate a sharp pulse. The Schmitt trigger arrangement

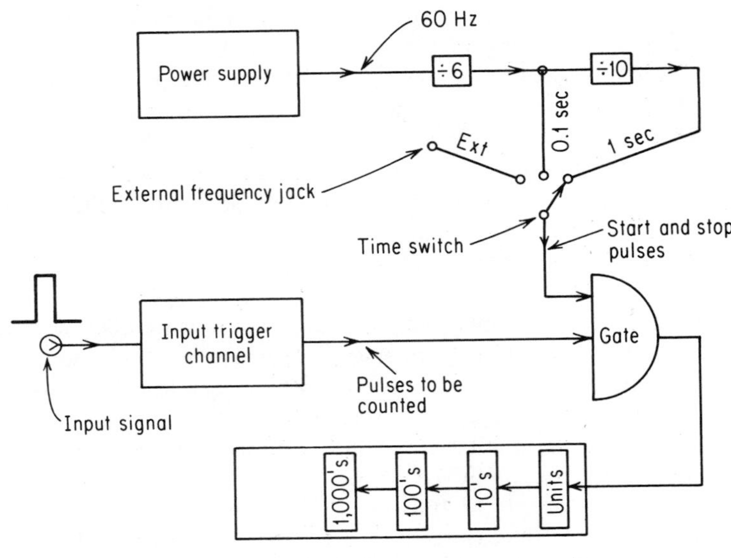

Figure 17-8. Block diagram of four-digit counter.

will trigger on the positive-going or negative-going portion of the signal, as desired, at a sensitivity at the millivolt level. These pulses are applied to the gate and are counted during the precise period that the gate is opened by the 60-Hz line supply. The time switch is shown positioned for a 1-sec gate, but may be switched to a 0.1-sec gate. (Line-frequency stability is generally expected to be 1 part in 10^3; greater stability than this can be obtained from the connection for a crystal-controlled oscillator.)

The display may be set to remain visible, before changing, for variable lengths of time by manual adjustment.

17-8. DIGITAL MULTIMETERS

The advantages of digital readout that won ready acceptance for digital counters also have been readily adopted for digital voltmeters and multimeters. In addition to the obvious benefit of the clear and definite *numerical display*, the digital multimeters offer a variety of flexible versions for *improved accuracy* of the readings.

For example, in comparison with a conventional pointer type of panel meter for dc volts with a $\pm 1\%$ tolerance, a digital panel meter provides a basic accuracy of $\pm 0.1\% \pm$ one digit, thus improving the tolerance figure by at *least* 10 times, without incurring any possible additional inaccuracies due to human error in making the reading.

The *digital voltmeter* (*DVM*), and the *digital multimeter* (*DMM*) extend this improvement in accuracy by the increased number of digits in the readout. Many advanced models of digital multimeters are available with four or five digits, providing an accuracy of $\pm 0.01\%$ or better. These advanced multimeters also offer a *combination of functions* beyond the customary DC VOLTS and OHMS of the VTVM; many provide additional functions of AC VOLTS, RATIO, and DC CURRENT. Moreover, an important feature in some models is the provision for a *binary-coded output* to make them compatible for interfacing with a digital computer.

We can thus summarize the salient advantages of digital multimeters as offering highly desirable features in clearer readings and improved accuracy; and in cases where the increased cost warrants, they also offer a very flexible combination of functions and a binary-coded output that acts as an analog-to-digital converter for input to a computer.

The digital volt–ohmmeter of Fig. 17-9 (*Fairchild model 7050*) is an example of the so-called "$3\frac{1}{2}$-digit" meters, where the fourth digit lights up only as 1; this provides a 50% over-range, extending the regular 999 reading on any given range to 1500.

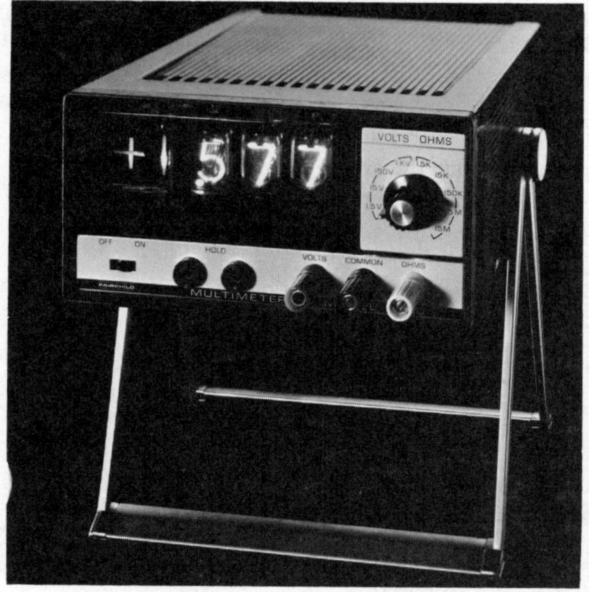

Figure 17-9. Digital volt-ohmmeter of the $3\frac{1}{2}$-digit type; the fourth digit on the left lights up only as a 1 when the value of 999 is exceeded, thus providing a 50% over-range for readings up to 1500. (*Fairchild model 7050*, manufactured by Systron-Donner.)

The ranges of the $3\frac{1}{2}$-digit volt–ohmmeter, which uses integrated circuits to achieve its compactness, are as follows:

Direct-Current Volts: four voltage ranges, manually selected with automatic polarity indication: ±1.500 V, ±15.00 V, ±150.0 V, and ±1 kV full-scale.

Accuracy: ±0.1% of reading, ±1 digit.

Resolution: 1 mV on 1.500-V range.

Response Time: Less than 5 sec.

Input Impedance: 10 MΩ, and greater than 10 MΩ on 1500-V range.

Ohms: five ranges, manually selected, from 1500 kΩ to 15.00 MΩ, full-scale.

Direct Current (option): five precision shunt resistances available as plug-ins on the 1.5-V range, to provide five full-scale current ranges, from 150.0 μA to 1.500 A, with resolution of 0.1 μA, ±0.2% of reading, ±1 digit.

17-9. OPERATING PRINCIPLES OF DIGITAL VOLTMETERS

The fundamental process involved in digital measuring instruments involves an analog-to-digital conversion of the incoming signal variation. This conversion can best be shown by the process of changing a dc voltage input to a digital readout of dc volts, as in a digital voltmeter (DVM). With the digital DC VOLTS function as a basis, modified versions of this basic unit can then be fashioned into other functions, such as DC CURRENT, OHMS, AC VOLTS, and RATIO for incorporation into the single multimeter instrument.

Of the many methods of accomplishing the A–D (analog-to-digital) conversion in the DVM, three main principles may be distinguished, as follows:

1. *Comparison principle*, using a successive approximation method
2. *Counting principle*, using a voltage-to-frequency method
3. *Integrating principle*, using a ramp to accomplish a voltage-to-time method

A fourth method, the so-called "*dual-slope*" *method* is a popular modification of the third method (the single ramp approach), which retains the noise-cancelling advantage of the integrating principle, while lessening the possible errors that might be incurred in the voltage-to-time measurement.

Comparison Principle

The *successive-approximation* method employs the potentiometric comparison technique and is the fastest and one of the most stable of the four A–D conversions mentioned. A block diagram of one of the early models is shown in Fig. 17-10. The voltage-comparison block produces an error voltage that is applied to a program-switching arrangement. This may take various forms; an oil-filled step-

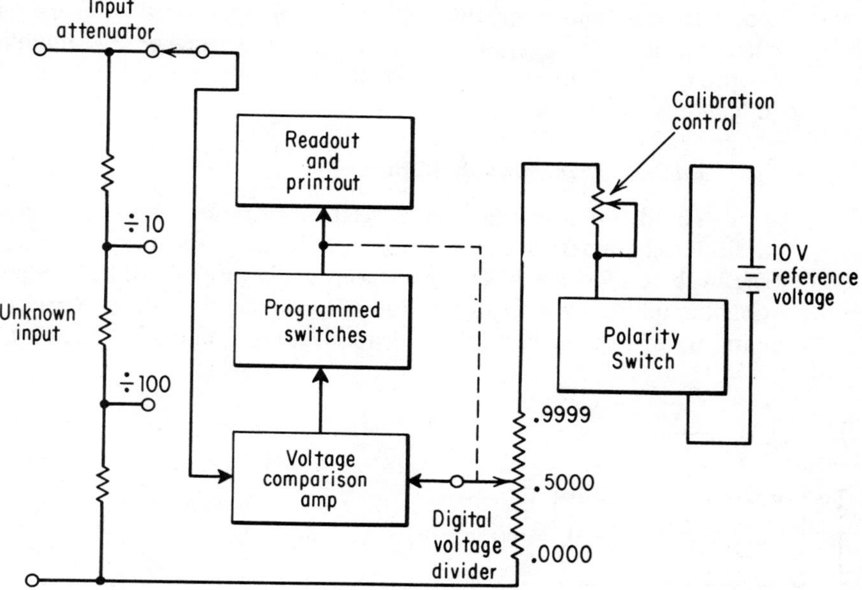

Figure 17-10. Block diagram of DVM, using a "successive approximation" method; it employs the potentiometric comparison principle of providing a digital readout of DC VOLTS.

ping-switch arrangement is the one used in the DVM of Fig. 17-10, while other models use reed relays or all-electronic switching methods.

The switching arrangement presents comparison voltages that become progressively closer to the input voltage, until a zero error exists. Through the common coupling between the switching arrangement and the comparison voltage, a digital readout is obtained, which can also be employed for print-out or other forms of data processing.

Counting Principle: Voltage-to-Frequency Conversion

In cases where a counting function is already provided in the instrument, a *voltage-controlled oscillator* (*VCO*) can provide voltage readings by a voltage-to-frequency conversion. Although this method has the disadvantage that the stability and accuracy of this method is heavily dependent on the VCO, it offers an effective method for averaging out 60-Hz line noise, without the use of noise filters that would reduce reading speed.

Integrating Principle: Voltage-to-Time Conversion

When the input voltage is integrated to produce a single ramp, the value of the unknown voltage is related to the *RC* time constant and a known time. In this method, the integrating process has the great advantage of minimizing noise. For good accuracy, however, the single ramp requires excellent characteristics in both the linearity of the ramp and in the time measurement. A modification of this method, offered in the dual-slope technique, serves to overcome both of these difficulties.

Dual-Slope Integration Method

The dual-slope technique is relatively simple, without sacrificing the highly desirable characteristics of stability and accuracy. The block diagram of Fig. 17-11 shows the dual-slope A–D converter as consisting of five basic blocks—an operational amplifier used as an integrator in conjuction with its R and C, a level comparator, a basic clock, a set of decimal counters, and a block of logic circuitry.

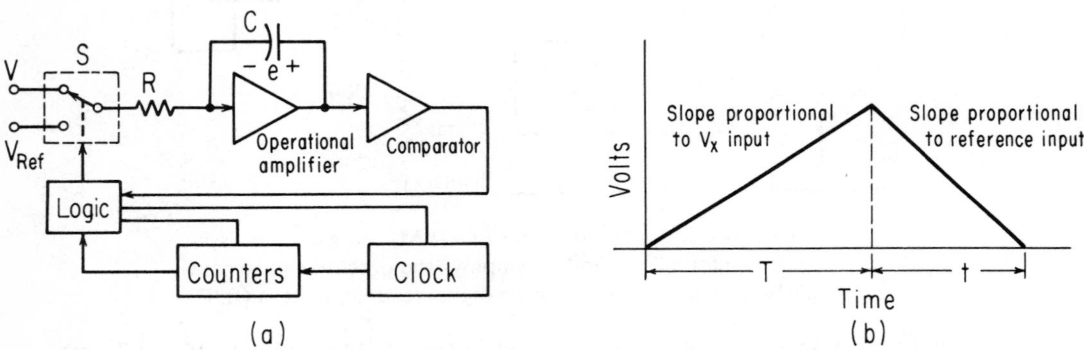

Figure 17-11. Operation of dual-slope DVM: (a) block arrangement where the unknown voltage V_x is compared to the known reference voltage V_{ref} for a variable time; (b) V_x is integrated over known time T and then over the proportional time t.

The conversion is made by connecting the integrator input to the unknown voltage through a switch for a *known* period of time T. The known period of time is usually determined by counting the clock frequency in the decimal counters. As shown in Fig. 17-11(b), a charge is stored in the integrating capacitor that is proportional to the input signal V_x during the known time interval T.

At the end of time interval T, the input is switched from the unknown signal input to a reference voltage of opposite polarity. This causes a current to flow in the integrator in such a manner that the charge begins to decrease with time and results in a downward linear ramp voltage. During this second period of time a *known* voltage is then observed for an unknown time. The *unknown* time t is determined by again counting the clock pulses until the voltage across the capacitor reaches its basic reference level. (In the example, the reference is ground, although any basic reference level may be used.) Mathematically the process may be written as a simple proportion:

$$\frac{V_x}{T(\text{fixed})} = \frac{V_{\text{ref}}}{t}; \qquad \text{thus} \quad t = \frac{V_{\text{ref}}}{V_x}T$$

where T = fixed integration time
V_x = unknown voltage
V_{ref} = reference voltage
t = variable discharge time

17-10. REPRESENTATIVE DIGITAL VOLTMETERS

Readout Forms of Digital Display

There is a wide choice among the many methods that may be used for displaying the numeral or alphanumeric characters constituting the readout of a digital instrument. This is so, even if we exclude the much more complex methods, such as the *character-producing CRTs* (that are employed to display a substantially large amount of information on the scope screen), or the highly flexible *graphic displays* (that employ a plotter interface to generate complicated graphic forms). If we confine ourselves only to the area of displaying a reasonably modest number of either digits or alphanumeric characters, we still find a variety that includes the following methods and their variations:

1. The old "*Nixie*" *tube type* that employs a stack of neon-type gas-discharge tubes, each of which is energized by the logic switching of relatively high voltages (100–200 V);

2. The *incandescent type*, that forms the desired character by the light which is successively directed through various transparent masks to form each character;

3. The *light-emitting-diode* (*LED*) *type* of semiconductor (usually gallium arsenide or phosphide) that produces either red or green visible light obtained from a relatively small amount of forward current (usually 30–50 mA) at a low voltage. This method has in its favor the indefinitely long-life characteristic, inherent in diode semiconductors, as a very strong maintenance advantage. It is often used in two forms; one form is that of a *seven-segment form* that produces the number or letter by the selective lighting of various combinations of the seven semiconductor bars, or, in the other case, in the form of a *5 × 7 matrix of dots* that form the character from a logic selection of the 35 semiconducter dots making up the matrix. These LEDs also find use in the digital processing of punched cards; for such use, they are in the form of photocoupled pairs, where the emission of the pinpoints of visible light (or infrared) from the LED is sensed by a photodetector (or PIN diode) that is very accurately aligned with the emitted light.

QUESTIONS

Q17-1. Discuss the reasons that make binary arithmetic especially suitable for counting and for digital computation.

Q17-2. *Distinguish* between the following types of pulse circuits, and give *one example* of the use of each type of circuit:
(a) Bistable.
(b) Monostable.
(c) Astable.

Q17-3. Explain why a pulse suitable for a flip-flop trigger must not be too short or too long in duration.

Q17-4. State how a square-wave signal can be shaped to be more suitable for a trigger pulse.

Q17-5. For the purpose of indicating the condition of a bistable unit, compare a neon-tube with an incandescent-bulb indicator or an LED as to:
(a) Voltage and current required for operation.
(b) Suitability in tube or transistor circuits.

Q17-6. For decimal readout in a counter, compare the *operating principles* and *advantages* of each of the following digital-display indicators:
(a) Column-of-10 neon numerals.
(b) "Nixie" indicator.
(c) In-line, in-plane readout indicator.
(d) LED indicator.

Q17-7. State three advantages possessed by digital instruments over the corresponding deflection meters.

Q17-8. In a digital voltmeter, explain the principle of operation of each of the following types:
(a) Stepping-switch type.
(b) All-electronic type.

Q17-9. Explain one method for obtaining analog-to-digital conversion.

Q17-10. State the main factors favoring the use *semiconductor* diodes, transistors, and tunnel diodes over the corresponding *vacuum-tube* types in a digital computer.

PROBLEMS

P17-1. Express the following binary numbers in their *decimal* forms:
(a) 0011 (b) 0101 (c) 0111 (d) 1111

P17-2. Express the following decimal numbers in their *binary* forms:
(a) 2 (b) 15 (c) 16 (d) 32

P17-3. A square-wave signal is to be shaped to make it more suitable as a trigger for a bistable multivibrator (or flip-flop) tube circuit, used for binary counting purposes. *State* which of the circuits, (a), (b), or (c), shown in the Fig. P17-3, is better for this purpose and *explain* reasons for your choice, based on sketches of the resulting output e_{out} in each case.

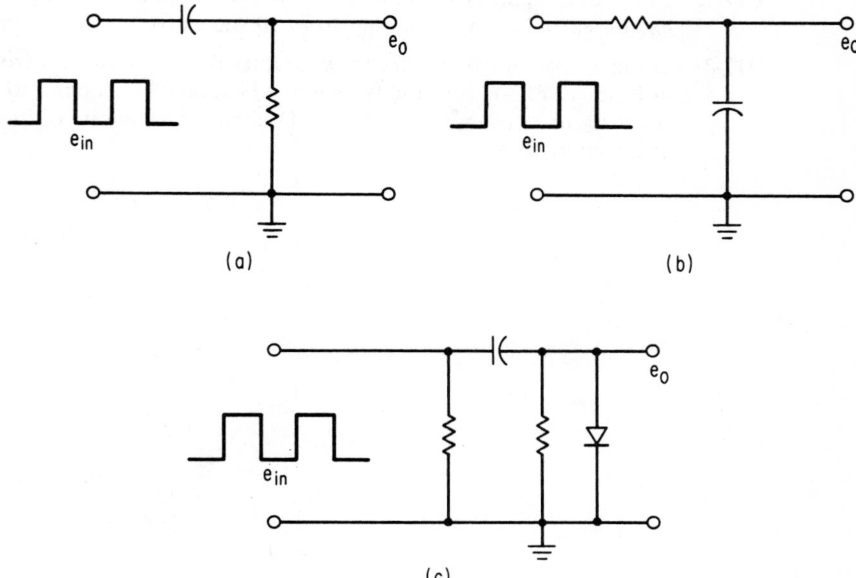

Figure P17-3

P17-4. The circuit shown in Fig. P17-4 is to be used to supply a manual trigger for a tube flip-flop, each time the pushbutton switch is pressed. Draw the output wave-form resulting from successive actions in pressing and releasing the switch.

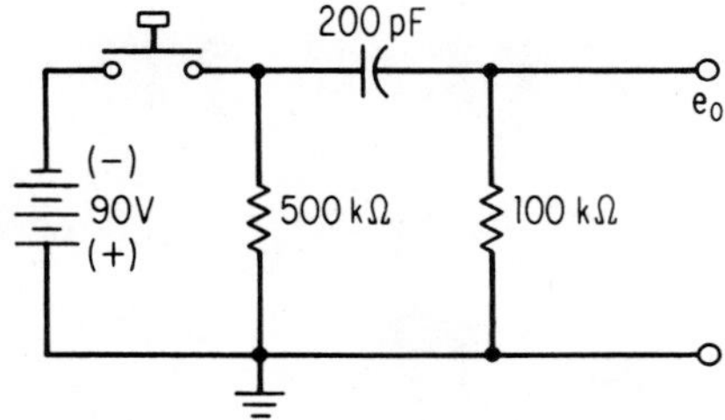

Figure P17-4

P17-5. When the manual trigger circuit of Fig. P17-4 is used with the values shown:

(a) Estimate the approximate *duration of the negative output* pulse at 50% amplitude.

(b) State the maximum *amplitude of the positive output* pulse that results when the pushbutton is released.

P17-6. *Draw* a scale-of-16 counter, showing the four flip-flops in symbolic form, and *indicate* the feedback arrangement to modify it into a decimal counting unit (DCU).

P17-7. In table form, show the binary display (indicated by a lit neon lamp, signifying *one* on each flip-flop) resulting from 10 successive counts in Problem P17-6.

P17-8. Using a symbolic diagram, show another method (different from the one used in Problem P17-7) for modifying a binary scale-of-16 counter to act as a decimal counting unit. (*Suggestion:* An AND circuit, having one of its inputs triggered at the count of 8, may be used.)

18

Analog Computers

18-1. ANALOG VERSUS DIGITAL COMPUTERS

While computers, strictly speaking, are more properly regarded as systems rather than electronic instruments, they do contain basic elements that deserve introductory attention as important parts of the instrumentation process.

The role of the *analog computer*, in particular, plays a vital role in the simulation of an actual dynamic situation and provides a significant readout of the results of the various simulated inputs; this readout is generally in the form of an *x–y* graph. With the advent of compact and relatively inexpensive integrated-circuit operational amplifiers, the great flexibility of even a desktop analog computer represents a powerful instrumental tool.

In discussing computers, we generally distinguish between two main types: analog and digital. As mentioned in the previous chapter, the *digital computer handles binary digits* (or bits) representing numbers. As an added feature, alphabetic characters or other information may be coded into numerical form, and the digital computer then handles so-called "alphanumeric" characters. On the other hand, the *analog computer represents variables in a given situation* (*or their analogous functions*) *by corresponding electrical or electromechanical quantities* (called "machine" variables to distinguish them from the actual problem variables). Such machine variables might be standard or derived dc voltages or shaft positions producing ac voltages at various phase angles. In effect, an analog computer simulates various system components (such as masses, springs, motors, and the like) and the forces acting in the actual system by their corresponding (or analogous) electrical components (such as resistors, inductors, and capacitors) and the voltages to drive them. Thus the investigator works with easily managed blocks of electrical components interconnected to produce electrical results corresponding to the results that would be obtained in some massive or otherwise complex actual dynamical system. The analog computer accordingly possesses great usefulness "*as a direct aid to the designer's thinking process, rather than just as a numerical calculator.*"[1]

A simple example—oversimplified to a degree for purposes of clarification—can be given to show the fundamental difference between the analog and digital computers. The problem presented to each computer is to add the quantities 6, 1, and 5. In the digital method discussed in Chapter 17, one rudimentary way of doing this would be to count successively the pulses in binary form (reading from the right), to yield a binary output of 1100 (or $2^3 + 2^2$) = 12, as shown in Fig. 18-1(a), where the corresponding indicators are lit in the 1–2–4–8 code.

For the analog computer, a rudimentary method, employing only passive components, is shown in Fig. 18-1(b). Here the voltage inputs, analogous to the numbers, are obtained from the individual potentiometers connected across the 10-V supply and are applied to a *summing circuit*, to produce a voltage output on the voltmeter V. As will be shown later in Fig. 18-4, where this summing circuit is analyzed, the output of the network is $(6 + 1 + 5)/4$ or 3 V. By arranging a special scale for the voltmeter V, on which the indicated readings would be four times as great as their true values, we would read the number 12 on our voltmeter, thus obtaining the required answer. Any subsequent settings of the potentiometers would then yield the desired sum on this special scale. It can be noted, from this example, that a numerical output is obtained in the case of the digital computer and, by contrast, a variation of an electrical quantity is the output obtained from the analog computer.

Granting that the method used in either case seems to be a long way around to achieve a simple result, nevertheless each method becomes a powerful computational tool when used in its more highly developed practical from. Thus, the *digital computer evolves from the simple binary counting function* illustrated, into a device possessing full adding, multiplying, and other arithmetic functions, and in its final form it performs enormous tasks of data processing, information storage, and automatic control functions. Likewise, the *analog computer evolves through the use of operational amplifiers* into a device capable of handling differential and higher-order equations by virtue of its integrating circuits, and in its final form it performs very significant tasks in solving highly complex equations and simulating intricate dynamical processes.

[1]G. A. Korn and T. M. Korn, *Electronic Analog Computers* (New York: McGraw-Hill, 1964).

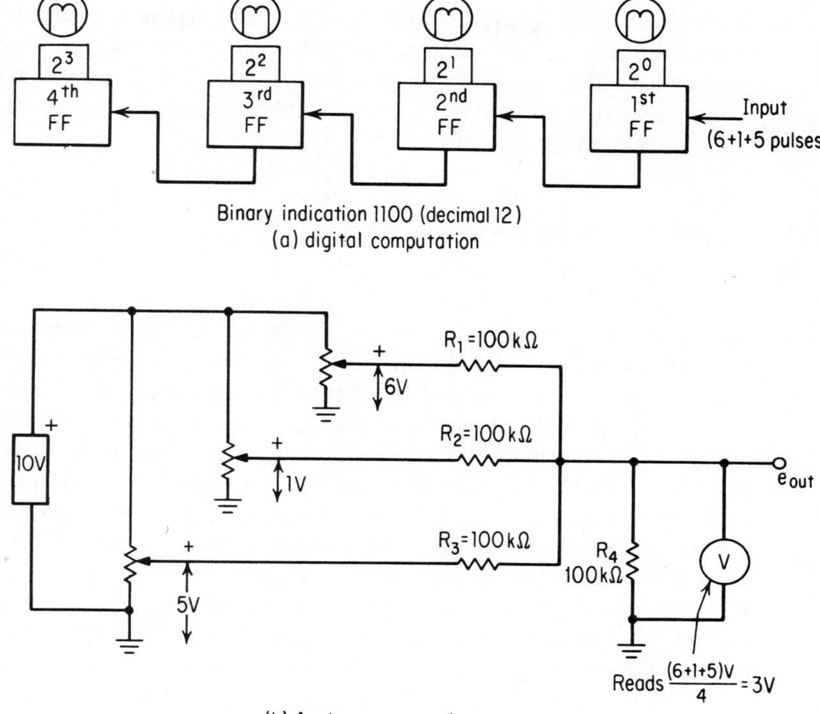

Figure 18-1. Digital versus analog methods in a simple computation ($6 + 1 + 5 = 12$): (a) digital counting by binaries; (b) analog summation of voltages.

Comparative Advantages

In many situations that are applicable to both analog and digital computers, the digital computer approach can generally be expected to be more accurate (and also considerably more expensive). For the analog computer, it can be said that accuracies of 1 to 3% are obtained at relatively lower cost and in a form where the effects of parameter changes can be more effectively observed. Higher order of accuracies (around 0.1 or 0.2%) can be obtained in analog computers at the cost of greatly increased complexity.[2]

Combinations of analog and digital computers called hybrid computers, are also employed when the situation warrants the inclusion of the advantages of

[2]A concise but very perceptive evaluation of the relative merits (and disadvantages) of each computer type is given by B. M. Oliver, "Digital Display of Measurements in Instrumentation," *Proceedings of the I.R.E.*, 50th anniversary issue, May 1962.

each type. Often converter units are employed in which a digital output display is provided for an analog instrument (as in the digital voltmeter), and even more frequently, use is made of the analog-to-digital converter for processing analog data by a digital computer. In the converse sense, digital computers may include digital-to-analog conversion for convenient continual display of changing data in the form of a graph record.

18-2. ELECTRONIC COMPONENTS OF ANALOG COMPUTERS

Electronic amplifiers form an essential part of analog computing systems. The dc amplifier is employed particularly for performing the mathematical operations of summing, multiplication, and integration. When used in this way, the dc amplifier is called an *operational amplifier*. The basic operational amplifier is a dc feedback amplifier, in which feedback (of the shunt type) can be arranged externally to provide various mathematical operations as required by the governing physical equation being solved. Comparatively recent developments have resulted in such amplifiers having large and stable gains. The original "differential analyzer" (which had been known scores of years ago as a highly complex device having a limited specialized usefulness) has been transformed into a modern instrument having great flexibility, able to be set up in a straightforward fashion to solve very simple as well as highly complex problem situations.

18-3. THE OPERATIONAL AMPLIFIER

The properties of the basic operational amplifier are shown in block form in Fig. 18-2(a) and in symbolic form in part (b). The dc amplifier is of the *direct-coupled* type discussed in Chapters 7 and 13 and may be employed in any one of a variety of forms producing a *linear, high-gain, feedback amplifier*. It should have an odd number of stages (or equivalent 180° phase shift), so that provision for overall negative feedback may be brought out to external terminals. In this form, the amplifier can handle steady dc or slowly varying ac components and can then be made suitable for its computation function by modifying the input impedance Z_{in}, the feedback impedance Z_f, or both.

Basic General Equation

The functional representation in Fig. 18-2(a) is generalized to cover any instantaneous input e_1 and any combination of resistive or reactive impedance (Z_{in} or Z_f), for the input and feedback impedances. The nominal gain of the amplifier $|A|$ is the magnitude of the gain without feedback (or open-loop gain) and is represented by the ratio of the instantaneous output e_2 to the net input e', which is the actual

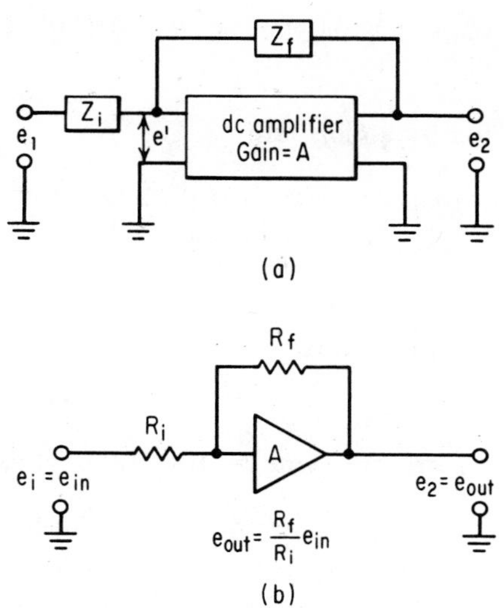

Figure 18-2. Basic operational amplifier: (a) block diagram; (b) symbol form.

input e_{in} (or e_1) reduced by the opposing feedback voltage:

$$A = -\frac{e_2}{e'}, \quad \text{where } A = \text{open-loop gain, } A_{VOL}$$

the negative sign indicating the inherent phase reversal of an odd number of stages. In the most general sense, the instantaneous values are most readily expressed by use of the p-operator notation, with p defined by d/dt; the operational impedances then become $Z_{in}(p)$ and $Z_f(p)$. Utilizing this notation, it can be shown[3] that when an amplifier with a large nominal gain A is used [i.e., when $A \gg 1 + Z_f(p)/Z_{in}(p)$], the output e_2 is related to the actual input e_1 by the fundamental relation:

$$e_2 = -\frac{Z_f(p)}{Z_{in}(p)} e_1$$

For our present purposes, it will be simpler to consider only resistive impedances (until the topic of integration is reached), and so we can rewrite the above *basic equation of the operational amplifier* in its simpler, limited form shown symbolically in Fig. 18-2(b), as

$$e_{out} = -\frac{R_f}{R_i} e_{in}$$

[3]J. D. Ryder, *Engineering Electronics, with Industrial Applications* (New York: McGraw-Hill, 1958), p. 227.

(the assumption of large gain needed to make this relation reasonably exact can easily be realized).

Multiplication by a Constant

The symbolic diagram, Fig. 18-2(b), can be used in the form shown to multiply by a constant. Thus, the general equation

$$y = ax$$

can be solved in the form

$$e_2 = -ae_1$$

by the circuit arrangements of Fig. 18-3. Thus in part (a) of the figure, if the

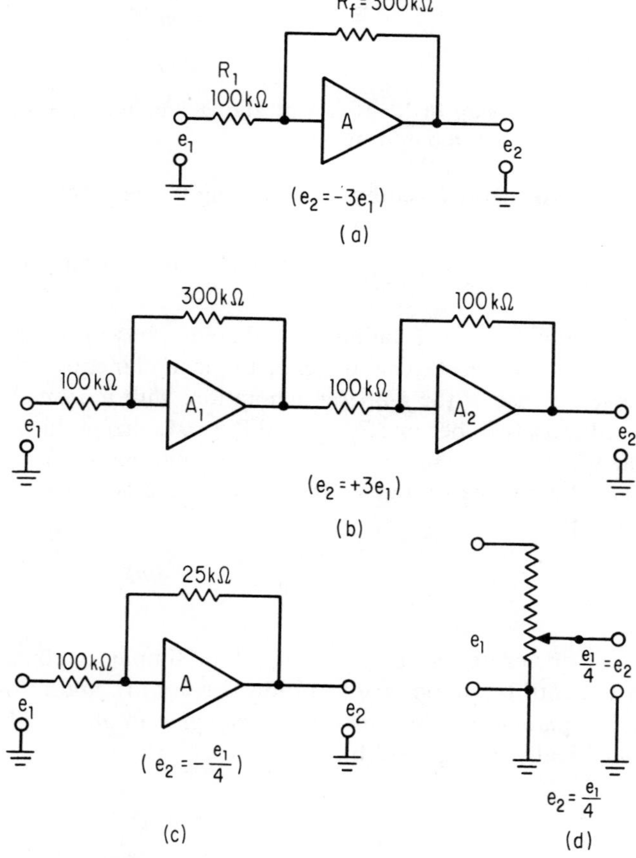

Figure 18-3. Multiplying by a constant; constant is (a) -3, (b) $+3$, (c) $\frac{1}{4}$, (d) $\frac{1}{4}$, without use of an operational amplifier.

multiplier a is to be -3, then R_f is made three times as large as R_{in}. If it is desired to produce a positive output three times as large as the input, the circuit in part (b) shows an extra *unity-gain amplifier* inserted to accomplish the *change of sign* and produce $e_2 = 3e_1$.

To multiply by a fraction, R_f is made smaller than R_{in} as in part (c), where $e_2 = -e_1/4$. (This can also be accomplished by making R_{in} correspondingly four times as large as R_f.)

Of course, multiplication by a fraction can be done by ordinary voltage-divider action, as in part (d) of the figure, without using an operational amplifier. However, it should be noted that the use of an operational amplifier makes it additionally possible to multiply by numbers larger than 1 and also enables sign reversals to be made.

18-4. SUMMING AMPLIFIER

The operational amplifier can accept a number of inputs and give the negative of their sum as its output. It is also possible, if desired, to produce an output proportional to the sum of the individual inputs by a *summing network* [as shown in Fig. 18-4(a)], without using an operational amplifier, but again, the use of the active *summing amplifier*, shown in Fig. 18-4(b), has its advantages over the passive summing network, as will be shown.

The network of resistors in Fig. 18-4(a) is most easily solved by the Millman theorem,[4] which expresses each resistance R in terms of its admittance $Y = 1/R$ (in this case its conductance) and yields the expression

$$e_{\text{out}} = \frac{e_1Y_1 + e_2Y_2 + e_3Y_3}{Y_1 + Y_2 + Y_3 + Y_4}$$

When all four resistors are equal, as in Fig. 18-4(a), the output becomes

$$e_{\text{out}} = \frac{Y(e_1 + e_2 + e_3)}{4Y}$$

or

$$e_{\text{out}} = \frac{e_1 + e_2 + e_3}{4}$$

Thus, in our example the output e_{out} is

$$\frac{6 + 1 + 5}{4} = 3\text{ V}$$

indicating that the output is equal to one-fourth of the sum of the inputs.

[4]S. Seely, *Electron-Tube Circuits*, 2nd ed. (New York: McGraw-Hill, 1958), p. 251.

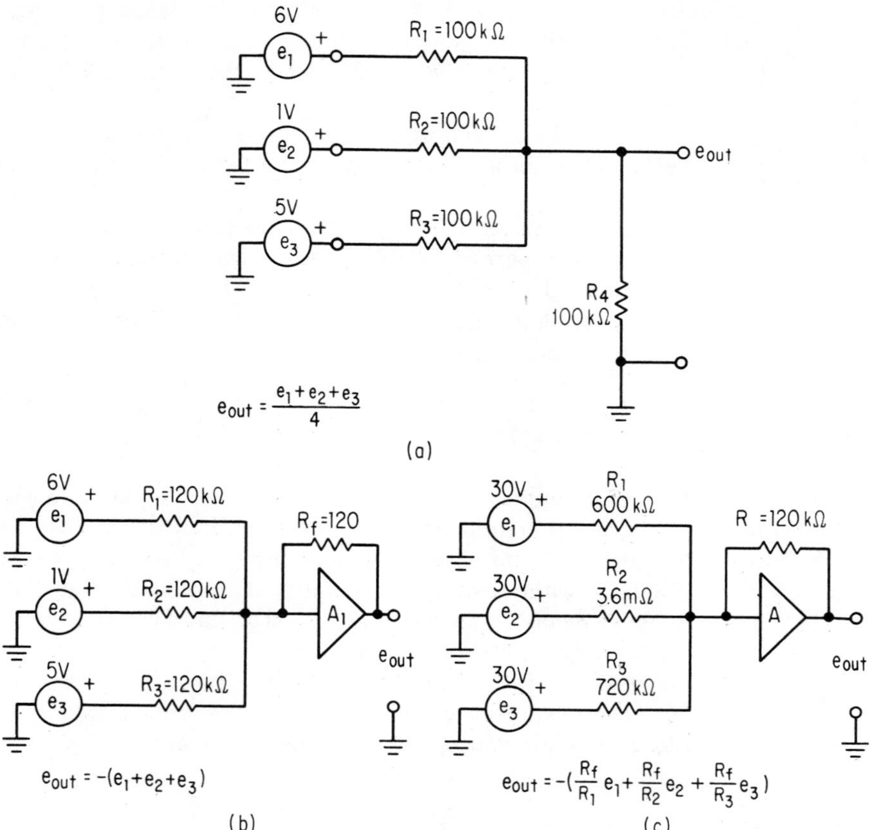

Figure 18-4. Summing network and summing amplifier circuits: (a) equal-resistor summing network; (b) equal-resistor summing amplifier; (c) summing amplifier with unequal resistances.

When the *summing amplifier* circuit is used, the basic expression for the operational amplifier becomes

$$e_{\text{out}} = -\left(\frac{R_f}{R_1}e_1 + \frac{R_f}{R_2}e_2 + \frac{R_f}{R_3}e_3\right)$$

In the example shown in Fig. 18-4(b), equal resistors are used again, and the expression for e_{out} yields

$$\begin{aligned} e_{\text{out}} &= -(e_1 + e_2 + e_3) \\ &= -(6\text{ V} + 1\text{ V} + 5\text{ V}) = -12\text{ V} \end{aligned}$$

thus obtaining the sum without attenuation. In Fig. 18-4(c), where unequal resistors are used as multiplying factors, the expression for e_{out} yields

$$\begin{aligned} e_{out} &= -\left(\frac{120}{600}30\text{ V} + \frac{120}{3600}30\text{ V} + \frac{120}{720}30\text{ V}\right) \\ &= -(6\text{ V} + 1\text{ V} + 5\text{ V}) = -12\text{ V} \end{aligned}$$

It can be seen from the examples above that the summing *amplifier* makes it possible either to obtain the sum without any attenuation or, where necessary, to apply various multiplying factors to each input to be added. It is important to note that the output of the operational amplifier is obtained as the *negative of the sum of the inputs*. If a positive value is desired, an additional amplifier may be used as a sign-changer or inverter. The solution, in the form of the output, is generally read as a voltage on the self-contained zero-center voltmeter of the computer, in this case of constant input.

The manner in which the analog computer elements are arranged (or programmed) to solve equations is based on a form of *implicit* reasoning that first assumes a solution to be obtained as an output, if the proper values are fed into a summing amplifier and then uses this implied output as a real input value wherever required. This facility, plus the ability to handle differentials very easily by *successive integrations*, allows for the solution of both simple and highly complex equation situations. Two examples will be given to illustrate this method, one for solving a set of simple simultaneous linear equations to obtain a numerical value for x and y, and the other for solving a familiar type of differential equation to obtain a graph form of the solution.

18-5. SOLVING SIMULTANEOUS EQUATIONS

For illustrative purposes, we again deliberately choose a simple example of two simultaneous equations in two unknowns, whose answer is fairly obvious, so that the basic method can be seen more clearly.

Let the two simultaneous equations be

$$\begin{aligned} 2x + 3y &= 28 \\ 2x - y &= 4 \end{aligned}$$

(Simple subtraction yields the answer by inspection that $4y = 24$ and $y = 6$, making x equal to $\frac{10}{2}$ or 5.)

We proceed first to arrange the two simultaneous equations so that each unknown is at the left side:

$$2x = -3y + 28$$

$$x = -\frac{3y}{2} + 14$$

$$y = 2x - 4$$

Next, the conditions necessary to generate x are set up as the input to one summing amplifier, and the corresponding conditions for generating y are set up as the input to another (so far independent) summing amplifier, as shown by the solid lines of Fig. 18-5. It will be noted that we insert the negative of the values given by the preceding equations in order to generate a positive output value, because of the inherent phase reversal of the summing amplifier. Thus for summing amplifier A_1, inputs of -14 and $-(-3y/2)$ [or $(+3y/2)$] are required to generate $+x$; similarly for A_2, inputs of $-(2x)$ and $-(-4)$ or $+4$, are required to generate $+y$. The constant values that are needed are tapped off the potentiometer voltage dividers connected to the proper negative or positive polarity of the ±100-V supply and are fed into their respective amplifiers on a 1:1 ratio basis. The required value of $-(-3y/2)$ [or $(+3y/2)$] is obtained from the $+y$ output of A_2 and is fed into the R' input resistor of A_1, thus simultaneously producing a multiplication of y by the R_f/R' ratio of 120:80 or 3:2, and so producing the desired $+3y/2$ (dashed line).

The negative of the $2x$ input needed for feeding into A_2 is obtained from $+x$ output of A_1 and is sent through an inverting amplifier A_3, which produces a *negative* x output, multiplied by the ratio R_f/R'', equal to $-(120x/60)$ or $-2x$, as required (as shown by dashed-line connection).

The solutions for x and y are then obtained by successively switching the indicating zero-center voltmeter to the points at which x and y are expected to be generated and by reading each value. As expected from the solution by inspection, the computer solution should show that the voltmeter at the x output reads $+5$ V while the voltmeter at y reads $+6$ V.

The following points may be noted from this simple example, which would apply equally to other cases with sets of equations involving three or more unknowns:

1. The inputs required to generate a desired output are obtained from voltage division of the power-supply voltages for the constants and by interconnections between the separate amplifiers that are assumed to generate each unknown.
2. The inputs required to generate an unknown, such as x, are the negatives of the values on the right side of the equation that add up to x.
3. Coefficients of an unknown may be obtained from the R_f/R_{in} ratio of the input resistors of the summing and inverting amplifiers.

In certain instances, it will also be found that a number of different ways of interconnecting the operational amplifiers will produce the same result.

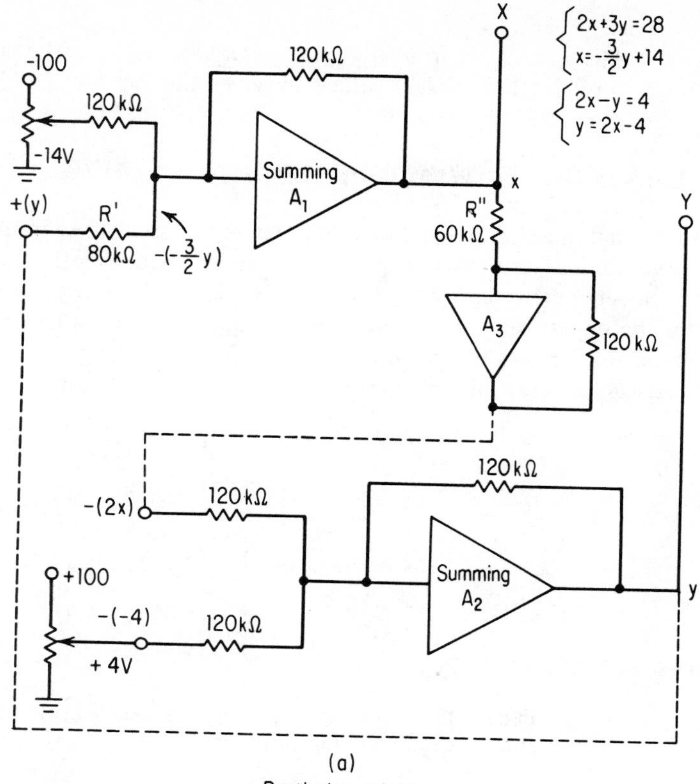

(a)
Particular solution

$$a_1x + b_1y + c_1 = 0, \quad \text{or} \quad y = -\frac{a_1}{b_1}x - \frac{c_1}{b_1}$$

$$a_2x + b_2y + c_2 = 0, \quad \text{or} \quad x = -\frac{b_2}{a_2}y - \frac{c_2}{a_2}$$

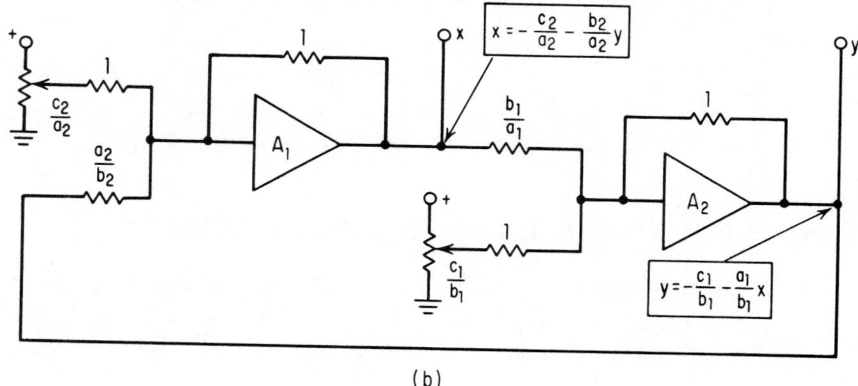

(b)
General solution

Figure 18-5. Analog computer solution of simultaneous equations: (a) particular solution; (b) general solution.

The *general solution for two equations in two unknowns* is shown in Fig. 18-5(b), where the resistor values are given in megohms.

18-6. THE INTEGRATING AMPLIFIER

The analog computer realizes its greatest potential in solving the *differential equations* that describe actual dynamic systems. It solves these differential equations by a series of integrations through the use of the integrating-circuit arrangement of the operational amplifier. This arrangement is obtained by the simple use of a capacitor as the feedback element Z_f, while the input impedance Z_i remains a resistor, as shown in Fig. 18-6.

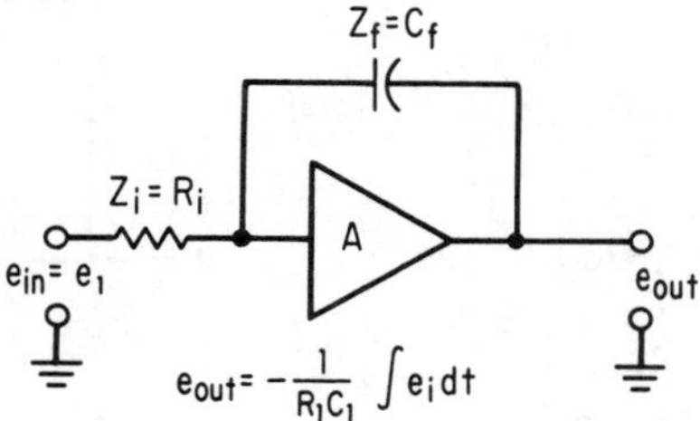

Figure 18-6. Integrating circuit produced by operational amplifier in an analog computer.

We apply the basic formula in the *p-operator* form to this integrating circuit, where $Z_i(p) = R_{\text{in}}$, and $Z_f(p) = 1/C_f(p)$, yielding

$$e_{\text{out}} = \frac{-x}{R_1 C_f(p)}$$

Here $1/p$ indicates integration with respect to time, so that

$$e_{\text{out}} = -\frac{1}{R_1 C_f} \int e_1 \, dt$$

or the output will be a *negative constant times the time integral of the input.*

18-7. SOLVING A SIMPLE DIFFERENTIAL EQUATION

A simple example of the solution of a differential equation by the analog computer can be given in terms of the equation of motion of a falling body.[5] Suppose that the constant acceleration of gravity acts as a decelerating force (negative value of

[5]Adapted from an example in ibid., p. 265.

-32 ft/sec²) on an object hurled vertically upward with an initial velocity (positive value) of 128 ft/sec. The instantaneous displacement y is given by the differential equation for constant acceleration:

$$\frac{d^2y}{dt^2} = -32 \text{ ft/sec}^2$$

Values for vertical displacement as a function of time, $y(t)$, will be traced by the computer in the form of a curve, as the solution for y; also, if desired, values for instantaneous velocity dy/dt can be traced by the same computer solution by connecting to the corresponding point. The ground-zero level is taken as the reference point, and the following initial conditions will be assumed: $y = 0$ at $t = 0$; and $dy/dt = +128$ ft/sec at $t = 0$, signifying initial velocity in the upward direction.

(It should be kept in mind that this example, like previous digital computer examples, is deliberately made simple enough to allow the answer to be easily checked by ordinary computation and to make it easy to concentrate on the method involved.)

The simplified setup to solve this differential equation is shown in Fig. 18-7.

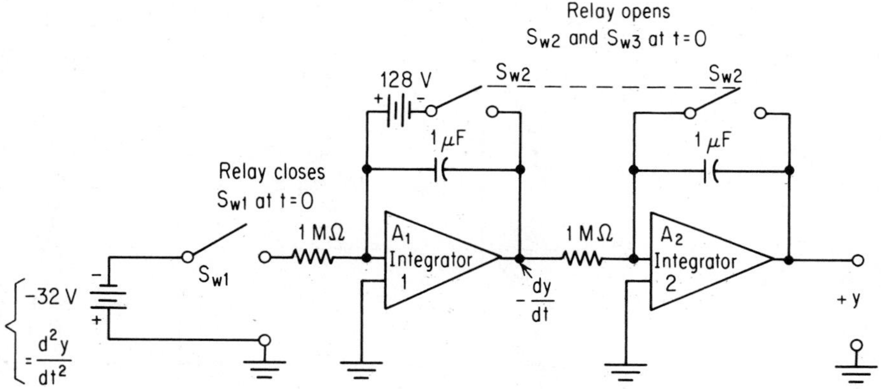

Figure 18-7. Computer setup to solve for $y(t)$ in the differential equation $d^2y/dt^2 = -32$ ft/sec².

Since the highest-order derivative is already on the left side of the given equation:

$$\frac{d^2y}{dt^2} = -32 \text{ ft/sec}^2$$

the value corresponding to d^2y/dt^2 is fed into the input of the first integrator A_1 in the form of negative 32 V (corresponding to the relation $d^2y/d^2t = -32$ ft/sec²). The RC value of each of the integrators is made equal to unity, by choosing $R =$ 1 MΩ and $C = 1$ μF. The result of the first integration is therefore labeled the negative of the integral of the input (d^2y/dt^2) or $-dy/dt$. By feeding this value into the second integrator, the resulting output will be the solution $+y$ as a function of time $y(t)$, which can be traced out as a curve in real time by an appropriate recorder connected to this output point.

To set in the initial conditions, a voltage corresponding to $dy/dt = 128$ ft/sec is applied to initially charge the capacitor of the first integrator to the value of 128 V (if the voltage scale factor is 1). This is accomplished in the proper polarity by arranging that the negative terminal of the 128-V supply is connected to the point at which the negative of dy/dt appears; in other words, $-dy/dt$ is made equal to -128 V. Switch SW_2 is in a closed position, as an initial condition before the start of computer operation. Similarly the initial condition that $y = 0$ at $t = 0$ is arranged by switch SW_3, which keeps the capacitor across the second integrator short-circuited until the computer operation starts. At the start of operation,

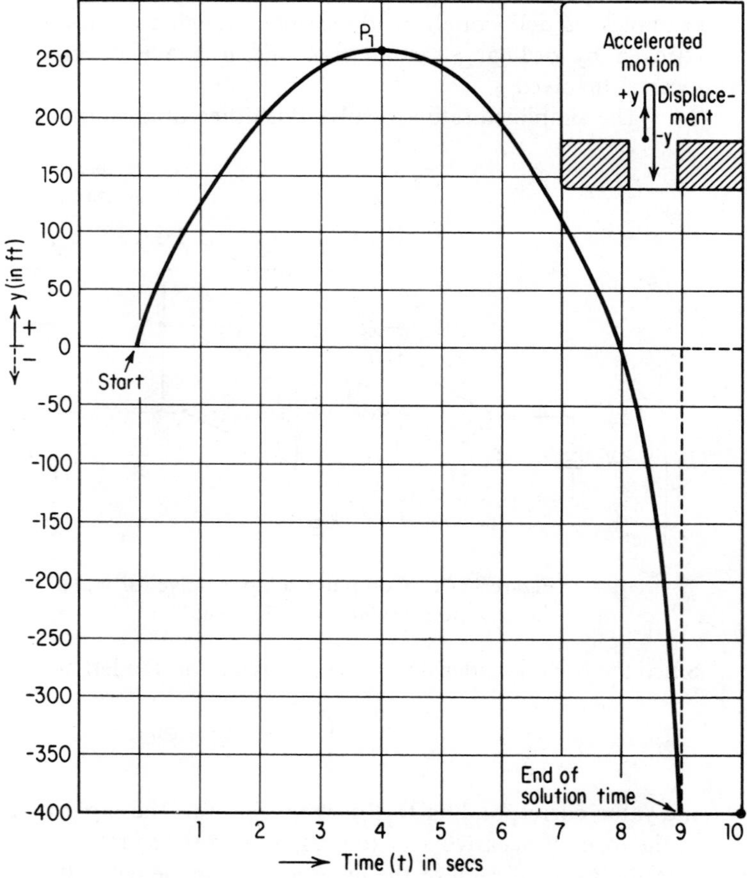

Figure 18-8. Output curve obtained as a computer solution of $y(t)$ in moving-body problem of Fig. 18-7.

relays operate to close SW_1, applying the negative 32-V input and to open simultaneously SW_2 and SW_3 to allow the integrations to take place at the time when the capacitors each have the proper charge.

The resulting output, obtained as a curve on a plotter (or equivalent recorder) in real time, is shown in Fig. 18-8.

The resulting plot can easily be checked at any desired points by applying the familiar equations in physics for the distance covered by a body acted upon by gravity. For example, the time required for the body to lose its upward velocity, obtained from $V = at$ (being the same as the time required by a falling body to acquire a velocity of 128 ft/sec), is

$$t = \frac{V}{a} = \frac{128 \text{ ft/sec}}{32 \text{ ft/sec}^2} = 4 \text{ sec}$$

The height y attained at the moment of zero velocity is similarly obtained at $t =$ 4 sec from $y = \frac{1}{2}at^2 = \frac{1}{2}(32)(4)^2 = 256$ ft. This checks with point P_1 on the curve, where the horizontal real time divisions are 1 division/sec, and the vertical y distances are 50 ft/division. It will be noted that although the height y has fallen to zero at the end of 8 sec, the computer continues the solution through 9 sec (since it was so programmed), even though the physical interpretation is valid only for the first 8 sec. Programmed relay action returns the computer to zero at the end of the programmed period of 9 sec, in this case.

18-8. "BUILDING-BLOCK" USES OF THE OPERATIONAL AMPLIFIER

As was pointed out in Chapter 13, the operational amplifier serves as the heart of the analog computer, because it posseses the widely useful ability to *provide a high value of precisely controlled amplification.* In addition to the use in the computer, its capabilities offer many possibilities for using the op amp as a basic building block, around which to fashion numerous circuit functions. Before entering upon further details of the analog computer, a few of these building-block uses will be presented. Any of these uses, incidentally, are equally valid, whether the op amp is employed in its integrated-circuit form to conserve space, or whether it is used in its discrete form as a plug-in device, where its slightly larger package is still compact enough for most applications.[6]

Referring to the six selected applications in Fig. 18-9(a), a *precision voltage-control circuit* shows the standard symbolic representation of the two-input (differential) operational amplifier, where the minus input is generally called the *inverting input* (since it produces a positive output voltage e_{out} for a negative input, and vice versa) and where the plus input stands for the noninverting input. The feedback resistor R_f is shown as variable, to control the amount of amplification of the

[6]See M. Kahn, *The Versatile Op-Amp* (New York: Holt, Rinehart and Winston, 1970).

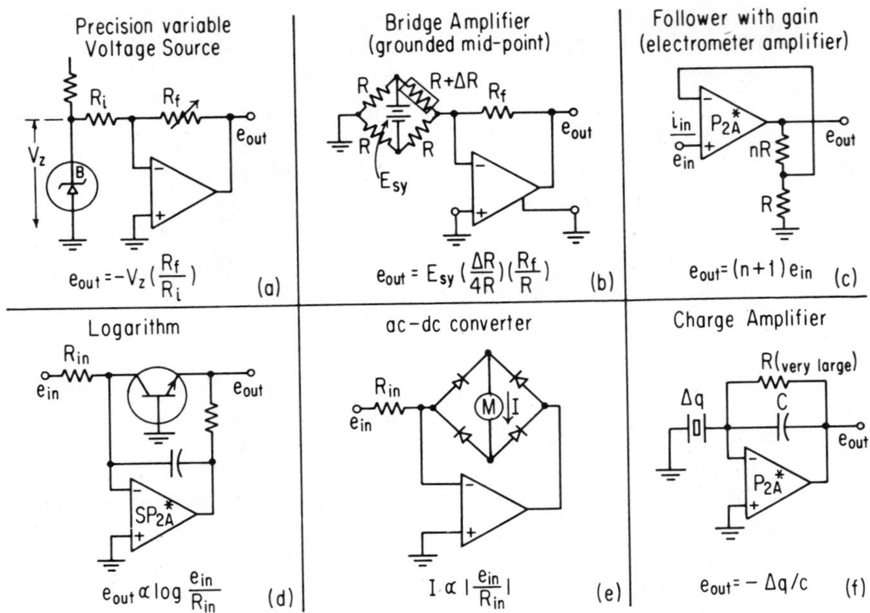

Figure 18-9. Using the operational amplifier as a building block. These six examples illustrate some of the functions (in addition to summing and integrating) that may be fashioned from the op amp. (Model numbers adapted from Philbrick/Nexus Research, Division of Teledyne.)

precision dc voltage V_z across the Zener diode. Hence, the resulting function is

$$e_{\text{out}} = -V_z(R_f/R_i)$$

A very useful *amplifier for bridge unbalance* is shown in part (b) of the figure, which allows a single-ended output to ground e_{out} and also allows grounding of the midpoint of the bridge circuit. The bridge-unbalance voltage is amplified by the familiar factor of R_f/R_i, where R_i, in this case, is practically equal to the R of the equal-arm bridge. Hence, for a given supply voltage E_{sy} (to the same approximation previously given for small unbalance in the equal-arm bridge):

$$e_{\text{out}} = E_{sy}(\Delta R/4R)(R_f/R)$$

The *follower circuit* in part (c) of the figure not only presents a very high input impedance (input current i_{in} is less than 0.01 mμA, or nA), as is usually the function of follower circuits, but also provides amplifier gain greater than 1, making it especially useful as an electrometer amplifier:

$$e_{\text{out}} = (n + 1)e_{\text{in}}$$

An output proportional to the *logarithm* of the input voltage e_{in} is obtained in part (d) of the figure, with the aid of a common-base transistor configuration and an *RC* combination, in addition to input resistor R_{in}:

$$e_{out} \text{ varies as } \log (e_{in}/R_{in})$$

The *ac-to-dc converter* of part (e) of the figure provides the meter M with dc current I:

$$I \text{ varies as magnitude of } |e_{in}/R_{in}|$$

A *charge amplifier* is shown in the final part (f) of the figure. The output e_{out} of this arrangement (when using a capacitor of negligible leakage) is equal to the negative of the change in charge (Δq) divided by capacitor C:

$$e_{out} = -\Delta q/C$$

In listing the six miscellaneous examples that have been given above, in addition to the fundamental operations of summing and integration, we might further add the important circuit arrangements of *comparators* and *oscillators* that may also be built up from the basic op amp. And so, it can readily be seen that the operational amplifier has many uses as a versatile building block, besides its wide use in performing mathematical operations in the analog computer.

18-9. COMPUTER OPERATIONAL MODES

The result of an analog computation is generally displayed as a plot traced out on an X–Y recorder, with appropriate scaling factors for amplitude and time. (Constant dc output voltages, of course, can be read off from the self-contained voltmeter.)

Figure 18-10 shows a *desktop type of analog computer*, using solid-state construction with a ±100-V computing range (*Systron-Donner model 3300*). Its visual-problem-board panel layout is particularly aimed at student use, and its 10-amplifier capacity is suitable for tutorial problems. In addition to the ten 10-turn potentiometers that form a row at the bottom, there are five more coefficient potentiometers (for initial conditions), forming a vertical column at the left.

Repetitive Mode of Operation

In another mode of operation, the output may be viewed on an oscilloscope, thus providing a method that is particularly effective in cases where it is desirable to observe the effect of parameter changes immediately. Figure 18-11 shows a transistorized general-purpose analog computer (*Pace model TR-10*) connected to a dc oscilloscope for such continuous viewing.

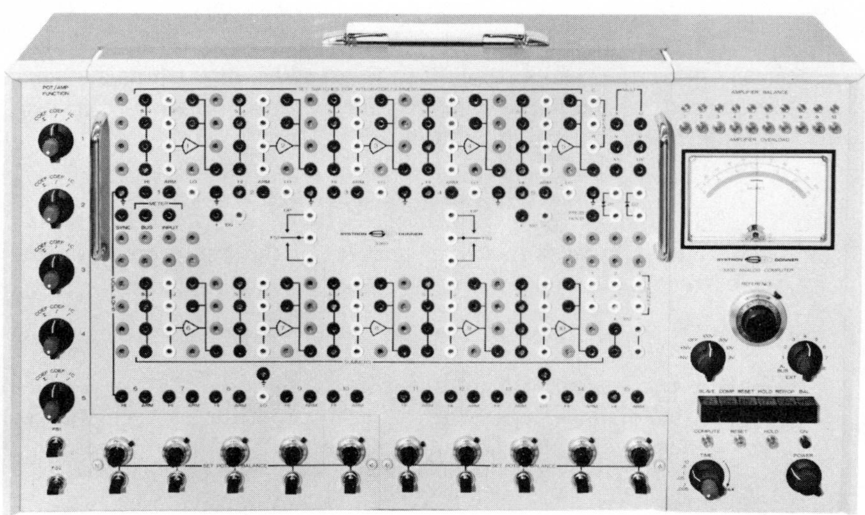

Figure 18-10. Desktop analog computer, having detachable problem board with visual outlines of computer circuits for interconnections. (*Systron-Donner model 3300.*)

Figure 18-11. Transistorized desktop analog computer with oscilloscope connected to output in high-speed repetitive mode. (*Electronic Associates, Inc., Pace model TR10.*)

The oscilloscope display is obtained by a *repetitive mode of operation* of the computer. The problem solution time is speeded up, so that the output may be displayed many times per second, generally around 16–20 times/sec, for a continuous display without objectionable flicker. In the model shown, for example, provision is made for a 100:1 reduction in problem solution time. Hence, a solution normally programmed for a 5-sec operation would be achieved in 50 msec, and consequently, in this case, would be repeated 20 times.

The repetitive mode is achieved by control and timing circuits that operate high-speed electromechanical (or equivalent) relays between the OPERATE and RESET positions of the computer.

18-10. HYBRID COMPUTERS

The hybrid computer incorporates both analog and digital techniques; as a result, it takes advantage of the conveniently speedy output displayed in graphical form from the analog portion and of the high order of computational accuracy from the digital portion.

An example of the greatly enhanced capability of the hybrid form may be cited from a simulation of a space mission.[7] In this case, the complex calculations governing the dynamic motions of the vehicle in three-dimensional coordinates are obtained from the digital-computer portion, while this information is fed to the analog-computer section for display of the system simulation. For similar reasons the hybrid computer is being increasingly used in complex industrial-process situations.

The combination of analog and digital functions in a hybrid computer necessitates interfaces of analog-to-digital (A–D) conversion and similarly digital-to-analog (D–A) conversion, as discussed below.

Conversion between Analog and Digital Outputs

The convenience of a digital display has been mentioned in the previous paragraphs, and the digital voltmeter was given as an example of an analog-to-digital conversion method. Some units are specially designed for such conversion, as for example the *Fischer & Porter Analog-to-Digital Converter* (*model 50 DC 1000*), which accepts the analog voltages and converts them to any of the conventional digital output codes, such as a 1–2–2–4 binary-decimal or straight binary code, and provides both serial and parallel outputs to a digital computer, while displaying the numerical values. Allowing a reading time of 130 μsec, it performs 6000 complete conversions per second.

Conversely, *digital-to-analog converters* are available for transmitting data from the digital computer in an analog form either for curve tracing or for further processing by an analog computer.

[7] A simple diagram of such a simulation is given in R. C. Weyrick, *Fundamentals of Analog Computers* (Englewood Cliffs, N.J.: Prentice-Hall, 1969).

18-11. COMMERCIAL ANALOG COMPUTER CHARACTERISTICS

As in the case of the digital computer, it can be seen that the characteristics of commercial analog computers will vary considerably according to the complexity of the particular application in which one is involved. Both tube and transistor types are available.

The cost relative to the application is frequently a deciding factor.

Figure 18-12 shows a large desktop analog computer with 60-coefficient potentiometers (*Electronics Associates, Inc., model TR-48*). The control panel on

Figure 18-12. Versatile desktop analog computer. It has provision for adding a console for hybrid operation, control and readout bay on left, removable prepatch at center, and 60-coefficient-potentiometer bay at right. (*Electronic Associates, Inc., model TR-48.*)

the left sets the operating modes. The readout on this panel comprises a digital voltmeter, a deflection-type null meter, and a plug-in scope for displaying repetitive operation (an X–Y recorder may also be plugged in for external readout). The removable prepatch panel is seen in the center.

A digital expansion system (DES-30) is available for mounting at the side of the desktop computer to provide hybrid capability.

QUESTIONS

Q18-1. Explain the difference between an analog and a digital computer with respect to:
(a) Type of results obtained.
(b) Type of electronic apparatus used in each case.

Q18-2. What requirements must be met by an amplifier to be suitable for use as an operational amplifier in an analog system, with respect to:
(a) Type of coupling?
(b) Number of stages?
(c) Gain without feedback?

Q18-3. Explain the advantage gained by the use of chopper-stabilized, rather than ordinary dc amplifiers, for the operational amplifiers.

Q18-4. If there are two inputs to an integrating amplifier, does the amplifier act as a summer, an integrator, or both? Explain.

Q18-5. Give *two examples* of situations where mechanical values may by simulated by voltages, to produce an analog presentation of a mechanical problem.

Q18-6. Explain how scaling factors are introduced in an analog computer for:
(a) Magnitude values.
(b) Time values.

Q18-7. Explain what method may be used in an analog computer to:
(a) Multiply by a constant k.
(b) Multiply by a variable z.

Q18-8. Explain the principle involved in using diodes to generate nonlinear functions (diode-function generator).

Q18-9. Describe the action required in an analog computer, if the output is to be viewed on an oscilloscope.

PROBLEMS

P18-1. For the summing circuit [shown in Fig. 18-1(b)] to produce an output e_{out} that is closely equal to the actual sum $(e_1 + e_2 + e_3)$, what must be the gain of a linear amplifier;
(a) When R_4 is 100 kΩ?
(b) When R_4 is 3.3 MΩ?

P18-2. In Fig. 18-1, the two widely used symbol representations of summing amplifiers shown in parts (a) and (b) are equivalent. Express the output:
(a) At point p.
(b) At point q.

P18-3. In the circuit for Problem P18-1, express the final output at the point e_{out}.

P18-4. The symbol representations shown in Fig. P18-4 are equivalent and represent multiplication by a constant k. Express the output available at point p.

P18-5. Find the output e_{out} of the analog set up in the Fig. P18-4.

P18-6. Draw the analog setup to obtain the solution for x and y in the simultaneous equations:

$$3x - 4 = 3$$

$$x + y = 8$$

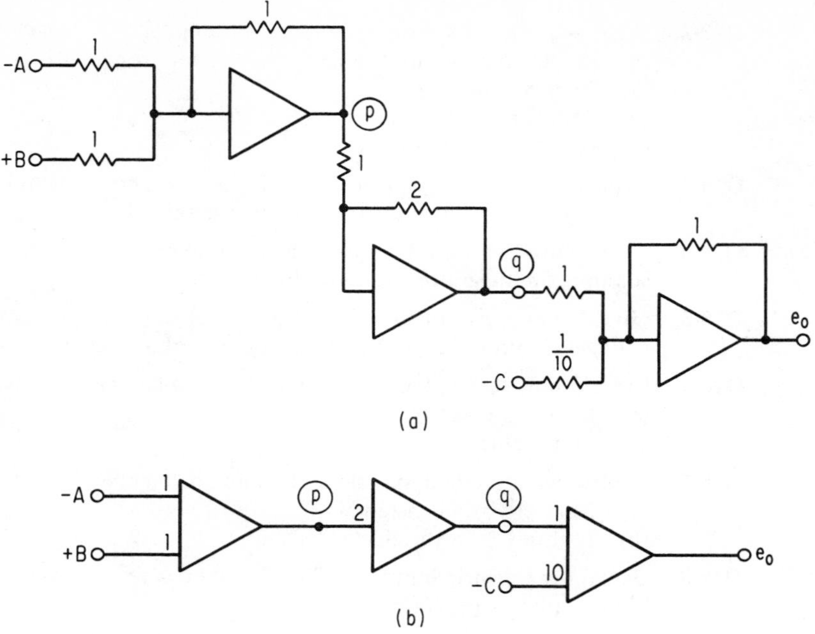

Figure P18-1

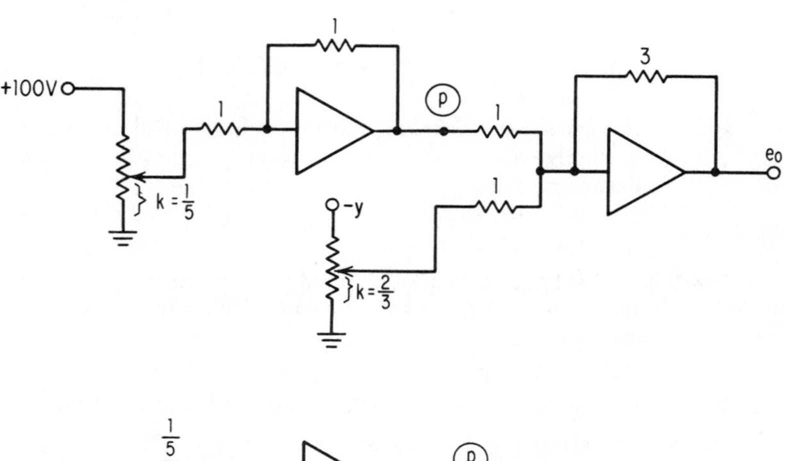

Figure P18-4

P18-7. Two equivalent representations of an analog computer setup requiring two integrations of a second derivative (such as d^2x/dt^2) to solve for the unknown x are shown in Fig. P18-7. Find the equation represented by this setup.

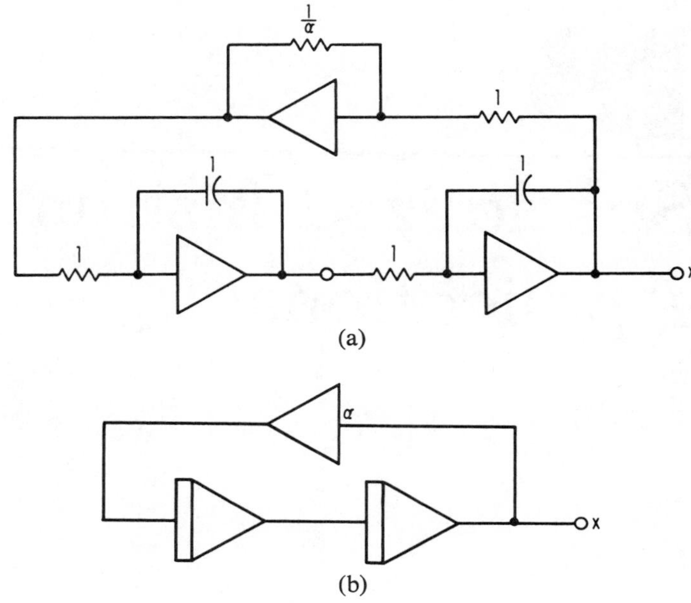

Figure P18-7

P18-8. Draw the analog computer setup to solve the equation

$$A\frac{d^2x}{dt^2} + \frac{dx}{dt} + x = 20$$

P18-9. Find the equation represented by the analog setup shown in the Fig. P18-9.

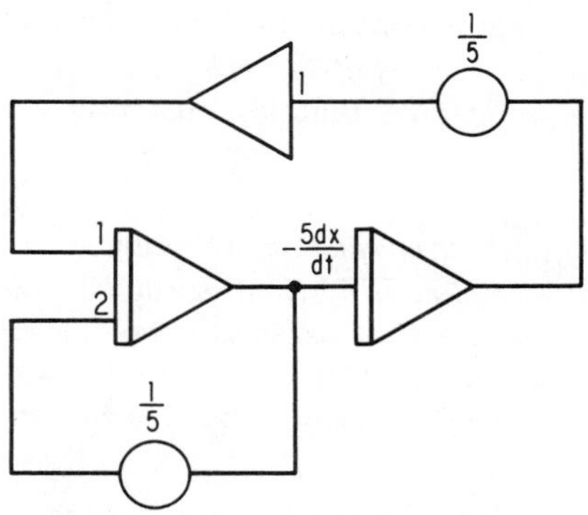

Figure P18-9

19

Specialized Instrument Applications

19-1. APPLICATION FIELDS OF SPECIALIZED INSTRUMENTS

If we try to visualize a bird's-eye view of the overall instrumentation field, we quickly come to realize how very extensive a field it covers. It is helpful to think of this broad territory as divided into three main areas: the first two (the *service/maintenance test instruments* and the *general-purpose laboratory test instruments*) have been the major concern of the previous chapters. The third field is the *specialized area of industrial/research instruments*, which constitutes a large and rapidly developing area that is constantly extending the "state of the art" in instrumentation. Some of these applications are:

1. *Analysis instrumentation*, including:
   (a) *Chemical analysis*, using such instruments as conductivity/pH meters, gas chromatographs, spectrophotometers, atomic absorption meters, neutron activation analysis, and air- and water-quality analyzers.[1]
   (b) *Biomedical analysis*, using such instruments as electrocardiographs (ECG

[1]Current literature in this expanding field is listed in "New Literature Digest" section, *Environmental Science and Technology* magazine, American Chemical Society, Washington, D.C.

or EKG), cardiac pace meters and defibrillators, electroencephalographs (EEG), and patient-monitoring (or intensive-care) systems.

(c) *Electrophysical analysis*, including vibration and acoustic analyzers, nuclear-magnetic resonance (NMR) and electron-spin resonance (ESR) detectors, and lasers (with their associated holograph systems).

2. *Industrial control systems*, including process monitors, automated process control, direct-digital control (DDC or numerical-control) instrumentation, and automatic-testing consoles.
3. *Precision measurement* (*metrology*), including primary voltage and resistance standards (traceable to NBS), secondary power supply and frequency references, and working calibration instruments.
4. *Aerospace/telemetry*, including body-function transducers, multiplexers, and FM/FM transmitters.

Advanced Instrument Forms: Noise Reduction—Improving Signal-to-Noise Ratio

Common to all of the four listed groups of advanced systems is the necessity for *extracting weak signals from noise*. This has become an increasingly serious requirement when dealing with signals at low-microvolt (and lower) levels, where the presence of ordinarily negligible noise becomes a predominant factor and where, consequently, the signal-to-noise ratio (S/N) progressively deteriorates. The current trend in tackling this difficult noise problem includes three forms of attack: *signal-averaging*, *phase-locking*, and *correlation techniques*.

The *signal-averaging* process takes advantage of the fact that random noise tends to average out to zero. By storing the result of successive averagings and displaying the stored result, the signal display will show the signal increasingly predominant over the noise.[2]

The *phase-locking* process takes advantage of another characteristic of random noise, i.e., the fact that random noise tends to contain a whole spectrum of frequencies (so-called "white noise"), thus allowing the frequency of interest to be emphasized. This is done by forcing the amplification to favor the phase (and therefore the frequency) to which the amplifier is "locked," by reference to the desired frequency.[3]

Correlation, as applied to wave-forms, is a real-time method for detecting periodic signals buried in noise. It accomplishes this by testing for the similarity between two wave-forms (one desired and one random) by a complicated mathematical process that produces a derived *autocorrelation function*, which has a period the same as the wave-form, operated on, with a time-shifted version of

[2]The *signal-averaging* method for noise reduction is described in the *Hewlett-Packard Journal* of Apr. 1968, in connection with their digital *model 5480A signal analyzer*.

[3]The *phase-locking* method of noise reduction is described in the bulletins for the *Keithley phase-sensitive detector Brookdeal model 822*, *Hewlett-Packard ac microvolter model 3410A*, and *Princeton Applied Research Technical Bulletin 109* for their *model JB-5 lock-in amplifier*.

itself. By taking advantage of the fact that the autocorrelation function of random noise is very much smaller than that of the periodic signal, the desired signal can be greatly enhanced, even when it seems completely hidden by the noise. Although known mathematically for a long time, it is only recently that the complicated mathematical manipulation of the signal in the correlation technique has become practical, through the use of digital techniques and the corresponding integrated circuits.[4]

Guarded circuits are an additional technique, used to protect against unwanted *leakage currents*. Connection to the *guard terminal* diverts such leakage currents (or ground-loop currents), so that they do not affect the meter indication.

Instruments featuring these advanced noise-reduction methods—although beyond the scope of this book—have become increasingly important, especially in the biomedical and aerospace instrumentation fields. The footnotes given in this section for the manufacturers' literature contain very good introductory material for each of these methods.[5]

Application Fields Emphasized in the Text

A consideration of the four specialized areas previously listed indicates that the fields of *industrial control systems* and *metrology* are beyond the scope of this book. Accordingly, within appropriate space limitations, this concluding chapter will concentrate on representative aspects of *analysis instrumentation* (*chemical-bio-chemical*, including pollution-monitoring) and of *aerospace telemetry*; the last-named area in particular lends itself to a helpful summary of the salient aspects of electronic instrumentation.

19-2. CHEMICAL INSTRUMENTAL ANALYSIS

Instrumental analysis in the chemical field has advanced greatly by means of developments in more sensitive and selective detection. Detection systems have been developed that respond with great selectivity to the progressive presence of "foreign" molecules in *gas chromatography* or to selective absorption of specific wavelengths in *absorption spectroscopy*. These newer methods of detection add to or supplement older methods of *conductivity-sensing* and *pH determination*, to mention just a few. When such a variety of sensors are combined with the sensitive measurement capabilities of electronic amplifiers and indicators, the resulting instruments become powerful tools in the field of chemical analysis.

[4]Correlation techniques for noise reduction are discussed and available in the *Hewlett-Packard Journal* of Nov. 1969, in connection with the *H-P model 3721-A correlator*.

[5]See also Letzter and Webster, "Noise in Amplifiers," *IEEE Spectrum*, Aug. 1970.

Conductivity Measurements

Since the ability of a solution to conduct current depends on the presence of ions in the solution, the measurement of the electrical resistance (or its inverse, the conductivity) of a solution frequently offers a simple means for checking on the purity of distilled water or various solutions involved in chemical processing.

The electrical resistance of a solution is measured by what is basically a Wheatstone-bridge circuit, arranged as shown in Fig. 19-1, so that a conductivity cell becomes one arm of the bridge. This cell contains electrodes having a known

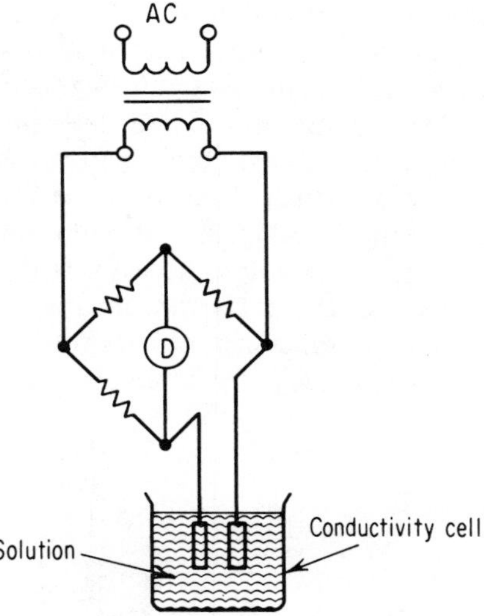

Figure 19-1. Measurement of solution conductivity.

area and separation, so that a *cell constant* can be used to relate the actual measured value of resistance to a standard value, giving the ratio of specific conductance to measured conductance. *Specific conductance*, as the standard unit of measurement, is defined as the conductance (in siemens) of 1 cm^3 of solution measured between two electrodes, 1 cm apart, at standard temperature (25°C).

It will be noted that the bridge uses an ac exciting potential, to avoid one-way electrolytic action and consequent polarization of the electrodes. Accordingly, the null detector *D* must also be an ac indicator, generally of the electronic-voltmeter type. Although the illustration shows the electrodes immersed in a stationary conductivity cell, the instrumentation can also be arranged to handle the flow of solutions through pipes A similar method is used in *oceanography* to measure the *salinity of seawater*.

19-3. HYDROGEN-ION CONCENTRATION (pH) DETERMINATION

An important element in chemical process control (and also in biomedical studies) is the measurement of the acid or alkaline properties of a solution, as determined by the pH or hydrogen-ion concentration of the unknown. In addition to its familiar use in determining the exact point at which an acid is neutralized by a base (or vice versa) by titration procedures, it is also very useful in tracing the progress of a wide variety of reduction-oxidation (redox) reactions.

The unit of pH measurement is based upon the negative logarithm of hydrogen-ion concentration. Applied to water solutions, it compares the degree of acidity or alkalinity of solutions on a scale of 0–14 pH, pure water having a pH of 7. As the solution departs from a neutral character, the values less than 7 progressively indicate greater acidity and values greater than 7 indicate greater alkalinity.

The pH of a solution may be determined by suitably measuring the dc voltage developed across a prepared pair of electrode combinations immersed in the solution being measured. Each of the electrode combinations produces a voltage by electrochemical action, in much the same way as in the familiar lead-cell storage battery that uses sulphuric acid as the electrolyte. The pair of electrode combinations used in the pH arrangement, shown in Fig. 19-2, is made up of a *reference-electrode* combination, which produces a constant potential, independent of the pH, and a *measuring-electrode* combination, whose potential varies with the

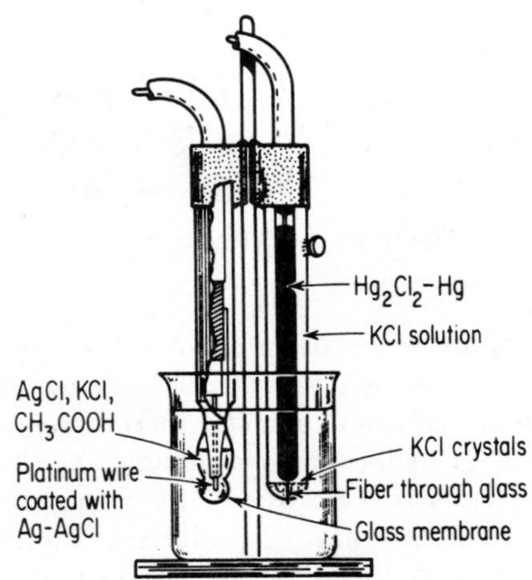

Figure 19-2. Electrode-pair assembly for pH determination: potential of glass electrode (on left) varies with pH, while constant potential of calomel electrode (on right) serves as a reference. [From F. Daniels et al., *Experimental Physical Chemistry* (New York: McGraw-Hill, 1970).]

pH of the unknown solution. The reference electrode combination commonly used is called the *calomel electrode* and consists of an assembly containing liquid mercury and mercurous chloride (calomel) in a reservoir filled with a saturated solution of potassium chloride (KCl), all enclosed in a glass tube. This forms a self-contained cell assembly producing a constant voltage as a reference. It makes electrical contact with the unknown solution by means of an opening forming a "salt bridge" (shown as a fiber through the glass in Fig. 19-2).

The measuring element is generally the so-called *"glass" electrode* assembly, which consists of a glass tube with a special glass membrane at its lower end. The thin membrane of soft glass encloses a dilute buffer solution, in which is immersed a platinum wire coated with silver–silver chloride combination. The glass-electrode potential varies with the hydrogen-ion concentration of the unknown solution, changing 59 mV/pH unit (at 25°C).

Since each electrode is in conductive contact with the solution, the output of the cell is the algebraic sum of the two electrode potentials. With the reference potential constant, the voltage produced at the junction of the glass membrane and the test solution is the only potential that varies. This potential variation of 59 mV/pH unit can provide a direct-reading indication of pH values on a suitable millivoltmeter of the high-impedance electrometer type.

pH Indicating System

The pH indicating system must be capable of accepting the millivolt output signal of the cell from an equivalent source resistance of order of $10^8\ \Omega$ (around 100 MΩ), with a correspondingly low value of current flow in the 10^{-12} (or picoampere) range. This calls for preamplification by an electrometer before the indication can be obtained by either an electronic voltmeter, or by a potentiometer instrument, such as the type previously discussed in Chapters 7 and 8.

In effect, the output of the pH cell is the difference in voltage between the constant value of the reference cell and the particular value produced in the measurement cell by the test solution. The reference potential E_r of the saturated calomel reference electrode at 25°C is constant. The potential of the measuring glass electrode E_m is given by

$$E_m = E_0 + 0.059\ \text{pH}$$

where E_0 varies with the individual electrode. Thus, expressing all the potentials in millivolts, the pH cell output voltage E_{out} is

$$E_{out} = E_r - [E_0 + 59\ \text{mV (pH)}]$$

If it is desired to calibrate an electrometer-type voltmeter to read the pH directly on a 0–14-pH scale, using a zero-center meter as the indicator, the calibration would be arranged so that 7 on the scale would correspond to 0 V output on the pH scale. If this were done, then the value of pH = 14 for the right end of the scale would correspond to 7 × (59 mV), or a positive 413 mV.

Since the indicating system, whether of the direct-reading or the potentiometer type, is essentially a millivoltmeter, many instruments also provide scales for using the pH meter as a millivoltmeter, generally with a full-scale range (zero at left) of 0–1400 mV.

A typical example of a pH meter used in routine analysis is shown in Fig. 19-3 (*Beckman model* 76). It is of the electrometer type and transistorized through-

Figure 19-3. pH meter, having standard 0–14-pH scale, plus an expanded scale for a 2-pH unit span, and also a 0–1400-mV range. (*Beckman Instruments, Inc., model 76.*)

out, with the exception of the electrometer tube. The transistor amplifier is of the chopper type to eliminate drift. Once standardized, it allows pushbutton determination of the pH on a 0–14-pH range. An expanded scale with a 2.0-pH full-scale span can be selected by the pushbuttons to provide a span of 2 pH anywhere in the 0–14 range, such as 2–4, 10–12, or any other span of 2 pH units. It also includes millivolt readings, either on a 0–1400-mV range, or on an expanded scale over any 200-mV span. The millivolt output is available at an outlet for connection to a standard potentiometer recorder.

In a manner basically similar to the pH electrode selecting the hydrogen cation H^+, *ion-selective electrodes* are used to measure cation or anion activity, such as in Cl^-, Ca^{2+}, F^-, or any other ion.[6]

19-4. GAS CHROMATOGRAPHY

As an analytical technique in chemistry, gas chromatography is one of the most versatile methods for detecting and measuring individual constituents in combinations and mixtures containing many chemically similar complex organic compounds. It is highly effective in separating organic compounds that exist in two or more forms, known as isomers.

The name *chromatograph* given to the recording produced by this method is a misleading one, since there is nothing involving color properties (or chromatic) about the method or its results. The name comes from the original method of separating colored constituents of a solution by color bands, obtained as compounds migrate at different speeds through a filter column or across a sheet of porous paper.

Chromatograph-Detection Principle

In gas chromatography, the fact that the individual compounds can be caused to migrate at different speeds is utilized by employing an inert gas to carry the mixture along through a tube (or column) that has been packed in a special way (Fig. 19-4). This long column exerts its fractionating effect by means of a nonvolatile liquid (called a partitioner), mixed with inert material to pack the column. Some more recent instruments employ a *Golay* column, developed for *Perkin-Elmer*, as the fractionating column. This is a capillary tube, about 0.01 in. in diameter, having a length ranging from 150 to 1000 ft, coiled into a compact helix. By the use of various columns and partitioners, combined with suitable detectors, the gas chromatograph is able to analyze complex organic mixtures, such as those found in petroleum products, flavors, and other substances of biological origin.

The important aspect of the chromatographic analyzer, from the standpoint of electronic instrumentation, is the *detector*, which must respond definitely and quickly to the separated components, as they emerge from the end of the fractionating column. The job of the detector is not to identify the emerging compound, but merely to signal when the output gas is carrying foreign molecules and when it is not. The detector response is interpreted by comparison with known samples.

Of the three forms of detectors widely used, one employs the effect of *resistance variation* (*or thermal conductivity*) and the other two utilize a *change in ionization*. The resistance-variation type is illustrated in the example that follows.

[6] R. K. Kaminski, "The Basics of Ion-Selective Electrodes," *Instrumentation Technology*, Sept. 1969.

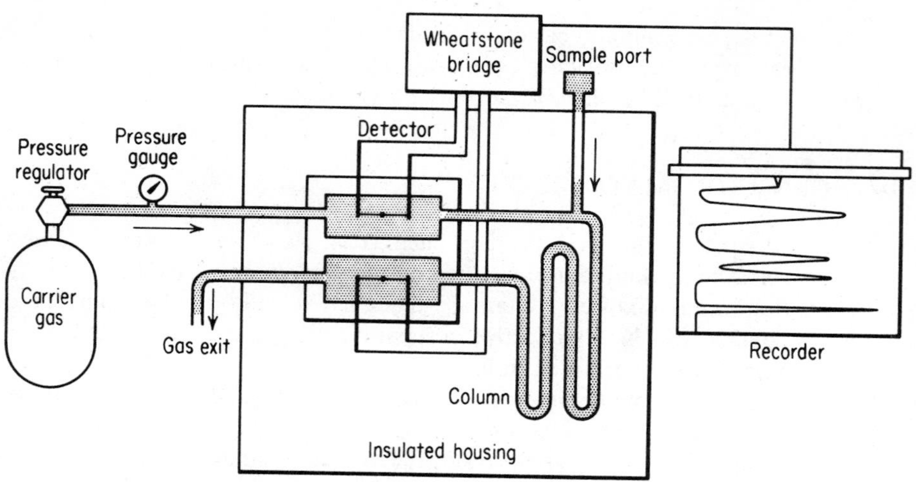

Figure 19-4. Gas chromatography system: the sample is swept by the carrier gas through a packed column, through which various components of the sample migrate at different speeds and are detected as they emerge separately (From R. A. Keller, "Gas Chromatography," *Scientific American*, Oct. 1961.)

Typical Example

A simplified functional diagram of a gas chromatograph system is shown in Fig. 19-4. This shows one form of detector as a Wheatstone bridge. Each of the active arms consists of a heated wire, whose resistance varies with its temperature. When a gas of constant composition and flow rate is allowed to pass over the heated wire, the wire will be cooled by a constant amount and so register a constant resistance. If a gas of different thermal conductivity appears in the stream striking the wire, the change in temperature, and therefore resistance, will be recorded on the strip-chart monitoring the output of the detector. The sensitivity of the detector is sufficient to recognize the very slight changes in the thermal conductivity of the various fractions as they emerge from the column. These output readings can then be compared with the reference charts obtained by feeding samples of known composition into the instrument.

Of the ionization detectors, one form employs a *hydrogen flame* to break up the chemical compounds emerging from the column into electrically charged fragments or ions. By determining the change in ionization, the detector responds to the changing fractions of the sample being investigated. An instrument employing this form of detection is illustrated in Fig. 19-5.

Many models of specialized gas chromatograph instruments are manufactured. In the excellent exposition by Keller,[7] the following results are summarized:

[7]R. A. Keller, "Gas Chromatography," *Scientific American*, Oct. 1961, p. 58.

Figure 19-5. Gas chromatography with hydrogen-flame ionization detector, a high-temperature manual-programming model. (*Wilkens Instruments, Aerograph model A-600.*)

"Samples containing as many as 76 different substances have been successfully analyzed in one pass. Analysis time is typically a few minutes, and sometimes only a few seconds. Some instruments can handle samples weighing not more than a millionth of a gram, and in such samples, they can detect the presence of substances that weigh no more than a trillionth of a gram—about the weight of a single bacterium."

A sample chart, showing the "fractograms" resulting from analyzing a mixture of seven very similar forms of the four-carbon butane compounds, is reproduced from Keller in Fig. 19-6(a).

Since the rate of separation of the constituents in the column is quite dependent on temperature, in addition to the length of the column, a temperature that rises at known rate greatly extends the range of mixtures that can be separated. This control of "temperature programming" is an added instrumentation feature of the advanced chromatographs.

An example of the broad range of separation that can be obtained with temperature programming is shown in the chromatograph in Fig. 19-6(b), which was obtained in experiments with *linear-programmed chromatography to 500°C.*[8]

In comparison with the graph in Fig. 19-6(a), for the separation of the

[8]Preprint of paper by Martin, Bennett, and Martinez, F & M Scientific Corp., New Castle, Del., presented at Edinburgh, Scotland Symposium, June 1960.

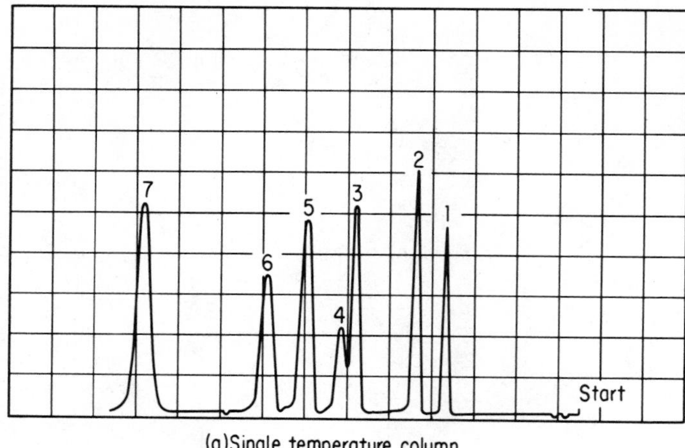

(a) Single temperature column

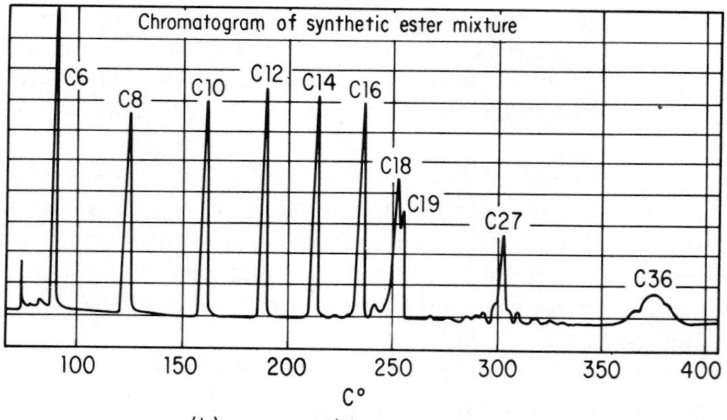

(b) programmed temperature column

Figure 19-6. Chromatograms: (a) separation of seven isomers of four-carbon butane in column operated at 0°C (From R. A. Keller, "Gas Chromatography," *Scientific American*, Oct 1961); (b) separation of a synthetic ester mixture of methyl esters of C-6 to C-36 fatty acids, in a column programmed from 100° to 400°C (From preprint of "Linear-Programmed Temperature Gas Chromatography," paper by Martin, Bennett, and Martinez, F & M Scientific Corp., New Castle, Del.)

varieties of four-carbon isomers whose boiling points were encompassed within a restricted span of 20° C, Fig. 19-6(b) illustrates how the higher-number carbon compounds (methyl esters of fatty acids) separate out from a C-6 to C-36 mixture, at the higher temperatures. Martin's paper also gives a very useful bibliography of some 30 modern reference papers on the details of gas chromatography.

19-5. ABSORPTION SPECTROPHOTOMETER

The technique of analysis by an absorption spectrophotometer (or spectrometer) system involves successively passing radiation of different wavelengths through a substance and measuring the degree of absorption at each of the exciting wavelengths, to determine the material's identity, concentration, and molecular structure.

Ultraviolet, Visible, and Infrared Spectrum

The portions of the electromagnetic spectrum as they are commonly designated are shown in Fig. 19-7(a). Although, theoretically, all sections of the complete spectrum can be used for analytical work, the analysis of molecular structure for chemical identification is best done by using wavelengths in the ultraviolet, visible, and infrared regions, as shown in Fig. 19-7(b). It will be seen here that the *ultraviolet and visible energy have their greatest effect on electron shifts;* accordingly the detection of wavelengths in the UV region (0.2–0.4 μm) and in the visible region (0.4 μm for violet to 0.7 μm for red) are usually lumped together in one instrument, using selected photocells as detectors. The *infrared region, which is effective in causing molecular vibration,* is quite extensive and is divided into three subdivisions, the near infrared, the fundamental infrared, and the far infrared. The IR region most useful in chemical analysis is from 2.5 to 15 μm in the fundamental infrared subdivision, and it uses a thermopile for detection. There are, of course, no sharp boundaries between these regions, and there is therefore quite an overlap in the area utilized by many of the absorption spectrometer systems.

Principle of Wavelength Separation

The value of the absorption spectrometer system lies in the fact that matter absorbs radiation very selectively with respect to wavelength. Specific to the molecular structure of each substance, the atoms of each molecule vibrate at a definite frequency. If radiation of a given frequency (or wavelength) strikes a compound that has the same vibration frequency, this specific wavelength will be absorbed by the molecule, increasing its natural vibration. Stated simply, *the molecule absorbs most at frequencies to which it can resonate.* Other frequencies pass through the molecule with comparatively little change in radiant energy. Since the molecular vibrational frequencies of many substances fall in the infrared region, the response

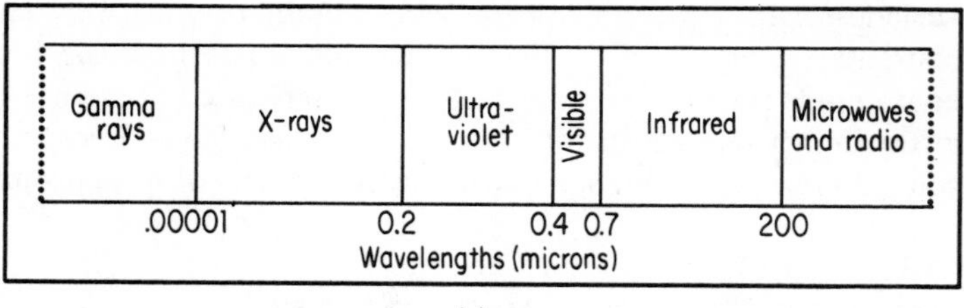

(a)

(b)

Figure 19-7. Electromagnetic spectrum: (a) extent of spectrum in microns, with infrared at 0.7 μm = 7000 Å; (b) wavelength regions, showing effects on atoms in sample. (From R. Sawyer, "Absorption Spectrophotometry," *ICS*, Nov. 1961.)

to IR radiation, as a particular example, can be effectively used to characterize the molecular structure. Pursuing this IR example further (but also remembering that similar statements may be made for other portions of the spectrum), the instrument problem becomes one of introducing the various wavelengths (or bands of wavelengths) so that they will pass progressively through the sample and then indicating the response of the material to the individual wavelengths by the output of a thermopile detector.

A simplified sketch of the basic principles involved in the absorption spectrophotometer is shown in Fig. 19-8. Shown here is the separation of the introduced wavelengths (in an extreme simplification) by refraction of the rays from the infrared source by means of a prism; the wavelength desired is selected by being lined up with a narrow slit. The optical system needed for such fine separation is, of course, much more refined in an actual instrument than shown in the simplified diagram.

Either prisms or diffraction gratings can be used to separate the energy into its spectral components. Gratings provide higher resolution over a limited portion of the spectrum. Often more than one grating is used, each being employed in the

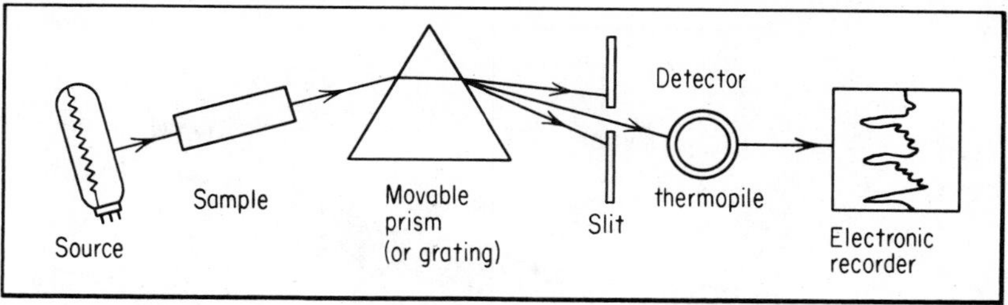

Figure 19-8. Principle of the absorption spectrophotometer (infrared).

most efficient order of the grating. Various schemes are also designed to lengthen the optical path and produce a greater amount of dispersion. The need for very fine separation of wavelengths can be appreciated from the units of length that are used, as follows:

$$1\ \mu\text{m} = 10^{-6}\ \text{m (or } 10^{-3}\ \text{mm)}$$
$$1\ \text{m}\mu\text{m (nm)} = 10^{-9}\ \text{m (or } 10^{-6}\ \text{mm)}$$
$$1\ \text{Å} = 10^{-10}\ \text{m (or } 10^{-7}\ \text{mm)}$$

Thus

$$1\ \text{Å} = 0.1\ \text{m}\mu\text{m (or 1 nm} = 10^{-9}\ \text{m)}$$
$$1\ \text{m}\mu\text{m (nm)} = 10\ \text{Å}$$

Frequency (f or ν) is related to wavelength (l or λ) by the fundamental relation that holds throughout the entire electromagnetic spectrum:

$$\begin{aligned} lf\,(\text{or } \lambda\nu) &= \text{velocity of light } (c) \\ &= 2.998 \times 10^{10}\ \text{cm/sec (or } 299.8 \times 10^{6}\ \text{m/sec)} \end{aligned}$$

Instrument Description

As a simple illustration, without entering into the complexities of the optical system, Fig. 19-9(b) shows how the controls of a basic spectrophotometer Fig. 19-9(a) might operate (*Beckman model B*), for the region from 325 to 1000 mμ. The specific group of wavelengths of the spectrum range desired is selected by a fine-control positioning of the dispersing or refracting element—in this case, a prism. The slit adjustment and sensitivity controls are set to some value to be considered as a reference (or 100% transmission). This reference value would then be indicated by the meter, corresponding to the output of the detector (a phototube in this case), responding to the visible (and some infrared) energies of the light source. The sample cells include solutions, some of which would be used either empty or filled with only the solvent, to assist in the calibration of the instrument. Typical spectra showing the detection of four elements in one solution are shown in Fig. 19-10.

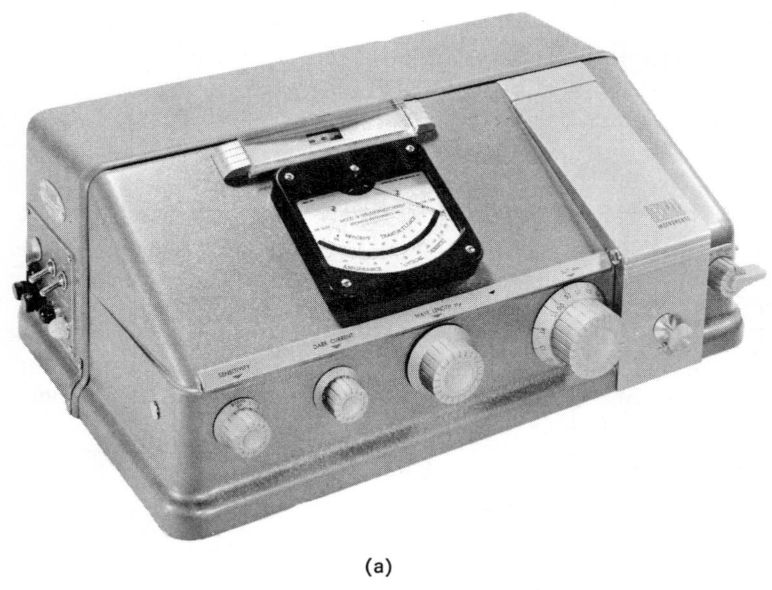

(a)

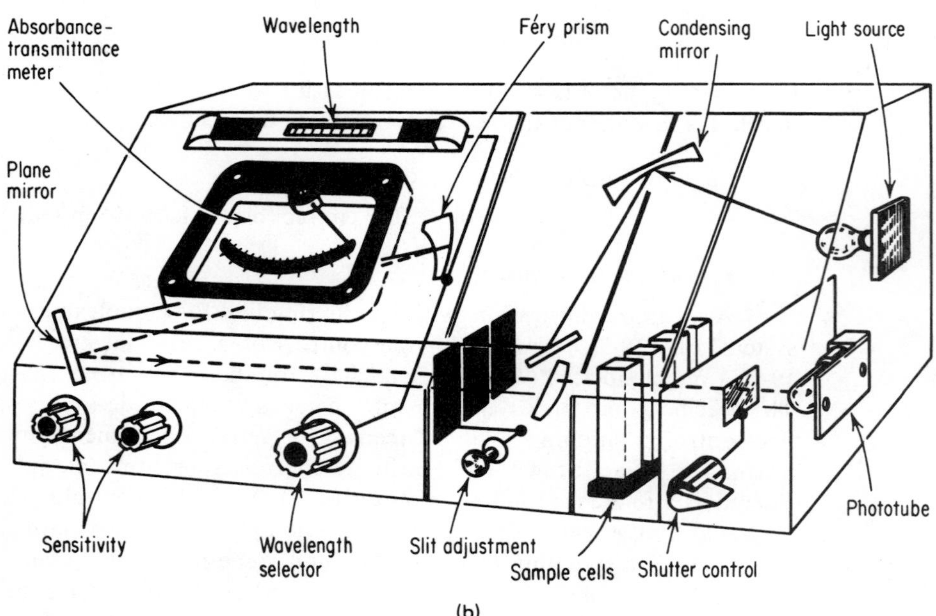

(b)

Figure 19-9. (a) Spectrophotometer; (b) controls of spectrophotometer. (*Beckman Instruments, Inc., model B.*)

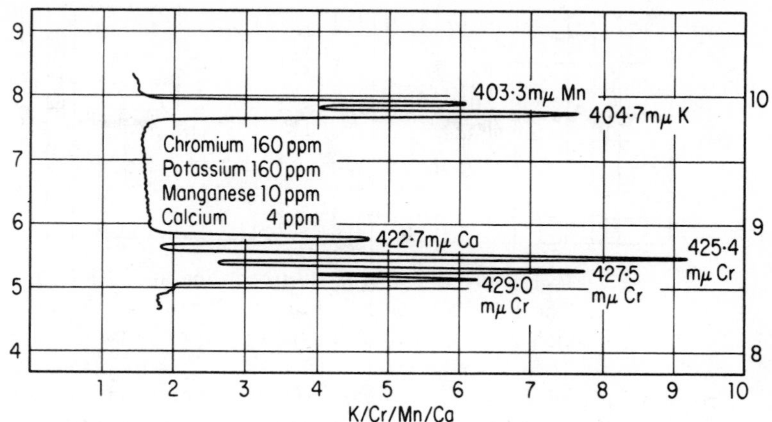

Figure 19-10. Typical spectra in the blue-violet region of the spectrum (403–429 mμ) detect the presence of four elements (Mn, K, Ca, and Cr) in one solution in a flame spectrophotometer. (*Unicam Instruments, Cambridge, England, model SP900.*)

This *optical absorption spectrophotometer* as an analytical instrument should not be confused with the *mass spectrometer*, which depends upon the difference in atomic weights of the ionized atoms. Other useful (albeit highly specialized) analysis methods include *nuclear-magnetic resonance* (*NMR*), *electron spin resonance* (*ESR*), *atomic absorption* (*AA*), and *nuclear activation analysis* (*NAA*).

19-6. BIOLOGICAL/MEDICAL INSTRUMENTATION

The spread of electronic instrumentation into biological fields is closely allied to its employment for medical purposes, and both fields can conveniently be discussed together. (Some writers have suggested the name *bionics* for this bio/medical/electronic area of investigation.)

Recording of Heart-Action Potentials (Electrocardiograph)

A long-familiar example of a medical record obtained through electronic means is the *electrocardiograph* (variously abbreviated as EKG and ECG), where the electrical potentials generated by the action of the heart are tapped off by contact with various parts of the body and after amplification are recorded by direct writing oscillographs. A sample of such an electrocardiogram is shown in Fig. 19-11. The various portions of the heart-action sequence are identified by letters referring to voltage waves produced as different muscles of the heart pump go into action. (This generation of electrical potentials produced by muscular contraction is not

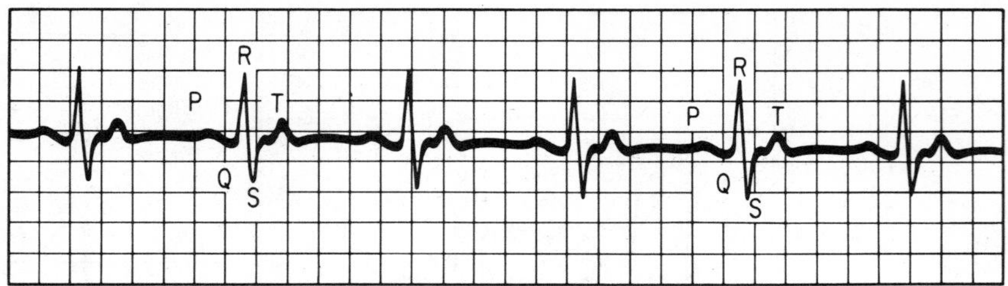

Figure 19-11. Sample electrocardiograph (EKG) recording. The *PQ*, *RST* waves indicate production of electrical potentials by heart action. (Limascope, New York, Chart 106.)

to be confused with the "*lub-dub*" *sound* of heart action, to which the physician listens with his stethoscope, for interpretation of heart-valve opening and closing sequences.)

The electronic aspect in the production of an electrocardiogram is based on obtaining stable (and closely repeatable) amplification of the generated potentials, which are then in a form suitable for application to the galvanometer pen motor of the direct-writing oscillograph. The slow rate of repetition of the heart wave (50–90 beats/min) produces a slowly varying signal that fits very well into the frequency capability of the galvanometer type of recording instrument. As a familiar instance of the instrument technician's responsibility, one can see the significance of a careful calibration of the electrocardiograph instrument by the medical-instrument technician. It is surely important that no element of electronic malfunctioning or even miscalibration be interpreted by the physician as abnormal functioning of the patient's heart.

The Electroencephalograph

In spite of the tongue-twisting names so dear to the medical profession, the basic principle of the instrument for recording brain waves, *the electroencephalograph* (EEG), is of the same simple nature as that already mentioned, namely, the recording of potentials produced by a part of the body—in this case, by the nerve impulses to and from the brain. To the medical-instrument technician, however, the application of the electrodes is far from simple. Current practice calls for contacting the scalp by as many as 17 different connections, each having a meaning of its own, and switching in various combinations of these 17 wire leads, as the long process of obtaining the EEG record proceeds.

From the electronic standpoint, the brain-wave signals applied to the recording instrument are at a much lower level than the heart-action signals of the ECG and hence call for much more sensitive amplifiers, with the accompanying care necessary to prevent false signals caused by *electrical noise*. The noise level, of course, is an important factor that must be considered with any form of low-level instrumentation.

A good medical-instrument technician must also have skills of a psychological nature. Consider the patient trying to achieve a relaxed state of mind, with 17 pins stuck in his head, and being told to close his eyes while he hears the ominous racket of relays closing in sequence. It takes quite a bit of doing to convince the patient that he is not in danger of being shocked (or even electrocuted), even though the technician knows that there is no possible danger from this source, not only because the signals are at such low voltage levels, but also because the energy flow goes *out from the patient* rather than into him. Knowing this is one thing, but assuring the naturally alarmed patient may be quite another!

Electronic Monitoring of Body Functions

In monitoring a patient's condition for temperature, respiration rate, and the like, there is a growing tendency, especially in hospital recovery rooms where up to 12 patients are recovering from major surgery, to obtain a continuous running record of the patient's condition, displayed on a central panel. By electronic instrumentation, the output of each of the individual monitoring instruments can be conveniently displayed at such a central point.

An arrangement such as shown in Fig. 19-12 can provide much better surveillance in cases where it is vitally important to know, at a glance, how the patient is progressing. The recording chart for each patient, in the *Honeywell body-function recorder* shown in Fig. 19-13, displays continual sample readings of temperature, blood pressure (generally both systolic and diastolic), respiration rate, and pulse, as major body functions. This is a print-wheel type of readout, with a different letter printed for each variable.

Biomedical Transducers

In the *Honeywell* system illustrated, the method used for monitoring the pulsing heart action makes use of a photodiode, which detects the rhythmic change in light transmission properties through an ear lobe, as the blood in the thin capillaries of the ear follows the pulse rate.

The outputs of various transducers are amplified and fed as recurrent samples to the recording system. The readout in this system is in the form of a graph of each of the variables being monitored, produced, in the example cited, by means of a multiple print-wheel.

The readouts for the various parameters in this example require the following calibration of the various biomedical transducers:

Respiration rate (per minute)
Heart rate (per minute)
Temperature (°C)
Systolic pressure (mmHg)
Diastolic pressure (mmHg)

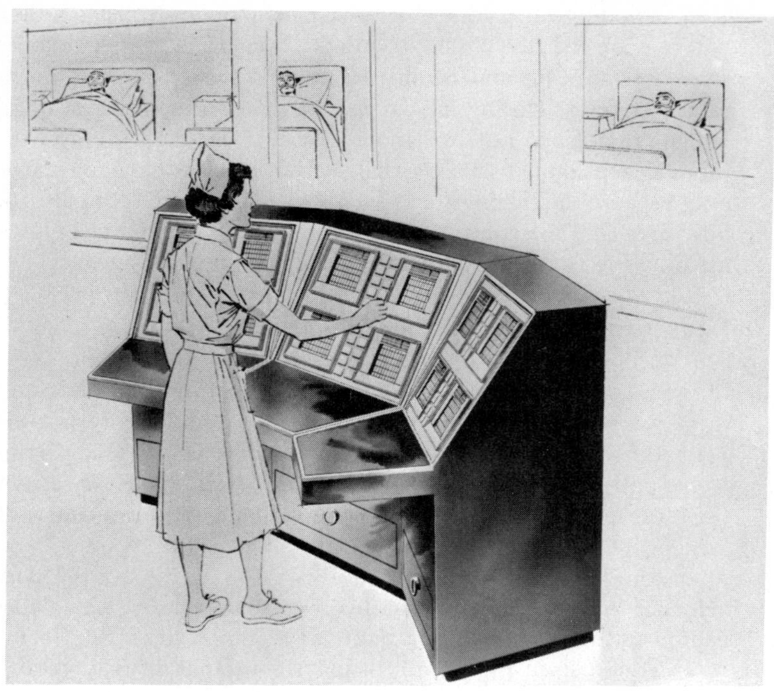

Figure 19-12. Electronic body-function monitor system, for central monitoring of patients' conditions. For each of 12 patients, the pulse rate, respiration rate, temperature, and diastolic and systolic blood pressures are recorded. (*Honeywell body-function recorder.*)

Oscilloscope presentation in a patient-monitoring unit generally also provides a capability for viewing electrocardiograph (ECG) and other desired waveforms.

19-7. POLLUTION-MONITORING INSTRUMENTS

A growing awareness and national concern regarding pollution developed and spread across the country in the 1960s. Along with the clearer realization of the potential dangers inherent in the deterioration of our air, water and land environments, there came intensified efforts to develop effective instruments to monitor this pollution.

The development of this instrumentation proceeded simultaneously in two major directions: first, toward *standardizing existing pollution-monitoring instruments*, in connection with the promulgation of effective regulations specifying

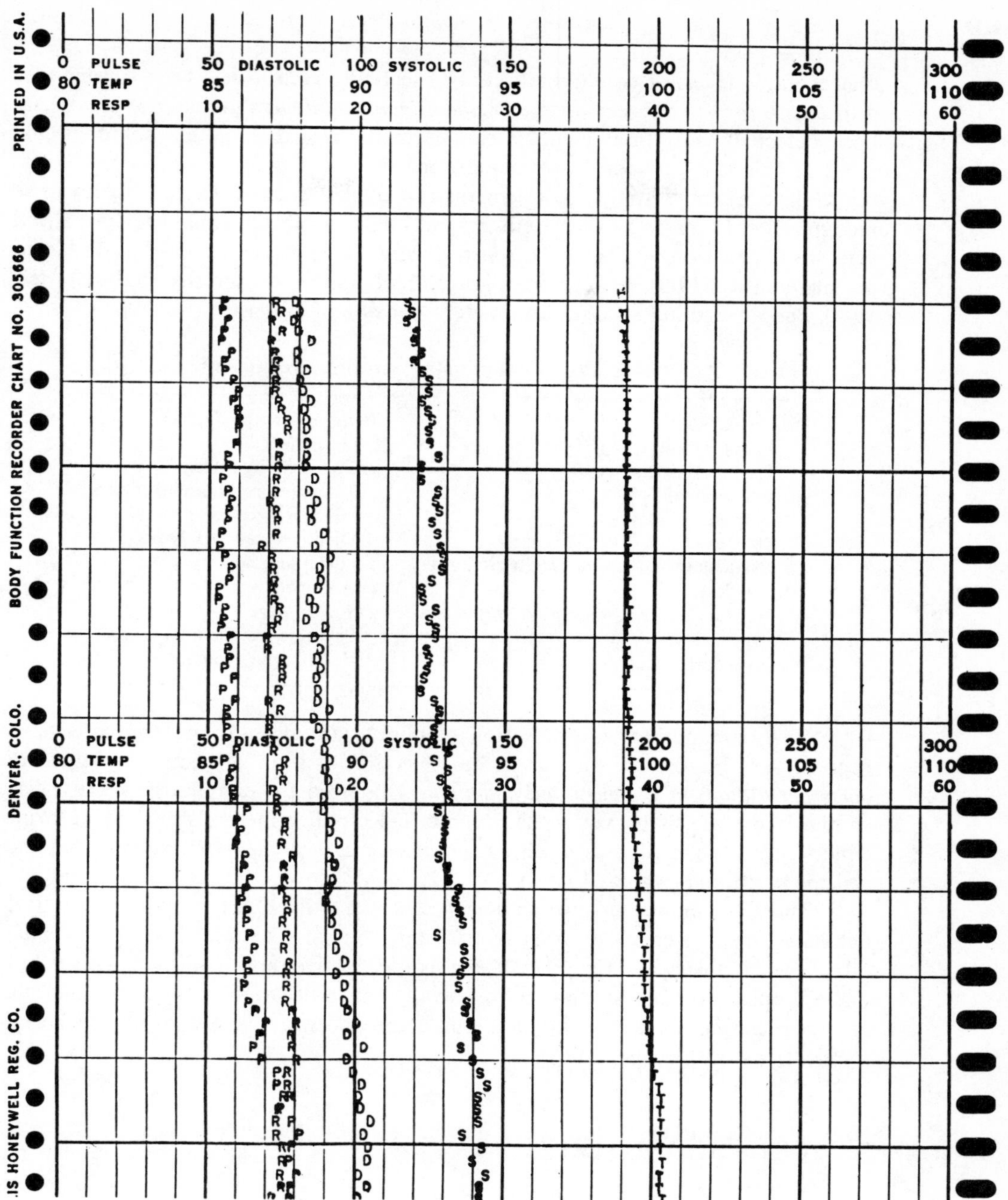

Figure 19-13. Typical strip-chart record, produced by the Honeywell body-function recorder that monitors five physiological variables for patients requiring intensive hospital care. Measurements are printed out consecutively in multicolored ink to allow easy identification. Each printed point includes the letters *P* (pulse), *R* (respiration), *D* (diastolic), *S* (systolic), and *T* (temperature).

unacceptable levels of contamination. The second direction has been toward *new instruments for detecting and measuring additional pollutants* that had not previously been detected. Only brief mention can be made here to highlight some specific areas of this fast growing instrumentation field.

The areas of existing monitoring instruments are given in Table 19-1. It will be recognized that many of the monitors and analyzers in the table are specialized versions of general-purpose *analysis instruments*. Thus, preceding sections discussed pH meters (Sec. 19-3), gas chromatographs (soil and food impurity analysis) (Sec. 19-4), and absorption spectrophotometers (sensing of all gases) (Sec. 19-5).

TABLE 19-1. Pollution-Monitoring Instruments

Air-quality sensors:	*Water-quality analyzers:*
Carbon monoxide (CO)	pH meters
Carbon dioxide (CO_2)	Conductivity meters
Sulfur dioxide (SO_2)	Temperature (thermal) meters
Nitrogen oxides	Dissolved-oxygen meters
Ozone	Dissolved-salts analyzers
Particulate matter (soot and visible particles)	Bacterial analyzers
	Land-quality monitors:
Aerosol matter (nonvisible suspended particles)	Soil analyzers
	Pesticide-residue analyzers
Hydrocarbon analysis	Food-impurity analyzers

New demands for the detection of minute amounts of contaminants have become very important, such as in the detection of trace amounts (*in fractional parts per billion*) *of mercury and cadmium*. This need for extreme sensitivity, has emphasized analysis by very sensitive but less familiar methods employing principles from atomic physics, such as the *atomic-absorption* (*AA*) method and the *neutron-activation analysis* (*NAA*).

The vital matter of *environmental control* involves not only the technical details of specialized monitoring instruments, but also cuts across the *economic interests* of industry and the *social concern* of legislative representatives. This is but one example of the broader and more socially-concerned view required of scientific and engineering workers who, hopefully, are increasingly aware of the social implications of their technology.

19-8. TELEMETRY AND INSTRUMENTATION SUMMARY

Since telemetering is basically the process of providing *remote measuring* or *monitoring*, a brief discussion of the various methods used can also serve as a useful summary of an instrumentation system. Often, the telemetering system also includes remote control (frequently called supervisory control) and the term *telemetry* is used to encompass both functions of *measurement and supervisory control conducted at a distance.*

These systems comprise the familiar, conventional measuring and control systems discussed in this text (transducers, amplifiers, indicator/recorders, and the like), with the *additional element of a transmitter–receiver communication link to* relay data from one place to another.

For the transmission of information over the communication link, both analog and digital techniques are in common use, using a wide variety of signal-shaping methods to accomplish the desired communication objective.

Apart from the specialized methods called for in particular applications that may be concerned with aircraft, satellites, medical monitoring, and the like, it can be seen that a telemetering system combines many of the elements of electronic instrumentation already discussed. With respect to these common elements, such a system furnishes a good summary of basic practical instrumentation principles. With this thought in mind, a very brief explanatory statement of the *types of telemetering equipments* is included here as an appropriate summarizing topic for this text.

Types of Telemetering Equipment

The types mentioned are selected from a most useful survey[9] of available telemetering equipments.

Types of Signal Transmitted

The main *types of signals* transmitted are as follows:

1. *Current*, as in direct-writing oscillographs.
2. *Voltage*, for actuating potentiometric or bridge-type recorders, digital voltmeters, or analog computing circuits.
3. *Frequency*, including tone modulation, for actuating counters or frequency-selective circuits, including FM discriminators.
4. *Phase*, for actuating phase-sensitive detectors.
5. *Pulse-duration* (*or pulse-width*), resulting in pulse-duration-modulation (PDM), where the value of the signal determines the duration of the transmitted pulse.
6. *Pulse-amplitude modulation* (*PAM*), where the value of the signal determines the amplitude of the transmitted pulse.
7. *Pulse-code-modulation* (*PCM*), including pulse-count, often combined with digital and tone techniques.

The last-named illustrates the possibility of combining basic modulation techniques in various ways, to achieve methods that combine the specific advantages of each method. Tones and the frequency shift of FM are often combined

[9] R. C. Nelson, "Telemetering and Remote Control" (no. 6 in a series of articles on telemetering), *Instruments and Control Systems*, Feb. 1962.

with pulse techniques. Thus, methods suitable for data transmission over a telephone line using an ordinary handset have been developed via PCM and digital-tone techniques.

Medium of Transmission

The means for transmitting the information signals are of three kinds:

1. *Wire*, which can be the ordinary telegraph or telephone line, as indicated above.
2. *Carrier-current "wired-wireless,"* which usually imposes signals of about 20–200 kHz on power-distribution lines.
3. *Radio-frequency*, which includes many types, from short-distance transmission by miniature (perhaps 1-mW) crystal-controlled AM or FM transmitters in medical electronics, on through the high-frequency ranges for FM/FM transmitters (around 1 W or so) used in satellite communications in the megahertz region, and including much higher powers in the microwave relay systems operating in the kMHz or gigahertz (GHz) range.

Choice of Telemetering Systems

The choice of signal type and transmission medium generally includes three basic interrelated factors: (1) data rate, (2) accuracy desired, and (3) cost. The availability of a large number of precise analog-to-digital converters (including the selsyn type of shaft-angle converters and the electronic voltage converters), as well as high-speed multiplexing and scanning equipments, has increased the interest in digital communication links.

Summarizing Telemetry Developments

In this penultimate section, a summary of significant developments in telemetry can also serve appropriately as a summary of how the main elements of electronic instrumentation are being developed. It will be recalled that these three main elements were identified in the first chapter as, first, the *transducer* that converts the quantity to be monitored into an equivalent electrical voltage; second, the *signal-modifier* that shapes and amplifies the electrical signal; and finally, the *indicating system* that presents the results in the form of a display of the variable being monitored.

Examples of expanding developments in these elements have appeared, often in dramatic form, not only in the biomedical field, but also in *aerospace* and *oceanographic* instrumentation. Thus, one thinks of the continuous monitoring of heart activity in intensive-care patients by both passive and active implanted transducers, or similarly, in the process of landing astronauts on the moon. One's attention is again directed to the high state of the development of *sensitive transducers* for obtaining electrical signals that respond to their physical condition, as shown by cardiac activity and blood-pressure changes.

The body-function signals from the sensors are then processed by *stable high-gain amplifiers*, often in IC form, which are able to build up these weak signals to a satisfactory amplitude for the desired display.

In the matter of indicating, storing, and/or displaying the monitored signal, there is a wide variety in the choice of methods. If a wire link is involved, as was illustrated in the case of a central display panel for nurses, the display may include a continuous *strip-chart recorder* and/or *oscilloscope*, to supplement a *digital readout*, with appropriate accessory indicators for warning signals to alert personnel when any output exceeds predetermined limits.

Alternately, in the case of radio-monitoring of signals, the body-function signals *modulate* a radio carrier wave (by means of such techniques as FM/FM or pulse modulation, with appropriate multiplexing) to impress these and other desired signals on the carrier. At the receiving end, the signals are *demodulated* and may then either be immediately displayed, as shown in Fig. 19-14 for a Bionic telemetering system, or in other cases, may be stored on *magnetic tape*, often in digital form. The reception of these extremely weak signals, which may by deeply buried in noise after their journey through the tremendous distances of space, involves highly advanced forms of *noise-reduction techniques*, as previously mentioned. Where the desired signals have been stored on tape, they may than be processed through *digital computers* to provide a comprehensive analysis.

In addition to these extensive monitoring functions, a brief reminder is in order to recognize the role of compact digital computers in *space guidance and control.* The very successful achievements in space travel have forcefully demonstrated the capabilities involved in the *highly accurate tracking* of a space capsule at fantastic distances, as well as in exerting *extremely delicate control* of its maneuvers and of its role in performing scientific experiments. These capabilities are made possible by advanced control techniques in combination with highly sophisticated *digital-computer programming.*

Progress in Instrumentation Systems

Seen in broad outline, the predominant direction in which advanced instrumentation is moving is increasingly toward the development of more integrated *instrumentation systems.* The employment of digital techniques (and their associated integrated circuits) has produced compact packages capable of performing highly flexible functions, such as were pointed out in the examples given for automated testing and telemetry. These functions can then be directed and controlled by computer programs, to form a complete unit or integrated system, for accomplishing the desired task with a minimum of manual intervention.

This growth of the *system approach* to instrumentation imposes an added opportunity for the progressive worker in the field of instrumentation. Going beyond the understanding of the functioning of the individual electronic instrument, he may then go on to the interconnections or interfaces required to feed the individual instrument output, to be in a compatible form with the other portions of the system. These other portions may provide such extended functions as multi-

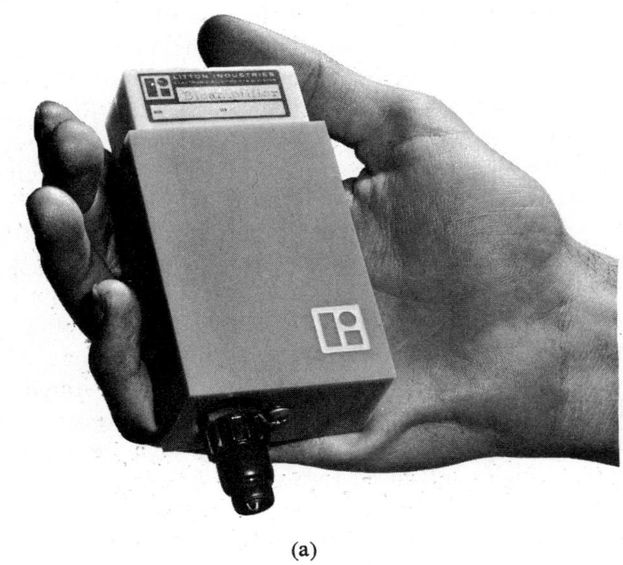

(a)

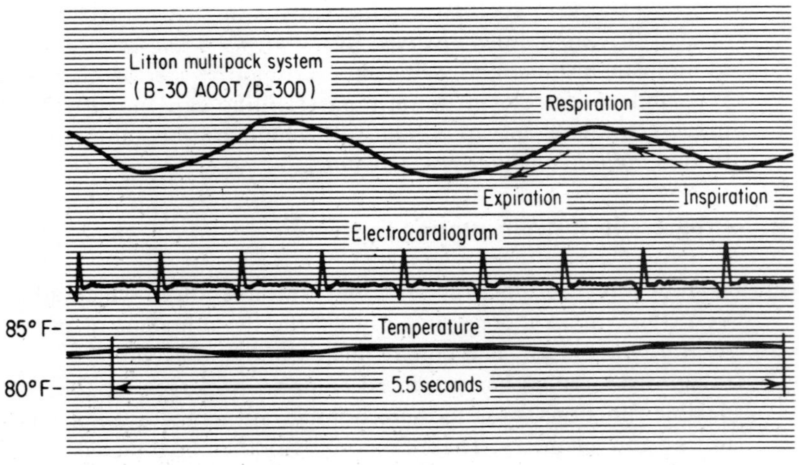

(b)

Figure 19-14. Bionic telemetering system: (a) Biopack modular unit that can be combined (multipack) to transmit three channels for display and recording; (b) sample of typical curves for respiration, electrocardiogram, and temperature record. (*Litton Systems Biopack model B-30A TP.*)

plexing with other instruments, computer direction of test operations, or various forms of graphic display.

QUESTIONS

Q19-(1 to 8). Starting with each electronic instrument given in the right-hand column (numbered 1–8), select a *matching process* from the left-hand column (letters a–h), and name the sensor (or detector) associated with the detection process (page 437).

Q19-(9 to 10). *Select two processes* (or applications) from the left-hand column given below, one from the chemical and one from the physical or biological field, and write a brief description of this application of a specialized electronic instrument in each of the two cases.

Table of Specialized Electronic Instruments

Process for Detection or Measurement	*Electronic Instrument*
a. Identifying chemical ingredients by detecting their passage through a packed column	1. Electrocardiograph
b. Determining acidity or alkalinity of chemical solution	2. Absorption spectrophotometer
c. Identification of chemical compounds by their response to selective infrared wavelengths	3. Electroencephalograph
d. Measurement of total radiation in the infrared region	4. Gas chromatograph
e. Recording of electrical potential produced by heart muscles	5. Body-function monitor
f. Receiving distant signals modulated by physical variables.	6. Telemetry
g. Recording of electrical potentials produced by nerve messages to brain centers	7. pH meter
h. Recording biological conditions in critical patients	8. Radiation pyrometer (temperature determination by radiation)

Appendices

Thévenin's Circuit Theorem[1]

Thévenin's theorem may be stated as follows: Any linear and bilateral network with two accessible terminals may be replaced by an EMF acting in series with an impedance—the EMF is that between the terminals when they are unconnected externally, and the impedance is that presented by the network to the terminals, when all sources of EMF in the network are replaced by their internal impedances.

This means, in effect, that for any element, such as a load resistor, that we wish to connect to two points in a circuit, the whole circuit, no matter how complicated, may be replaced by two simple elements, an equivalent EMF and an equivalent impedance in series with it. The EMF E_{TH} is the open circuit voltage across the two points, before the new element is connected. The impedance Z_{TH} is that which is seen looking into the circuit from the two points (all EMFs set equal to zero, but internal impedance of generators retained). First, we shall illustrate the theorem by the simple dc circuit shown in Fig. A-1(a). With S open:

$$I_0 = \frac{125 \text{ V}}{2 + 3 + 20\ \Omega} = 5 \text{ A}$$

[1]Adapted from M. B. Stout, *Basic Electrical Measurements*, 2nd ed. (Englewood Cliffs, N.J.: Prentice-Hall, 1960).

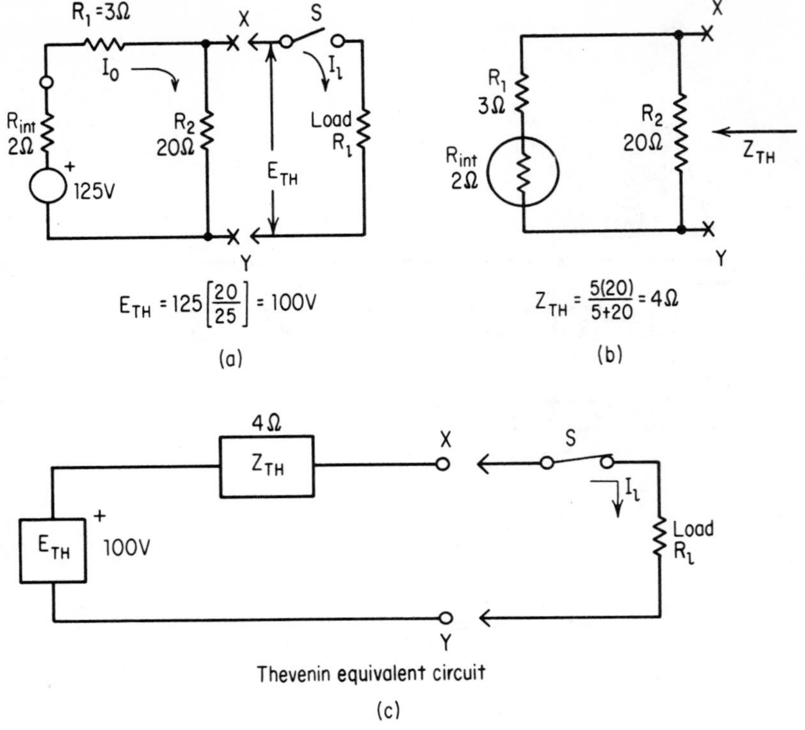

Figure A-1

$$E_{\mathrm{TH}} = 5 \times 20 = 100 \text{ V}$$

or

$$E_{\mathrm{TH}} = 125\left(\frac{20}{25}\right) = 100 \text{ V}$$

The resistance seen from terminals x–y, Fig. A-1(b), is the Thévenin equivalent impedance Z_{TH}:

$$Z_{\mathrm{TH}} = R_{xy} = \frac{5 \times 20}{25} = 4\ \Omega$$

Now, by Thévenin's theorem, we may replace the original circuit by the circuit of Fig. A-1(c), which is *equivalent for any load we wish to connect from* x to y. This may be tested by trying any convenient load, say, 6 Ω, connected first to the original circuit and then to the equivalent circuit. The resulting current in the 6-Ω load would then be 100 V/(4 + 6) = 10 A, and the corresponding voltage across the 6-Ω load would be 6 × (10) = 60 V.

It should be noted that the results obtained from the Thévenin equivalent circuit are valid *only* for the current in (and the voltage across) the element connected to the right of the breakpoints x–y, in this case the 6-Ω load resistor. The values of current in the hypothetical Z_{TH} to the left of the points x–y are not to be treated as actual values. In spite of this limitation, however, Thévenin's theorem often serves as an extremely useful simplification, as it would in this case if we wished to obtain new values of *load current* for many different load resistors, without changing the rest of the circuit.

B

Selected Bibliography

Note: The starred references are at the advanced undergraduate/graduate levels.

Alpert, N. L., Keiser, W. E., and Szymanski, H. A., *IR Theory and Practice of Infrared Spectroscopy*, 2nd ed. New York: Plenum Press, 1970.

*Baird, D. C., *Experimentation—An Introduction to Measurement Theory*. Englewood Cliffs, N.J.: Prentice-Hall, 1962.

Buckstein, E., *Basic Servomechanisms*. New York: Holt, 1963.

——— *Industrial Electronics Measurement and Control*. Indianapolis, Ind.: Howard W. Sams, 1961.

Cooper, W. D., *Electronic Instrumentation and Measurement Techniques*, 2nd ed. Englewood Cliffs, N.J.: Prentice-Hall, 1978.

Eimbinder, J., *Linear Integrated Circuits*. New York: Wiley-Interscience, 1968.

——— FET *Applications Handbook*. Blue Ridge Summit, Pa.: Tab Books, 1967.

Foster, A. R., and Wright, R. L., Jr., *Basic Nuclear Engineering*, 3rd ed. Boston: Allyn and Bacon, 1977.

Hill, D. W., *Principles of Electronics in Medical Research*. London: Butterworth, 1965.

Jackson, H. W., *Introduction to Electric Circuits*, 5th ed. Englewood Cliffs, N.J.: Prentice-Hall, 1981.

Liptak, B. G., ed., *Instrument Engineers Handbook*, Vol. 1, *Process Measurement*. Philadelphia: Chilton, 1970.

Malmstadt, H. V., and Enke, C. G., *Electronics for Scientists*. New York: W A. Benjamin, 1963.

——— *Digital Electronics for Scientists*. New York: W. A. Benjamin, 1969.

Malvino, A. P., *Electronic Instrumentation Fundamentals*. New York: McGraw-Hill, 1967.

Marcus, A., and Lenk, John D., *Measurements for Technicians*. Englewood Cliffs, N.J.: Prentice-Hall, 1971.

Morrison, R., *Grounding and Shielding Techniques in Instrumentation*. New York: Wiley, 1970.

Prensky, S. D., *Electronic Demonstration Manual*. Brooklyn, N.Y.: Radiolab Publ., 1943.

Rider, J. F., and Prensky, S. D., *How to Use Meters*, 2nd ed. New York: Hayden–Rider, 1960.

Roth, C., *Use of the Oscilloscope: A Programmed Text*. Englewood Cliffs, N.J.: Prentice-Hall, 1970.

Shunaman, Fred, *How to Use Test Instruments in Electronics Servicing*. Blue Ridge Summit, Pa.: Tab Books, 1970.

*Terman, F. E., and Pettit, J. M., *Electronic Measurements*, 2nd ed. New York: McGraw-Hill, 1952.

United Detector Technology, *Silicon Photodetector Design Manual*. Santa Monica, Calif.: United Detector Technology, 1970.

*Von Handel, P., *Electronic Computers (Digital and Analog)*. Englewood Cliffs, N.J.: Prentice-Hall, 1961.

Warner and Fordenwait, *Integrated Circuits*. Phoenix, Ariz.: Motorola, 1965.

Wedlock, B. D., and Roberge, J. K., *Electronic Components and Measurements*. Englewood Cliffs, N.J.: Prentice-Hall, 1969.

Weyrick, R. C., *Fundamentals of Analog Computers*. Englewood Cliffs, N.J.: Prentice-Hall, 1969.

Yeager, D. A., and Gourley, R. L. *Introduction to Electron and Electromechanical Devices*. Englewood Cliffs, N.J.: Prentice-Hall, 1976.

C

Answers to Odd-Numbered Problems

Chapter 2

P2-1. (a) 600 mV; (b) 195 mV; (c) 48.75 μA.

P2-3. (a) 24 kΩ; (b) 50 Ω; (c) the switch connects the shunt into the circuit only on the 0–2.5-mA range.

P2-5. (a) and (b) are diagrams.

P2-7. (a) 128.6 V, or over 14% error from nominal 150 V; (b) $\cong$ 150 V.

P2-9. (a) Loading of circuit by voltmeter; (b) R_1 = 300 kΩ, R_2 = 150 kΩ.

Chapter 3

P3-1. (a) 4.5 kΩ; (b) 450 Ω/V.

P3-3. (a) 2.2 kΩ; (b) 225 Ω/V; (c) 1.8 kΩ.

Chapter 4

P4-1. (a) 15 V; (b) E_{TH} = 15 V, Z_{TH} = 6 Ω; (c) 1 mA.

P4-3. (a) E_{TH} = 300 V, Z_{TH} = 10 kΩ; (b) 5 kΩ; (c) 100 V; (d) 32 mA.

P4-5. 10 Ω.

P4-7. 0.5 mA.

P4-9. (a) E_{TH} = 384 mV, $Z_{TH} \cong$ 880 Ω; (b) $\cong$ 0.4 mA.

Chapter 5

P5-1. 100 Ω.

P5-3. 160 pF (or $\mu\mu$F).

P5-5. (a) $R_x = 250$, $I_G = +2.22$ mA;
$R_x = 500$, $I_G = +0.85$ mA;
$R_x = 750$, $I_G = +0.30$ mA;
$R_x = 2$ kΩ; $I_G = -0.47$ mA;
$R_x = 3$ kΩ; $I_G = -0.64$ mA.

P5-7. (a) Estimate = 0.2 mA;
(b) exact value = 0.202 mA.

P5-9. 71.4 mV.

Chapter 6

P6-1. (a) $33\frac{1}{3}$ kΩ; (b) $16\frac{2}{3}$ kΩ;
(c) 10 kΩ.

P6-3. (a) 40–70 V; (b) 0–40 V.

P6-5. (b) About a 10-mA change, compared to about a 3-mA change where curves are more crowded; (c) $\cong -12$ V.

P6-7. 20.

P6-9. 8 kΩ.

P6-11. 2500 μ siemans.

P6-13. Load line passes through point (300 V, 0 mA) and point (0 V, 15 mA); it is parallel to load line shown on the figure.

P6-15. $I_b = 5$ mA, $E_b = 100$ V.

Chapter 7

P7-1. (a) See Fig. P7-2; (b) 2 mA.

P7-3. (a) 2.5 mA;
(b) I_m varies from 0 to 2.5 mA, as E_s varies from 0 to +5 V.

P7-5. VOM reads 2 V instead of 3 V, loading error = $\frac{1}{3}$ or $33\frac{1}{3}$%; VTVM reads practically 3 V, with negligible (much less than 1%) loading error.

P7-7. (a) 0.3 V dc; (b) 1.2 V dc;
(c) 0–50-μA meter, with 18 kΩ in series; or equivalent combination.

Chapter 9

P9-1. 2000 microstrain.

P9-3. (a) 10.3 mV; (b) 1.6 mV.

P9-5. 20 μA.

Chapter 10

P10-1. (a) 180 Hz; (b) 90 Hz; (c) 40 Hz.

P10-3. (a) 0°; (b) 30°; (c) 45°; (d) 150°.

P10-5. (a) 42.4 V rms;
(b) $E_{o2} \cong 0$; $E_{b2} = 110$ V;
(c) 100 p-p.

Chapter 12

P12-1. (a) 380 Ω.

P12-3. $R_s = 36$ Ω; $X_s = 48$ Ω.

P12-5. $C = 35$ pF; $Q = 60$.

Chapter 14

P14-1. (a) 21; (b) 7.4.

P14-3. (a) 14.1 V; (b) 20 V; (c) 100 V;
(d) 398 V; (e) 1,995 V.

P14-5. Amplifier B is down 1 dB at 400 kHz; amplifier A, therefore, has the wider frequency response.

Chapter 17

P17-1. (a) 3; (b) 5; (c) 7; (d) 15.

P17-3. Circuit of part (c) gives a suitable peaked negative pulse.

P17-5. (a) Approx. $2RC/3$ or 14 μsec; (b) 15.

P17-7.

	Output				*Input*
	Eight's 2^3	*Four's*, 2^2	*Two's*, 2^1	*Units*, 2^0	*Decimal Count*
	0	0	0	0	0
	0	0	0	1*	1
	0	0	1	0	2
	0	0	1	1	3
	0	1	0	0	4
	0	1	0	1	5
	0	1	1	0	6
	0	1	1	1	7
	1	0	0	0	8—no feedback
	1	1	1	0	8—with feedback
to 2^4	1	1	1	1	9
1	0	0	0	0	10

*A ONE is indicated by lighting of neon lamp.

Chapter 18

P18-1. (a) 3; (b) $\cong 2$.

P18-3. $2A - 2B + 10C$.

P18-5. $2y - 60$.

P18-7. $\dfrac{d^2x}{dt^2} - x = 0$.

P18-9. $\dfrac{5\,d^2x}{dt^2} + \dfrac{2\,dx}{dt} + x = 0$.

Index

E

J

K

L

S

W

X

Z